W0225808

Transforming Communications – Studies in Cross-Media Research

Series Editors
Uwe Hasebrink
Leibniz Institute for Media Research
Hans-Bredow-Institut
Hamburg, Germany

Andreas Hepp
ZeMKI
University of Bremen
Bremen, Germany

We live in times that are characterised by a multiplicity of media: Traditional media like television, radio and newspapers remain important, but have all undergone fundamental change in the wake of digitalization. New media have been emerging with an increasing speed: Internet platforms, mobile media and the many different software-based communication media we are recently confronted with as 'apps'. This process is experiencing yet another boost from the ongoing and increasingly fast sequence of technological media innovations. In our late modern social world, communication processes take place across a variety of media. As a consequence, we can no longer explain the influences of media by focusing on any one single medium, its content and possible effects. In order to explain how media changes are related to transformations in culture and society we have to consider the cross-media character of communications. Furthermore, today's digital media are not only means of communication, but also of continuous generation and processing of data.

In view of this, the book series 'Transforming Communications' is dedicated to cross-media communication research and related processes of datafication. It aims to support all kinds of research that are interested in processes of communication and datafication taking place across different kinds of media and that subsequently make media's transformative potential accessible. With this profile, the series addresses a wide range of different areas of study: media production, representation and appropriation as well as media technologies and their use, all from a current as well as a historical perspective. The series 'Transforming Communications' lends itself to different kinds of publication within a wide range of theoretical and methodological backgrounds. The idea is to stimulate academic engagement in cross-media issues by supporting the publication of rigorous scholarly work, text books, and thematically-focused volumes, whether theoretically or empirically oriented.

More information about this series at
https://link.springer.com/bookseries/15351

Andreas Hepp • Juliane Jarke
Leif Kramp
Editors

New Perspectives in Critical Data Studies

The Ambivalences of Data Power

Editors
Andreas Hepp
ZeMKI, Centre for Media, Communication and Information Research
Universität Bremen
Bremen, Germany

Leif Kramp
ZeMKI, Centre for Media, Communication and Information Research
Universität Bremen
Bremen, Germany

Juliane Jarke
ZeMKI, Centre for Media, Communication and Information Research
University of Bremen
Bremen, Germany

ifib, Institute for Information Management Bremen
University of Bremen
Bremen, Germany

ISSN 2730-9320 ISSN 2730-9339 (electronic)
Transforming Communications – Studies in Cross-Media Research
ISBN 978-3-030-96179-4 ISBN 978-3-030-96180-0 (eBook)
https://doi.org/10.1007/978-3-030-96180-0

© The Editor(s) (if applicable) and The Author(s) 2022. This book is an open access publication.

Open Access This book is licensed under the terms of the Creative Commons Attribution 4.0 International License (http://creativecommons.org/licenses/by/4.0/), which permits use, sharing, adaptation, distribution and reproduction in any medium or format, as long as you give appropriate credit to the original author(s) and the source, provide a link to the Creative Commons licence and indicate if changes were made.

The images or other third party material in this book are included in the book's Creative Commons licence, unless indicated otherwise in a credit line to the material. If material is not included in the book's Creative Commons licence and your intended use is not permitted by statutory regulation or exceeds the permitted use, you will need to obtain permission directly from the copyright holder.

The use of general descriptive names, registered names, trademarks, service marks, etc. in this publication does not imply, even in the absence of a specific statement, that such names are exempt from the relevant protective laws and regulations and therefore free for general use.

The publisher, the authors and the editors are safe to assume that the advice and information in this book are believed to be true and accurate at the date of publication. Neither the publisher nor the authors or the editors give a warranty, expressed or implied, with respect to the material contained herein or for any errors or omissions that may have been made. The publisher remains neutral with regard to jurisdictional claims in published maps and institutional affiliations.

Cover illustration: Beate C. Koehler

This Palgrave Macmillan imprint is published by the registered company Springer Nature Switzerland AG.

The registered company address is: Gewerbestrasse 11, 6330 Cham, Switzerland

Contents

Notes on Contributors

Katrin Amelang is a post-doctoral researcher and lecturer in the Department of Anthropology and Cultural Research at the University of Bremen, Germany. Her research is inspired by and situated at the intersection of cultural anthropology and (feminist) science and technology studies, particularly in the area of health and medicine. She has received her doctoral degree from the Humboldt University (HU) of Berlin for an ethnographic study about the production of everyday life and normality after organ transplantation. In her current work, she grapples with cultural processes of digital transformation, especially related to algorithms and data as well as body-technology relations.

Jo Bates is a senior lecturer at the University of Sheffield, UK. Her main research interests are around the socio-material dynamics that influence the production and use of data, data cultures and practices, and algorithmic bias and transparency.

Nancy Baym is a senior principal research manager at Microsoft Research. Her research concerns social dynamics of new communication technologies in personal and professional relationships. Her books include *Twitter: A Biography* (co-authored with Jean Burgess, 2020), *Playing to the Crowd: Musicians, Audiences, and the Intimate Work of Connection* (2018), *Personal Connections in the Digital Age* (2010, Second Edition 2014), *Internet Inquiry: Conversations About Method* (co-edited with Annette Markham, 2010), and *Tune In, Log On: Soaps, Fandom and Online Community* (2000).

Göran Bolin is Professor of Media and Communication Studies at Södertörn University, Sweden. His research focuses on datafication, commodification, and cultural production and consumption in digital markets. His most recent research is collected in *Value and the Media: Cultural Production and Consumption in Digital Markets* (2011), in the edited volume *Cultural Technologies: The Shaping of Culture in Media and Society* (2012), and *Media Generations: Experience, Identity and Mediatised Social Change* (2016). He is a member of the Executive Board of European Communication Research and Education Association (ECREA) and Chair of the section Film, Media and Visual Studies in *Academia Europaea.*

Jonathan Bonneau is a data scientist in research-creations since 2015, teaches in the Communication Faculty at the University of Quebec, Montreal, Canada, and coordinates research and events for the Research Laboratory on Social Media and Gamification.

Andreas Breiter is Full Professor of Informatics at the University of Bremen, Germany, member of the Centre for Media, Communication and Information Research (ZeMKI), and scientific director of the Institute for Information Management Bremen (ifib). His interdisciplinary research focuses on management of educational technologies, digital literacies, and learning analytics. In these fields, his work has been published in international and national journals, as well as three books on the mediatization of German education (2010, 2013, and 2015).

Alessandro Checco is an academic at the University of Roma La Sapienza, Rome, Italy. His main research interests are crowdsourcing, human computation, distributed private recommender systems, information retrieval, data privacy, and algorithmic bias.

Donna Cormack is from Kāi Tahu and Kāti Mamoe. She is a researcher and teacher with joint positions at Te Kupenga Hauora Māori, University of Auckland, and Te Rōpū Rangahau Hauora a Eru Pōmare, University of Otago, Aotearoa, New Zealand. Cormack's work focuses on the impacts of racism and colonialism on Māori health, Māori data sovereignty, and critical, decolonial research practices.

Lina Dencik is a reader in the School of Journalism, Media and Culture at Cardiff University, UK, and co-founder of the Data Justice Lab. She has written widely on digital media, resistance, and the politics of data and is the Principal Investigator of the DATAJUSTICE project funded by an European Research Council (ERC) Starting Grant. Her most recent publications include *Digital Citizenship in a Datafied Society* (with Arne Hintz and Karin Wahl-Jorgensen, 2018) and *The Media Manifesto* (with Natalie Fenton, Des Freedman, and Justin Schlosberg, 2020).

Claude Draude is Professor of Computer Science and one of the directors of the Research Center for Information System Design (ITeG) at the University of Kassel, Germany. She works at the intersection of gender studies and computer science. Her research covers exploring artistic research and critical design approaches to enhance diversity, participation, and inclusion in Information and Communication Technologies (ICT) and Human-Computer Interaction (HCI); employing social inequality research to explore the role of social bias in algorithmic systems; and developing frameworks, methodologies, models, and technological prototypes for critical, reflective computing. She has been serving as scientific expert for the Third Equal Opportunity Report of the German Federal Government on the topic of digitalization.

Elli Gerakopoulou is a researcher at the University of Sheffield Information School, UK, and in the Department of Organisation, Work and Technology at Lancaster University, UK. She is working in the Community-Led Open Publication Infrastructures for Monographs (COPIM) project focused on developing new consortial library funding models for open access (OA) book publishing. Her main research interests are around critical data studies, the future of work, big data, and society.

Lyndsay Grant is Lecturer in Education and Digital Technologies in the School of Education at the University of Bristol, UK. Drawing on critical data studies, science and technology studies, and socio-material approaches, her research critically examines how digital data practices reshape educational practice and policy. Her research explores the prediction and performance of educational futures through data and new critical and playful approaches to data literacies.

Laurence Grondin-Robillard is a doctorate student at the University of Quebec, Montreal, Canada, who specializes in social media communication and state-sponsored interferences in the public sphere during elections process.

Andreas Hepp is Professor of Media and Communications and Head of the Centre for Media, Communication and Information Research (ZeMKI) at the University of Bremen, Germany. He was a visiting researcher and professor at leading institutions such as the London School of Economics and Political

Science, Goldsmiths University of London, Université Paris II Panthéon ASSAS, and Stanford University. He is the author of 12 monographs including *The Mediated Construction of Reality* (with Nick Couldry, 2017), *Transcultural Communication* (2015), and *Cultures of Mediatization* (2013). His latest book is *Deep Mediatization* (2020).

Jenni Hokka is a post-doctoral researcher at the Aalto University, Finland. In her current project she investigates design processes of wearable technology and scrutinizes how datafication changes creative work. Previously, she has focused on public service media, including both drama and journalism, and analysed how the increased significance of social media platforms changes the conditions and methods of public service. In her other projects, she has examined platform politics and circulation of racism on social media and studied media coverage of religion. Her research interests include affectivity, platformization, and changes of cultural production in a data-driven society. Hokka holds a PhD in Media Studies and an MA in General History, both from Tampere University. She has written several articles on social media and television in journals such as *New Media & Society*, *European Journal of Communication*, and *International Journal of Digital Television*.

Gerrit Hornung is Full Professor of Public Law, IT Law, and Environmental Law at the University of Kassel, Germany, where he is also one of the directors of the university's interdisciplinary Research Centre for Information System Design (ITeG). His research interests cover legal issues of data protection, IT security, electronic government, and new surveillance technologies. Interdisciplinary research projects focus on legal criteria for IT design. He has written books on the legal problems of biometric ID cards and patient data cards and on fundamental rights innovations. He is also the co-editor of a comprehensive commentary on the General Data Protection Regulation (GDPR) and a handbook on IT security law.

Juliane Jarke is a senior researcher at the Institute for Information Management Bremen (ifib) and Centre for Media, Communication and Information Research (ZeMKI) at the University of Bremen, Germany. Prior to Bremen, she worked as a research associate at the Centre for the Study of Technology and Organisation (CSTO), Lancaster University. Her research focuses on public sector innovation, digital (in)equalities, user-centric design, and civic engagement. Jarke has co-edited *The*

Datafication of Education (with Andreas Breiter, 2019) and *Probes as Participatory Design Practice* (with Susanne Maaß, 2018). Her most recent book is *Co-creating Digital Public Services for an Ageing Society* (2020).

Rhianne Jones is research lead at BBC Research & Development. She oversees a programme of research that deals with critical questions arising from the increasing use of data in the media industry. She works with university and industry partners to conduct timely research that can inform technical and policy developments at the BBC.

Sigrid Kannengießer is Professor of Media and Communication Studies with a focus on media society at Centre for Media, Communication and Information Research (ZeMKI), University of Bremen, Germany. Her research focuses on media and sustainability; materiality of media technologies; digital, transcultural, and political communication; media ethics; and gender media studies. Her research is published in different journals such as *New Media & Society*, *Communication Theory*, *Convergence: The International Journal of Research into New Media Technology*, and *Feminist Media Studies.*

Helen Kennedy is Professor of Digital Society at the University of Sheffield, UK. Over 20+ years, she has researched how digital developments are experienced by citizens/publics/"ordinary people" and how these experiences can inform the work of digital media practitioners. She is interested in the datafication of everyday life. She is researching public attitudes to data mining and related issues such as trust in data, data and inequality, and what "good" data practice might look like. She is also interested in the role of visual representations of data in everyday life.

Heiko Kirschner was a member of the Centre for Media, Communication and Information Research (ZeMKI) Lab "Mediatization and Globalization" as a research assistant of the DFG-funded project "Pioneer Communities: The Quantified Self and Maker Movements as Collective Actors of Deep Mediatization". Previously, he was research assistant at the Human-Drone Interaction Lab, University of Southern Denmark (SDU), and researcher in the DFG-project "Mediatisierung als Geschäftsmodell III" (Priority Program 1505 Mediatized Worlds) at the University of Vienna. After completing his Master's Degree in Social Science Innovation Research at the Technical University (TU) Dortmund, he was a research assistant at the Chair of General Sociology at the TU Dortmund and assistant in the DFG-funded project "Scopic Media" (Priority Program 1505 Mediatized Worlds) at the University of Konstanz.

Goda Klumbytė is a research associate and PhD candidate at the Gender/Diversity in Informatics Systems group, University of Kassel, Germany. Her research engages feminist theory, science and technology studies, and critical computing, with focus on knowledge production in and through machine learning systems and critical epistemologies as tools for intervention. Her publications include co-authored articles on critical theory and computational practices in "Proceedings of the ACM nordiCHI conference 2020", and journals *Digital Creativity* and *Online Information Review*. She is also a co-editor of *More Posthuman Glossary* (2021) on key terms for contemporary posthumanist research and theory.

Leif Kramp is a post-doctoral media, communication, and history scholar and research coordinator of the Centre for Media, Communication and Information Research (ZeMKI) at the University of Bremen, Germany. He authored and edited various books and studies about the transformation of media and journalism. He is a founding member of the German Association of Media and Journalism Criticism (VfMJ/VOCER), a non-profit organization that supports journalists developing innovative and sustainable projects. Kramp has served as a jury member for the German Initiative News Enlightenment (INA) since 2011, for the #Netzwende Award since 2017, and as member of the nominating committee of the Grimme Online Awards since 2018.

Tahu Kukutai belongs to the Ngāti Tiipa, Ngāti Kinohaku, and Te Aupōuri tribes and is Professor of Demography at the National Institute of Demographic and Economic Analysis, University of Waikato, New Zealand. Kukutai specializes in Māori and Indigenous population research and has written extensively on issues of data sovereignty, Māori population change, official statistics, and ethnic-racial classification. She is a founding member of the Māori Data Sovereignty Network Te Mana Raraunga and the Global Indigenous Data Alliance. Kukutai co-edited the open access books *Indigenous Data Sovereignty: Toward an Agenda* (2016) and *Indigenous Data Sovereignty and Policy* (2020). She was previously a journalist.

Elena Maris is an assistant professor in the Department of Communication at the University of Illinois, Chicago, USA. Her research is concerned with the interactions and tensions between media/tech industries and users, particularly their uses of technology to understand and influence one another. She also studies how people experience and leverage their identities through their experiences with popular culture, technology, and the internet. Her work has been published in journals like *New Media & Society*, *Critical Studies in Media Communication*, *the European Journal of Cultural Studies*, and *Feminist Media Studies.*

Marc Ménard is an economist studying the commercial circulation of personal data. He is a professor at the Media School, University of Quebec, Montreal, Canada.

Stefania Milan (stefaniamilan.net) is Associate Professor of New Media and Digital Culture in the Department of Media Studies at the University of Amsterdam, the Netherlands. Her work explores the interplay between data infrastructure, political participation, and governance.

André Mondoux is a sociologist whose research is oriented towards societal (re)production and digital technologies. He is a professor at the Media School, University of Quebec, Montreal, Canada.

Iris Muis is coordinator of the Data Ethics Decision Aid (DEDA) at Utrecht Data School, the Netherlands. Utrecht Data School is a research and teaching platform within the Humanities Faculty at Utrecht University, the Netherlands. Her research focuses mainly on qualitatively studying the implementation of data ethics within Dutch government institutions. She has a background in law and international relations.

Haripriya Narasimhan is an associate professor in the Department of Liberal Arts at Indian Institute of Technology (IIT) Hyderabad, India. She is an anthropologist with research interests in three areas: medical anthropology, anthropology of the media, and, more recently, sustainability. She has worked extensively on manifestations of globalization processes in education and employment in India, and on caste. She is working on two research projects—an ethnographic study of the production and viewing of Hindi television soap operas in India, and on diabetes among the urban poor and middle classes in Chennai, South India.

Jack Linchuan Qiu is a professor and research director in the Department of Communications and New Media at National University of Singapore, Singapore. His research focuses on ICTs, digital labour, social class, activism, and globalization in the contexts of China and Asia. He is a co-editor of *Reimagining Utopias: Theory and Method for Educational Research in Post-Socialist Contexts* (2017). He has written more than 100 research articles and chapters and 10 books in English and Chinese including *Goodbye iSlave: A Manifesto for Digital Abolition* (2016), *World Factory in the Information Age* (2013), and *Working-Class Network Society* (2009). A recipient of the C. Edwin Baker Award for the Advancement of Scholarship on Media, Markets and Democracy, Qiu is the President of the Chinese Communication Association (CCA1.org).

Nimmi Rangaswamy is an associate professor at Kohli Centre on Intelligent Systems, International Institute of Information Technology (IIIT), Hyderabad, India. She brings an anthropological lens in understanding impacts of Artificial Intelligence (AI) research and praxis. She is also an adjunct professor at Indian Institute of Technology (IIT), Hyderabad. At both institutes, she teaches courses on human computer interaction (HCI) and science, technology, and society studies (STS). Her research approaches AI 'in the everyday' intersections of computational and social life. Using ethnographic methods, she studies human-machine interface to demystify the role and impact of digital technologies in ordinary life.

Mirko Tobias Schäfer is Associate Professor of Media and Culture Studies. His research focuses on the socio-political impact of media technology. He is the co-founder and project lead of the Utrecht Data School. He is the co-editor and co-author of *Digital Material: Tracing New Media in Everyday Life and Technology* (2009) and author of *Bastard Culture!: How User Participation Transforms Cultural Production* (2011), which was listed as best-seller in the computer science section of *The Library Journal*. His most recent publication is the edited volume (together with Karin van Es) *The Datafied Society: Studying Culture Through Data* (2017).

Anne Schmitz is a research associate at the Centre for Media, Communication and Information Research (ZeMKI), University of Bremen, Germany, in the project "Pioneer Communities: The Quantified Self und Maker Movements as Collective Actors of Deep Mediatization" funded by the German Research Association. She is part of the lab "Mediatization and Globalization" and the joint research network "Communicative Figurations" of the Universities of Bremen and the Leibniz Institute for Media Research Hans Bredow Institute. She previously completed her Master's Degree in Journalism and Communication Studies from the University of Hamburg and the University of Aarhus, Denmark.

Lotje Siffels is a PhD candidate of Practical Philosophy at Radboud University, the Netherlands. She has a research interest in the philosophy of technology, science and technology studies, and critical data studies. She is working on the Digital Good project, investigating the increasing involvement of big tech-corporations in the health domain. Before that she was a researcher investigating data ethics and the digitisation of local government at the Utrecht Data School. She has a background in history and philosophy.

Robin Steedman is Post-doctoral Researcher in Creative and Cultural Industries in Africa at Copenhagen Business School, Denmark. She is interested in global creative and cultural industries, and in questions of diversity and inequality in media production, distribution, and viewership.

David van den Berg is a researcher and teacher at the Utrecht Data School, the Netherlands, working on data ethics in government institutions with Data Ethics Decision Aid (DEDA), as well as investigating the effects of datafication and digital transformation on public servants in the research project *de Datawerkplaats*. He has a background in social studies as well as in applied ethics.

Irina Zakharova is a doctoral researcher at the Centre for Media, Communication and Information Research (ZeMKI), University of Bremen, Germany, and a research associate at the Institute for Information Management Bremen (ifib), Germany. Her doctoral thesis seeks to examine which kinds of methods are currently used to explore datafication processes and which kinds of knowledge about datafication do they help to produce in social science and in media studies.

List of Figures

The Value Dynamics of Data Capitalism: Cultural Production and Consumption in a Datafied World

Affinity Spaces as an Analytical Lens for Attending to Temporality in Critical Data Studies: The Case of COVID-19-Related, Educational Twitter Communication

Worker Perspectives on Designs for a Crowdwork Co-operative

List of Images

Public Values and Technological Change: Mapping how Municipalities Grapple with Data Ethics

List of Tables

New Perspectives in Critical Data Studies: The Ambivalences of Data Power—An Introduction

Andreas Hepp, Juliane Jarke, and Leif Kramp

Introduction

We live in a time of increasing global, national, and local insecurity. Despite the promises of an ever more connected world enabled through digital platforms and infrastructures, conflict zones are spreading, displacing millions of people that feel forgotten and disregarded by the rest of the world. Despite an increasing amount of concrete data about the causes and

A. Hepp (✉) • L. Kramp
ZeMKI, Centre for Media, Communication and Information Research, University of Bremen, Bremen, Germany
e-mail: andreas.hepp@uni-bremen.de; kramp@uni-bremen.de

J. Jarke
ZeMKI, Centre for Media, Communication and Information Research, University of Bremen, Bremen, Germany

ifib, Institute for Information Management Bremen, University of Bremen, Bremen, Germany
e-mail: jarke@uni-bremen.de

© The Author(s) 2022

A. Hepp et al. (eds.), *New Perspectives in Critical Data Studies*, Transforming Communications – Studies in Cross-Media Research, https://doi.org/10.1007/978-3-030-96180-0_1

consequences of climate change, policy actions have become less reliable, and the political will seems even less convincing. New analytic technologies promise a world in which practices become more personalised, yet the social world experiences newly formed inequalities, increasing the insecurity of different social actors. The idea of openness in the form of open government and open science is spreading globally, promising increased transparency, accountability, and participation, yet we see an unequal distribution of data ownership, published data sets, and civil society actors that actually engage with these data.

With increasingly globalised digital infrastructures and a global digital political economy, we face new concentrations of power, leading to new inequalities and insecurities with respect to data ownership, data geographies, and different forms of data-related practices. It is not only a concentration of power by a few corporations, but also a concentration of the availability in data on individual regions of the world. This includes (exerting) power over data (infra)structures and the processes of data creation, data collection, data access, data processing, data interpretation, data storing, and data visualisations.

Yet, data power is a highly ambivalent phenomenon. On the one hand—and this explains its "appeal"—digital data produces knowledge about society and social processes. For example, it is widely believed that digital data on urban mobility, energy consumption, and online shopping can help uncover patterns in human practice in order to make social processes "more sustainable", "more efficient", and "more reasonable". With such strident positivity, it is not only the utopia of the "Californian ideology" (Barbrook & Cameron, 1996, p. 44; Turner, 2006, p. 25) that resonates, according to which the use of digital technologies will "inevitably" (Kelly, 2016, p. 1) lead to a "better" life for all, digital data will allow for new ways of structuring social processes on the basis of self-organisation. On the other hand, we increasingly see the problems thrown up by digital data: It is used for surveillance (Andrejevic & Gates, 2014), on its basis new forms of capitalism become reality which are much more closely interwoven with everyday practices than earlier forms or stages of capitalism (Couldry & Mejias, 2019b; Zuboff, 2019), and inequalities and everyday racisms are reproduced in digital data (Eubanks, 2017; Noble, 2018), to name just a few of the most important points of discussion. The ambivalence of digital data can hardly be resolved, which is why we want to bring an argument to the fore with this book: It is crucial for critical data studies scholars and practitioners to address precisely such ambivalences if they want to develop new perspectives for their own research and a credible critique of data power.

In this edited volume, authors attend to these ambivalences in three areas and in so doing provide new perspectives in and for critical data studies: First, the *ambivalences between global infrastructures and local invisibilities*. These contributions challenge the grand narrative of the ephemeral nature of a global data infrastructure and instead make visible the local working and living conditions, resources, and arrangements required to operate and run them. Second is the *ambivalences between the state and data justice*. These contributions consider data justice vis-à-vis state surveillance and data capitalism and reflect the incongruities between an "entrepreneurial state" and a "welfare state". Third is the *ambivalences of everyday practices and collective action*, in which civil society groups, communities, and movements try to position the interests of people against the "big players" in the tech industry. It is such ambivalences from which the contributions in this volume develop future perspectives for critical data studies. With this introduction, we want to make this argument of seeing data power in terms of its irreducible ambivalences in a pointed way to provide an orientation to the chapters of this book. To this end, we first give a brief outline of the development of critical data studies. As part of this outline we also want to situate the series of data power conferences, the most recent of which this volume is based on. This will then serve as a basis for taking a closer look at three areas of data power's ambivalences.

Critical Data Studies as a Field: From Big Data to the Complexity of Digital Data and Data Infrastructures

The "transdisciplinary field" (Burns et al., 2019, p. 657) now called critical data studies has its origins in various disciplines of research on digital media and related infrastructures. In an incomplete list these include, among others, geography, media and communication studies, political science, science and technology studies, and sociology. The starting point for the emergence of critical data studies was the discussion about "big data": It was danah boyd and Kate Crawford (2012, p. 661) who raised critical questions against the increasing spread of sociotechnical imaginaries related to big data—imaginaries associated with a new "capacity to search, aggregate, and cross-reference large data sets". In their seminal article they drew attention to how big data is changing our understanding of knowledge, how misleading understandings of objectivity might be spread by big data, the quality isues big data can have, and why neglecting the context surrounding such data can lead to critical consequences. A broad discussion around these questions began to emerge across the disciplines mentioned above. This involved epistemological questions (Crawford et al., 2014), questions of the corporate interests of data production (Couldry & Turow, 2014), the forms of governance that filter through such data (Elmer, Langlois, & Redden, 2015b), and the role of infrastructures in generating these data (Mosco, 2014; Kitchin, 2014), as well as the myths that circulate around the subject of big data (Puschmann & Burgess, 2014). In essence, this discussion can be summed up as a critique of the implicit assumptions around big data in parts of the economy (i.e. Anderson, 2008): By contrast to what is said in public discourse and the economy, big data are not simply the "new oil" that just has to be extracted, it is neither "raw data" (Gitelman & Jackson, 2013, p. 1) nor in any other way just given. Rather, such data are always "cooked" (Bowker, 2005, p. xx), meaning that data processing always takes place through certain power structures (Beer, 2016).

In this wider context, it was Craig Dalton and Jim Thatcher (2014) who coined the term "critical data studies" in their online article "What Does a Critical Data Studies Look Like, and Why Do We Care?". Their aim was to make clear that technology is never something "neutral" and is the reason for their calling to avoid anything even approaching "technological determinism" and developing a critical attitude toward expectations around big data. The term "critical data studies" was quickly taken up in the same year by Rob Kitchin and Tracey Lauriault (2014), among others, who added important arguments to the original proclamatory call for a critical perspective on big data. In particular, they asked for an enhanced theoretical anchoring of the field of critical data studies: "Rather than produce an extensive list of questions, we want to conclude by calling for greater conceptual work and empirical research to underpin and flesh out critical data studies" (Kitchin & Lauriault, 2014, p. 14).

Looking at the discussion with the benefit of hindsight, there were two concepts in particular that were important for the theoretical foundation of critical data studies: Coming more sharply from geography as well as science and technology studies, the concept of *data assemblage*; and coming, again, more forthrightly from media and communication studies as well as sociology, that of *datafication*. It is worth taking a look at both concepts here to understand today's broad theoretical anchoring of critical data studies.

As an analytical term, assemblage was introduced by Gilles Deleuze and Felix Guattari to describe "complexes of lines" that build a "territoriality" (Deleuze & Guattari, 2004, p. 587). Assemblages in this sense are "wholes" characterised by the relations of exteriority. In terms that are closer at home in the social sciences, "social assemblage" refers to a "set of human bodies properly oriented (physically or psychologically) towards each other" (DeLanda, 2006, p. 12). As a concept, assemblage is widely used in actor-network theory, ultimately to capture the coming together of people and things in actor networks (i.e. Latour, 2007, pp. 16–17). It is against this broader context that the idea of ***data assemblage*** must be seen, a term that Kitchin (2014, pp. 24–26) in particular brought to the discussion and further developed with Tracey Lauriault. In essence, a "data assemblage" is defined as encompassing "technological, political, social and economic apparatuses and elements that constitutes and frames the generation, circulation and deployment of data" (Kitchin & Lauriault, 2014, p. 1). These include systems of thought, forms of knowledge, finance, political economy, governmentalities, and legalities, materialities and infrastructures, practices, organisations and institutions, subjectivities and communities, and (market-)places.

While data assemblage is a concept for describing certain sociotechnical relationships around data, ***datafication*** has not only a different origin, but also a different objective: It is about examining the processes associated with the rise and permeation of big data (logics). Critically reflecting on Mayer-Schönberger and Cukier's (2013) original arguments, José van Dijck (2014, p. 198) defined datafication as "the transformation of social action into online quantified data, thus allowing for real-time tracking and predictive analysis". This quote resonates with the double character of datafication's processuality. On the one hand, it is about the situated process of transformation, that is, about "translations" that take place when social processes are represented in data. These are complex, interest-driven processes that cannot be described simply as "digital reproduction" but, rather, as the sociomaterial construction of "data doubles" (Haggerty & Ericson, 2000; Ruppert, 2011) which must be understood as interest-driven technical articulations and not as 1:1 representations of people and their practices. Data do not provide a window on the social world and represent independently existing phenomena, the relationship with the social world they are meant to represent is recursive (see, e.g. Jarke &

Breiter, 2019). This recursivity may produce "new" and reproduce "old" inequalities or surveillance regimes but may also afford greater transparency and participation (Eubanks, 2017; Noble, 2018; D'Ignazio & Klein, 2020). On the other hand, then, datafication is about the transformation of society, how society changes when "online quantified data" become increasingly widespread (i.e. Iliadis, 2018, p. 219; Sadowski, 2019, p. 2). At this point, there exists a close connection with the discussion into the "deep mediatization" (Couldry & Hepp, 2017; Hepp, 2020) of society, an approach that critically describes the transformation of society with the increasing saturation by digital media and their infrastructures.

Anchored by both concepts—data assemblage and datafication—critical data studies is much more than just a reflection of the discourse around big data. Ultimately, critical data studies is concerned with the significance (and power) of digital data in contemporary society and how it relates to societal transformation. We can see this field of research as a response to the increasing spread of digital data and data infrastructures for decision- and meaning-making in various social domains. The fledgling history of critical data studies, then, is one of a broadening view—from big data in particular to digital data in general—and with it, the development of a sensitivity to the complexities and invisibilities of the sociomaterial figurations that operate global data infrastructures. This can be seen as the connecting line between the various current definitions of the field. Craig Dalton, Linnet Taylor, and Jim Thatcher argue that critical data studies "calls attention to subject formation within [...] data regimes, for a critical examination of where the interpellation of the individual emerges in algorithmic culture" (Dalton et al., 2016, p. 1). Annika Richterich (2018, p. 2) points out that research in critical data studies "deals with the societal embeddedness and constructedness of data", while Andrew Iliadis and Federica Russo (2016, p. 2) argue that critical data studies helps "define the questions that inform epistemological frameworks around social issues related to data" and are a "formal attempt at naming the types of research that interrogate all forms of potentially depoliticized data science and to track the ways in which data are generated, curated, and how they permeate and exert power on all manner of forms of life".

Ever since critical data studies emerged as a transdisciplinary field, the methodological reflection on *how* to critically examine data power was key. Stemming from its various disciplinary roots, critical data studies has by now developed and appropriated a rich body of methods for researching and challenging data power. Precisely because of their critical orientation, critical data studies have from the beginning opposed the naïve positivist methodology of many, especially commercial, data analyses (Couldry et al., 2016; Iliadis & Russo, 2016, p. 1). The idea was to set against such positivism a reflection of the epistemology behind it and a detailed description of people's data practices. In this sense, critical data studies called "for ethnographic and discursive work, for the thick description of data and the cultures around it, just as much as it relies on algorithmic analysis" (Dalton et al., 2016, p. 7). However, we would shorten the methodological discussion if we equated critical data studies with a particularly qualitative approach that positions itself "against" the quantifying idea of much "social analytics". At this point, it is well worth revisiting the original statement by Craig Dalton and Jim Thatcher (2014), because they already provided some insightful thoughts. In regard to the field of critical data studies, they argued for mixed methods approaches in which "big" and "small" data are "utilised in concert" (Dalton & Thatcher, 2014, p. 6). Even in this early reflection, it is not a question of positioning different methods against each other (Hepp et al., 2021) but, rather, of reflecting in an integrative way on which methods can contribute to a better, critical understanding of the construction of sociality by means of digital data. In addition to traditional qualitative research sensitivities, the roots of critical data studies in geography mean that many scholars brought their profound experience in analysing data relation to space along with critical approaches to spatial analysis such as counter-mapping (Dalton et al., 2016). Unsurprisingly, counter-mapping is also one of the examples that D'Ignazio and Klein (2020) provide in their book *Data Feminism*. Here, counter-mapping makes the *lack* of data on certain phenomena and groups of people visible and in so doing challenges dominant socio-political discourses.

Such a broad methodological orientation is also associated with adoption of "digital methods" (Rogers, 2013) and "computational social science" (Lazer et al., 2009) in critical data studies. Increasingly, it is a question of researching not only the social situatedness of data and data processing by means of qualitative methods, but also "digital traces" (Hedman et al., 2013) and digital data themselves. In doing so, critical data studies started to focus on the software and code, which is why so-called software studies (Fuller, 2008, p. 1) began to play an important role. Here a growing body of work builds on studies that explore how "software and code connect people, things, systems, places and events in a pervasive and sinuous fabric" (Mackenzie, 2013, p. 392). Scholars investigate "digital code and software from a wide range of perspectives—power, subjectivity, governmentality, urban life, surveillance and control, biopolitics or neoliberal capitalism" (Mackenzie & Vurdubakis, 2011, p. 3). The characteristic of critical data studies, however, remained one of scepticism and reflexivity towards a naïve implementation of the computational turn within social sciences research—their attitude remains one of "tool criticism" (van Es et al., 2021, p. 46) against the digital tools we use for research. In doing so, many of the arguments in favour of "putting digital traces in context" (Breiter & Hepp, 2018, p. 387) were anticipated: The aim was not to see digital traces and data generated online beyond or outside their contexts, but to triangulate methods in a critical analysis in such a way that the social bondage of the digital becomes accessible. Among all of these concerns, critical data studies have always had a connection to what is called "action research" (Bradbury-Huang, 2010; Wagemans & Witschge, 2019): Their proponents have not been completely outside the domains of their research, but have always been involved with people affected by digital data collection and processing. This is the point where methodological reflections are important in regard to how to communicate critical research back to the actors within the field of data science, or even—as Gina Neff et al. (2017) suggest—to integrate them into joint research.

The *Data Power Conferences*—the last of which this volume is based on—and the publications associated with them were fundamental to the emergence of critical data studies outlined so far. In contrast to conferences funded by big tech, the first Data Power Conference[1] in 2015 provided a physical space for the emerging interdisciplinary community to meet and critically discuss questions about the kinds of power that are "enacted when data are employed by governments and security agencies to monitor populations or by private corporations to accumulate knowledge about consumers".[2] They observed that emerging forms of data mining and data analytics allowed for "new, unaccountable and opaque forms of population management in a growing range of social realms" and argued that this required critical scholars to investigate data power in relation to control, discrimination, and social sorting. The conference resulted

[1] The first Data Power Conference was by Helen Kennedy, Jo Bates, and Ysabell Gerrard and hosted at the University of Sheffield (UK).

[2] The way the organisers described the conference can be seen at: http://datapowerconference.org/data-power-2015/about/ (accessed: 31.3.3021).

in a special issue of *Television and New Media* on *Data Power in Material Contexts* (Kennedy & Bates, 2017) that brought together media and communications scholarship concerned with datafication. The special issue featured five empirical studies that "ground the study of data power in specific, material contexts" and contributed to the overall aim of the first conference of bringing "together papers which analyze the operations of data power across a range of real-world domains" (ibid., p. 702). The research of the material contexts and everyday practices would allow for the questioning of social justice in a datafied world, data studies' "next phase", as the authors argued.

And indeed, the second Data Power Conference[3] in 2017 moved from a (stock-taking) analysis of increasing data power to questions about how agency and autonomy may be reclaimed in regimes of data power, how data may be mobilised for the common good.[4] The conference resulted in two special issues (Gerrard & Bates, 2019; Lauriault & Lim, 2019). Gerrard and Bates' collection attended to *tactics* for the opposition of data power (Lee, 2019; Currie et al., 2019), *access* to public data infrastructures for often marginalised social groups (Jarke, 2019; Scassa, 2019), and the social *shaping*, or *moulding*, of data (infrastructures) (Andrews, 2019; Iliadis, 2019; Mitchell, 2019). Lauriault and Lim's special issue followed the conference theme and focused on "the social and cultural consequences of data becoming increasingly pervasive in our lives" (Lauriault & Lim, 2019, p. 315), in particular on the "implications, biases, risks, and inequalities, as well as the counter-potential, of data practices and systems in various contexts" (ibid., p. 316).

In 2019, the third Data Power Conference took place at the University of Bremen.[5] The thematic focus of the conferences shifted again and considered the "global in/securities" of an ever-increasing data power.

[3] The second Data Power Conference was organised by Tracey P. Lauriault and Merlyna Lim in Ottawa, at Carleton University (Canada) in 2017 in collaboration with the previous organisers, Ganaele Langlois, Scott Dobson-Mitchell, and Jessi Ring. http://datapowerconference.org/data-power-2017/about/ (accessed: 31.3.3021).

[4] Again, see the conference website for this: http://datapowerconference.org/data-power-2017/about/ (accessed: 31.3.3021).

[5] The third Data Power Conference was organised by the editors of this volume at the ZeMKI, University of Bremen, in collaboration with Andreas Breiter, Monika Halkort, and the organisers from the conferences at Sheffield (Kennedy, Bates, Gerrard) and Ottawa (Lauriault, Lim). For more information, see http://datapowerconference.org/data-power-2019/about/ (accessed: 31.3.3021).

Dealing with in/securities focus on the above-mentioned ambivalences of data power: On the one hand, the availability of data seems to open up new securities, not only for companies and state authorities, but also for individuals. The desire for such social security through digital data and the associated phantasies and myths of accessibility, knowledge, and controllability was made apparent by the COVID-19 pandemic in 2020 and 2021. The course of the pandemic was presented to all of us in public discourse through "dashboards" with automatically updated data on the spread of the virus or later the vaccination programs that followed; digital tools such as the various tracking apps or sales platforms were hailed as a great hope in managing the pandemic. Again, with the benefit of hindsight, we can see that these ideas of security were imaginary. On the other hand, therefore, during the pandemic data power was always also associated with insecurities: Who has control over the data? How secure is it? Are ethical expectations regarding the handling of one's own data fulfilled? The scepticism about various forms of data visualisation in data journalism on the pandemic or the scepticism held by many against the various corona tracing apps can be understood as an expression of these insecurities at the individual level.

The joint work on this book made it clear that behind the question of global in/securities lies a larger theme which has now given the present volume its subtitle: *The ambivalences of data power*. Digital data and infrastructures may open up many potentials that can be emancipative; at the same time, however, this data power has many negative elements that should not go unnoticed. To put it succinctly, the main thesis that emerged throughout the conference and in the subsequent discussion with and among the authors is that, if we want to develop new perspectives for critical data studies, it would probably be expedient to realise them starting from the fundamental ambivalences of data power.

Perspectives in Critical Data Studies: The Ambivalences of Data Power

There are three particular areas of data power's ambivalences, which are not always easy to grasp, with which we are currently confronted. These are, first, the ambivalences that exist in the area of global infrastructures and local invisibilities, second, the ambivalences that emerge in the area of the state and data justice, and, third, the ambivalences that take rise in the area of individual everyday practices and collective action. Taking these ambivalences seriously opens up comprehensive perspectives for critical data studies.

The ambivalences in *global infrastructures and local invisibilities* are ultimately already embedded in the departure of big data as a social phenomenon. The kind of digital data we are dealing with today would not exist in its present form without the extensive engagement of the Big Five in Western companies (Amazon, Apple, Facebook, Google, Microsoft) or similar engagement by companies like Alibaba and Baidu in Asian countries (van Dijck et al., 2018, pp. 26–30; Hepp, 2020, pp. 19–30). For years, supported by an "entrepreneurial state" (Mazzucato, 2013)—which is itself interested in digital data for state surveillance (Greenwald, 2014; Lee, 2019)—these and other companies have built a globalised data infrastructure along their vested interests, serving a condition of "surveillance capitalism" (Zuboff, 2019) and "data colonialism" (Couldry & Mejias, 2019a). While these large companies are globally visible as "brands" and highly committed to emphasising their performance in building such a data infrastructure and the possibilities of exploiting this data for the "common good" of humanity (Webster, 2017), the local aspects of these infrastructures are sometimes decidedly invisible (i.e. Parks & Starosielski, 2015; Crawford, 2021). For example, Crawford and Joler's (2018) *Anatomy of an AI System* provides a detailed "anatomical map" of the human labour, data, and planetary resources required for the smooth operation of Amazon's Echo system. In her recent book, Crawford (2021) traces these networks or data assemblages in detail. Others have, likewise, pointed to the invisibility of the many workers in the Global South who operate global data infrastructures (Atanasoski & Vora, 2019; Gray & Suri, 2019; Qiu, 2016) and have argued how this invisibility increases the harms that the data industry inflicts upon them. In addition, scholars have argued the need to consider other invisibilities in the grand, global data infrastructure narrative: Namely, small businesses and initiatives that enable a connection to the globalised infrastructure (Arora, 2019), and the various local communities and forms of data activism (Chenou & Cepeda-Másmela, 2019). We are dealing with ambivalences of enablement through a global infrastructure, on the one hand, and the local invisibility, not only of relevant actors, but also of data power associated with this infrastructure on the other. These ambivalences can only be grasped beyond "data universalism" (Milan & Treré, 2019, pp. 319–322), that is, the assumption that digital data and infrastructures would be structured in an identical way throughout the world and could, therefore, also be recorded scientifically as such. Contributing to these, new perspectives in critical data studies, this volume also comprises stories of invisible labour in (the shadows of) data power.

In Part I of this book, contributors examine a variety of new perspectives on critical data studies that arise from these ambivalences. "Data Power and Counter-Power with Chinese Characteristics" by Jack Linchuan Qiu discusses the ambivalences of data power and counter-power with Chinese characteristics. Starting with China's internal conflicts and its relations with the external world, this chapter argues for a more holistic and historicising approach to critical data studies. Making some of the hidden labour (and counter-power) in the Chinese data power narrative visible, he recounts the dire working conditions in the emerging Chinese digital market which drive suicides among workers (996.ICU, anti-iSlave). This chapter is followed by "Transnational Networks of Influence: The Organisational Elites of the Quantified Self and Maker Movements on Twitter" by Anne Schmitz, Heiko Kirschner, and Andreas Hepp on the pioneer communities of the Maker and Quantified Self movements. Drawing on a Twitter analysis, they are able to show how an organisational elite based in Silicon Valley curates these apparently grassroots movements across countries—and in the course of doing so promotes certain imaginaries of data power, some of which are close to the Californian ideology. In "The Power of Data Science Ontogeny: Thick Data Studies on the Indian IT Skill Tutoring Microcosm", Nimmi Rangaswamy and Haripriya Narasimhan use an ethnographic approach to investigate the "Indian IT skill tutoring microcosm". This chapter emphasises the importance of a "thick" ethnographic description for the future development of critical data studies: Only through such an analysis can the invisibility of various actors in the Global South be overcome in favour of a more differentiated understanding of the globalised conditions of data power. They report from India's growing data science work force and describe the data economy's possibilities for upward mobility. Data science here is understood as enabling and facilitating livelihood. Jonathan Bonneau, Laurence Grondin-Robillard, Marc Ménard and André Mondoux's chapter "Fighting the 'System': A Pilot Project on the Opacity of Algorithms in Political Communication" reflects on the interrelations between algorithmic governmentality, identity, and political speech. The legitimacy of election processes and social media's contribution to the public sphere are now being questioned and it is important to document and analyse these new dynamics of political communication. In particular, they argue for the need to consider the role played by the automation of the production and circulation of political messages through the use of algorithms and artificial intelligence. Their chapter sets out a possible conceptual basis for such research. This first part of our book is concluded with "Indigenous Peoples, Data, and the Coloniality of Surveillance" by Donna Cormack and Tahu Kukutai examinig the relation of indigenous peoples, data, and the coloniality of surveillance. The authors explore the contemporary invisibilities of data colonialism from *within* indigenous frameworks of collective self-determination and collective rights. This includes, for example, resistance to surveillance through envisioning "data relations and data practices that are anti-colonial, relational and collective".

Part II of this book deals with the ambivalences of the *state and data justice*. As we have already seen, state or state agencies are in and of themselves highly ambivalent. It was the "entrepreneurial state" that made today's "surveillance capitalism" and "data colonialism" possible in the first place, and it has a vested state interest in digital data for surveillance purposes that only serve for their advancement. The Snowden affair in particular has shown how deeply involved the state is in current advances of surveillance (Greenwald, 2014; Lyon, 2014). On the other hand, the state also stands for the safeguarding of welfare, the balancing of interests, and public media, which, in the best-case scenario, can be a counterpart to the data power of globally operating technology corporations and a reference point for digital citizenship (Hintz et al., 2019). In such cases, the important questions surround the extent to which the state can secure and promote data justice as we move from the "entrepreneurial state" (in which neoliberal ideologies have a very contradictory position) to the "welfare state" and the new challenges it faces when it comes to data power (Dencik & Kaun, 2020).

Part II opens with "The Datafied Welfare State: A Perspective from the UK" by Lina Dencik which focuses on datafication and the welfare state. She advocates for a two-part argument about the ways in which data infrastructures are transforming state-citizen relations: On the one hand, by advancing an actuarial logic based on personalised risk and the individualisation of social problems (responsibilisation), and, on the other, by entrenching a dependency on an economic model that perpetuates the circulation of data accumulation (rentierism). In "The Value Dynamics of Data Capitalism: Cultural Production and Consumption in a Datafied World", Göran Bolin reflects on the value dynamics in data capitalism. He sees a need for analytical models to understand the ambivalent complexity, scale, and dynamics behind the datafication of social life. In so doing, he offers a perspective that focuses on data as a value and he presents an analytical model to study the dynamics of data capitalism as part of the process of datafication. This is followed by "Mapping Data Justice as a Multidimensional Concept Through Feminist and Legal Perspectives" by Claude Draude, Gerrit Hornung, and Goda Klumbytė on operationalising data justice in information systems. Here the authors contribute to new perspectives in critical data studies by showing that data justice can provide a multidimensional, conceptual ground that serves both the needs of legal formalisation and feminist imperatives of contextualisation and specificity. In chapter "Reconfiguring Education Through Data: How Data Practices Reconfigure Teacher Professionalism and Curriculum", Lyndsay Grant argues that in-depth explorations of how educational data practices work "on the ground" are needed to understand the ambivalences around how data power works in education. Lotje Siffels, David van den Berg, Mirko Tobias Schäfer, and Iris Muis in their chapter "Public Values and Technological Change: Mapping How Municipalities Grapple with Data Ethics" turn their attention to "action research" as discussed in the last section of this introduction, in their case realised in cooperation with public authorities. They developed DEDA, a tool that allows civil servants to critically reflect and engage with the ethical dimensions of a datafied public sector. "Welfare Data Society? Critical Evaluation of the Possibilities of Developing Data Infrastructure Literacy from User Data Workshops to Public Service Media" by Jenni Hokka brings us back to questions of data power and the welfare state, but in this case with a special focus on public service media. Her endeavour is to take part in finding solutions for the ambivalences surrounding datafication as discussed across the chapters in the second part of the book. She presents a study from Finland in which public service media improved the data literacy of citizens and in so doing increased digital equality.

Part III of this edited volume deals with the ambivalences of *everyday practices and collective action*. Datafication has a lot to do with everyday life: It is the quotidian use of digital media through which people leave online traces that then constitute comprehensive data sets that companies draw upon (Elmer, Langlois, & Redden, 2015a; Amoore & Piotukuh, 2016). But it is also everyday practices through which globalised data infrastructures and digital media are appropriated. On the one hand, there is an emancipatory potential here for people in their everyday lives—opportunities exist for individual empowerment through (self-generated) data (e.g. Gerhard & Hepp, 2018; Lupton, 2016; Neff & Nafus, 2016) or collective empowerment through open (government) data (e.g. Milan, 2017; Rajão & Jarke, 2018). On the other hand, these everyday practices are often associated with "resignation" (Draper & Turow, 2019, p. 1824), which results from the fact that the use of digital media is inevitably linked to the fact that companies and state authorities can use the data generated for their own purposes and that users can barely do anything about it. This resignation is not inevitable, however, because forms of "collective action" (Dolata & Schrape, 2015, p. 1) can emerge with reference to one's own everyday practices, which are directed against the hegemonic actors in the field such as "data activism", for example (Milan, 2017; Kennedy, 2018). The third and final part of this book deals with these ambivalences of individual everyday practices and collective action in relation to data power.

Part III opens with a contribution ("(Not) Safe to Use: Insecurities in Everyday Data Practices with Period-Tracking Apps") by Katrin Amelang on insecurities in intimate data practices relating to the everyday use of menstrual cycle apps. The ambivalence of this specific everyday data practice relates first to insecurities deriving from an endeavour to understand menstruating bodies with and through data (such as the trustworthiness of predictions). Second, the protection and privacy of data collected by period tracking apps are often insecure and wide open for third-party use. Amelang discusses the question of what "agential possibilities" datafication offers for people who menstruate from an everyday perspective. In their chapter, "Community Rankings and Affective Discipline: The Case of Fandometrics", Elena Maris and Nancy Baym argue that with platforms' increasing concentration of data power, critical data studies must attend to community-driven models of data and metrics. The fandom metrics phenomenon reflects larger anxieties about value, relevance, and power in increasingly metrified online spaces. Irina Zakharova, Juliane Jarke, and Andreas Breiter in "Affinity Spaces as an Analytical Lens for Attending to Temporality in Critical Data Studies: The Case of COVID-19-Related, Educational Twitter Communication", examine education-related Twitter communication during the COVID-19 pandemic through a hashtag predominantly used by German educators. They propose "affinity spaces" (Gee, 2005) as an analytical lens through which to attend to temporality in the analysis of Twitter communication. By following the changes of the affinity space in time, they are able to identify shifts in topics and actors central to the affinity space (and the associated collective) and trace the practices through which these shifts unfold. Rather than understanding Twitter as a site for content redistribution and stable data assemblage, they follow the dynamics of problematisation in times of crisis by attending to the reconfigurations of an affinity space that allow for collective action.

While such studies analyse the everyday level of datafication and outline perspectives of critical data studies in exploring the complexities involved, the following chapters emphasise the need for a differentiated engagement with collective action in relation to data and datafication. Sigrid Kannengießer, in her chapter "'Party Like It's December 31, 1983': Supporting Data Literacy at CryptoParties" examines how civil society initiatives such as CryptoParties provide revealing insights into how different actors critically reflect on the challenges of datafication and how they try to shape datafication. Through reconstructing the perspective of the actors involved, we not only learn about the challenges of datafication, such as different privacy risks in online communication, but we can also (critically) reflect on solutions that are developed and practised with the aim to create a more "data just" society. Robin Steedman, Helen Kennedy, and Rhianne Jones, in their chapter, "Researching Public Trust in Datafication: Reflections on the Deliberative Citizen Jury as Method" are interested in questions of public trust in data-driven systems through deliberative citizen juries. Through this example, they call for greater reflection into methods in the field of critical data studies. Jo Bates, Alessandro Checco, and Elli Gerakopoulou in, "Worker Perspectives on Designs for a Crowdwork Co-operative" draw attention to the labour of workers from crowdwork platforms and illuminate "structures of labour exploitation that many contemporary AI systems are dependent upon, and ask—with workers—how might these labour conditions be improved". The authors propose the idea of a crowdwork co-operative which would make workers more visible and collectively enforce better working conditions. This last part of the book concludes with "Counting, Debunking, Making, Witnessing, Shielding: What Critical Data Studies Can Learn from Data Activism During the Pandemic" by Stefania Milan on data activism in the post-COVID-19 world. This chapter explores data activism as a counterforce to the predominant state of data power, takes stock of its most recent evolutions, and identifies pathways for critical data studies in a post-pandemic world. It singles out three challenges for data activism in this world, namely the question of infrastructure, the diffusion of data poverty, and the scarcity of digital literacy.

As we have seen, issues of data power are highly ambivalent. They can open up opportunities, but they can also limit others; they are characterised by inequality, exclusion, and even exploitation. At its core, critical data studies is about addressing the ambivalences of data power in order to arrive at a better understanding of the role played by digital data and infrastructures in our societies today. The transdisciplinarity and openness of the field is certainly not a limitation but, rather, a great opportunity: It is precisely in this way that critical data studies can consistently succeed in integrating necessary knowledge from very different disciplines into critical engagement with digital data. In this way, critical data studies provides an extremely important contribution to the current social science discussion on today's transformation of society. We hope that with this volume we will be able to contribute to the continuation of this discussion.

* * *

Neither the Data Power conference on "Global In/Securities" nor the present volume would have been possible without diverse support, for which we would like to express our sincere thanks. Our thanks go first to Helen Kennedy, Jo Bates, and Ysabell Gerrard (the organisers of the first Data power Conference) as well as Tracey Lauriault and Merlyna Lim (the organisers of the second Data Power Conference) who supported the organisation of the third Data Power Conference in Bremen through their valuable experience, vision for the community, and its existing infrastructure. Furthermore, we thank Monika Halkort and Andreas Breiter, who realised the conference in Bremen together with us. Our heartfelt thanks also go to the ZeMKI (Centre for Media, Communication, and Information Research) at the University of Bremen for supporting the conference financially (which made travel grants for participants from the Global South possible) and for covering the open access costs for this volume. A large number of people helped us to organise the conference. We would like to thank Kerstin Biegemann, Matthias Franz, and Gabriele Köhn for their management and IT services, Dirk Vaihinger and Alexander Hillmann from the Centre for Multimedia in Education (ZMML) at the University of Bremen for producing videos about the conference, keynotes and some of the panels and, in particular, the highly motivated student assistants for their indispensable support before, during, and after the conference: Jona Andresen, Giulia Aureli, Arthur Belousov, Enna Gerhard, Helle de Haas, Hendrik Meyer, Paula Muche, Kiko Oorlog, Klara Pechtel, and Enqian Wu. We are also grateful for the help we received from the student assistants Lea Korte and Nicola Peters in the process of editing the book. Special thanks go to Marc Kushin for the careful language editing of the book's chapters and to Mala Sanghera-Warren and Felicity Plester from Palgrave for the careful handling in difficult times with the COVID-19 pandemic. However, this book was only made possible by the authors, who wrote their chapters in situations that were very difficult for some, for which we would also like to thank them very sincerely.

References

Amoore, L., & Piotukuh, V. (Eds.). (2016). *Algorithmic life. Calculative devices in the age of big data*. Routledge.

Anderson, C. (2008). The end of theory: The data deluge makes the scientific method obsolete. *Wired Magazine*, 23.06.08. http://www.uvm.edu/~cmplxsys/wordpress/wp-content/uploads/reading-group/pdfs/2008/anderson2008.pdf

Andrejevic, M., & Gates, K. (2014). Big data surveillance: Introduction. *Surveillance & Society, 12*(2), 185–196.

Andrews, P. (2019). The compliant environment. *Online Information Review, 43*(6), 1063–1079. https://doi.org/10.1108/oir-09-2018-0284

Arora, P. (2019). *The next billion users: Digital life beyond the West*. Harvard University Press.

Atanasoski, N., & Vora, K. (2019). *Surrogate humanity: Race, robots, and the politics of technological futures*. Duke University Press.

Barbrook, R., & Cameron, A. (1996). The Californian ideology. *Science as Culture, 6*(1), 44–72. http://www.comune.torino.it/gioart/big/bigguest/riflessioni/californian_engl.pdf

Beer, D. (2016). *Metric power*. Palgrave.

Bowker, G. C. (2005). *Memory and its 'other': Memory practices in the sciences*. MIT Press.

Boyd, D., & Crawford, K. (2012). Critical questions for big data: Provocations for a cultural, technological, and scholarly phenomenon. *Information, Communication & Society, 15*(5), 662–679.

Bradbury-Huang, H. (2010). What is good action research. *Action Research, 8*(1), 93–109.

Breiter, A., & Hepp, A. (2018). The complexity of datafication: Putting digital traces in context. In A. Hepp, A. Breiter, & U. Hasebrink (Eds.), *Communicative figurations: Transforming communications in times of deep mediatization* (pp. 387–406). Palgrave Macmillan.

Burns, R., Hawkins, B., Hoffmann, A. L., Iliadis, A., & Thatcher, J. (2019). Transdisciplinary approaches to critical data studies. *Proceedings of the Association for Information Science and Technology Computer Science, 55*(1). https://doi.org/10.1002/pra2.2018.14505501074

Chenou, J.-M., & Cepeda-Másmela, C. (2019). #NiUnaMenos: Data activism from the Global South. *Television & New Media, 20*(4), 396–411.

Couldry, N., & Hepp, A. (2017). *The mediated construction of reality.* Polity Press.

Couldry, N., & Mejias, U. (2019a). *The costs of connection. How data is colonizing human life and appropriating it for capitalism.* Stanford UP.

Couldry, N., & Mejias, U. A. (2019b). Data colonialism: Rethinking big data's relation to the contemporary subject. *Television & New Media, 20*(4), 336–349. https://journals.sagepub.com/doi/pdf/10.1177/1527476418796632

Couldry, N., & Turow, J. (2014). Advertising, big data and the clearance of the public realm: Marketers' new approaches to the content subsidy. *International Journal of Communication, 8*, 1710–1726. http://eprints.lse.ac.uk/57944/1/__lse.ac.uk_storage_LIBRARY_Secondary_libfile_shared_repository_Content_Couldry,N_Advertising big data_Couldry_Advertising big data_2014.pdf

Couldry, N., Fotopoulou, A., & Dickens, L. (2016). Real social analytics: A contribution towards a phenomenology of a digital world. *British Journal of Sociology, 67*(1), 118–137.

Crawford, K. (2021). *The Atlas of AI: Power, politics, and the planetary costs of artificial intelligence.* Yale University Press.

Crawford, K., & Joler, V. (2018). *Anatomy of an AI system: The Amazon Echo as an anatomical map of human labor, data and planetary resources.* https://anatomyof.ai

Crawford, K., Gray, M. L., & Miltner, K. (2014). Critiquing big data: Politics, ethics, epistemology. Special section introduction. *International Journal of Communication, 8*, 10. http://ijoc.org/index.php/ijoc/article/download/2167/1164

Currie, M. E., Paris, B. S., & Donovan, J. M. (2019). What difference do data make? Data management and social change. *Online Information Review, 43*(6), 971–985.

D'Ignazio, C., & Klein, L. F. (2020). *Data feminism.* MIT Press.

Dalton, C., & Thatcher, J. (2014). What does a critical data studies look like, and why do we care? Seven points for a critical approach to 'big data'. *Space and Society Open Site.* Retrieved September 6, 2016, from http://societyandspace.com/material/commentaries/craig-dalton-and-jim-thatcher-what-does-a-critical-data-studies-look-like-and-why-do-we-care-seven-points-for-a-critical-approach-to-big-data/

Dalton, C. M., Taylor, L., & Thatcher, J. (2016). Critical data studies: A dialog on data and space. *Big Data & Society, 3*(1), 1–9.

DeLanda, M. (2006). *A new philosophy of society: Assemblage theory and social complexity*. Continuum.

Deleuze, G., & Guattari, F. (2004). *A thousand plateaus: Capitalism and Schizophrenia [orig. 1980]* (New ed.). Continuum International Publishing.

Dencik, L., & Kaun, A. (2020). Datafication and the welfare state. *Global Perspectives, 1*(1). https://doi.org/10.1525/gp.2020.12912

Dolata, U., & Schrape, J.-F. (2015). Masses, crowds, communities, movements: Collective action in the internet age. *Social Movement Studies, 15*(1), 1–18.

Draper, N. A., & Turow, J. (2019). The corporate cultivation of digital resignation. *New Media & Society, 21*(8), 1824–1839.

Elmer, G., Langlois, G., & Redden, J. (Eds.). (2015a). *Compromised data: From social media to big data* (1st ed.). Bloomsbury Academic.

Elmer, G., Langlois, G., & Redden, J. (2015b). Introduction: Compromised data—From social media to big data. In G. Elmer, G. Langlois, & J. Redden (Eds.), *Compromised data: From social media to big data* (1st ed., pp. 11–13). Bloomsbury Academic.

Eubanks, V. (2017). *Automating inequality: How high-tech tools profile, police, and punish the poor*. St. Martin's Press.

Fuller, M. (2008). Introduction. In M. Fuller (Ed.), *Software studies: A lexicon* (pp. 1–14). MIT Press.

Gee, J. P. (2005). Semiotic social spaces and affinity spaces: From the age of mythology to today's schools. In D. Barton & K. Tusting (Eds.), *Beyond communities of practice: Language, power, and social context* (pp. 214–232). Cambridge University Press.

Gerhard, U., & Hepp, A. (2018). Appropriating digital traces of self-quantification: Contextualising 'pragmatic' and 'enthusiast' self-trackers. *International Journal of Communication, 11*, 683–700.

Gerrard, Y., & Bates, J. (2019). Introduction to the Data Power Special Issue: Tactics, access and shaping. *Online Information Review, 43*(6), 945–951.

Gitelman, L., & Jackson, V. (2013). Introduction. In L. Gitelman (Ed.), *'Raw Data' is an oxymoron* (pp. 1–13). MIT Press.

Gray, M. L., & Suri, S. (2019). *Ghost work: How to stop Silicon Valley from building a new global underclass*. Eamon Dolan Books.

Greenwald, G. (2014). *No place to hide: Edward Snowden, the NSA, and the US surveillance state*. Hamish Hamilton.

Haggerty, K. D., & Ericson, R. V. (2000). The surveillant assemblage. *British Journal of Sociology, 51*(4), 605–622.

Hedman, J., Srinivasan, N., & Lindgren, R. (2013). *Digital traces of information systems: Sociomateriality made researchable*. http://openarchive.cbs.dk/bitstream/handle/10398/8855/Hedman.pdf?sequence=1

Hepp, A. (2020). *Deep mediatization*. Routledge.

Hepp, A., Loosen, W., & Hasebrink, U. (2021). Jenseits des Computational Turn: Methodenentwicklung und Forschungssoftware in der Kommunikations- und Medienwissenschaft. *Medien & Kommunikationswissenschaft, 69*(1), 1–21.

Hintz, A., Dencik, L., & Wahl-Jorgensen, K. (2019). *Digital citizenship in a datafied society*. Polity.

Iliadis, A. (2018). Algorithms, ontology, and social progress. *Global Media and Communication*. https://journals.sagepub.com/doi/pdf/10.1177/1742766518776688

Iliadis, A. (2019). The Tower of Babel problem: Making data make sense with basic formal ontology. *Online Information Review, 43*(6), 1021–1045.

Iliadis, A., & Russo, F. (2016). Critical data studies: An introduction. *Big Data & Society, 3*(2). https://doi.org/10.1177/2053951716674238

Jarke, J. (2019). Open government for all? Co-creating digital public services for older adults through data walks. *Online Information Review, 43*(6), 1003–1020.

Jarke, J., & Breiter, A. (2019). Editorial: the datafication of education. *Learning, Media and Technology, 44*(1), 1–6.

Kelly, K. (2016). *The inevitable: Understanding the 12 technological forces that will shape our future*. Viking.

Kennedy, H. (2018). Living with data: Aligning data studies and data activism through a focus on everyday experiences of datafication. *Krisis: Journal for Contemporary Philosophy, 2018*(1), 18–13.

Kennedy, H., & Bates, J. (2017). Data power in material contexts: Introduction. *Television & New Media, 18*(8), 701–705. https://journals.sagepub.com/doi/full/10.1177/1527476417720034

Kitchin, R. (2014). *The data revolution: Big data, open data, data infrastructures and their consequences*. Sage.

Kitchin, R., & Lauriault, T. (2014). Towards critical data studies: Charting and unpacking data assemblages and their work. *papers.ssrn.com*. https://papers.ssrn.com/sol3/papers.cfm?Abstract_id=2474112

Latour, B. (2007). *Reassembling the social: An introduction to Actor-Network-Theory*. Oxford University Press.

Lauriault, T. P., & Lim, M. (2019). Special section on data power. *Canadian Journal of Communication, 44*(3), 1–2.

Lazer, D., Pentland, A., Adamic, L., Aral, S., Barabasi, A.-L., Brewer, D., & Gutmann, M. (2009). Life in the network: The coming age of computational social science. *Science, 323*(5915), 721–723. https://europepmc.org/articles/pmc2745217

Lee, C. S. (2019). Datafication, dataveillance, and the social credit system as China's new normal. *Online Information Review, 43*(6), 952–970.

Lupton, D. (2016). *The quantified self*. Polity Press.

Lyon, D. (2014). Surveillance, Snowden, and Big Data: Capacities, consequences, critique. *Big Data & Society, 1*(2), 1–13. https://doi.org/10.1177/2053951714541861

Mackenzie, A. (2013). Programming subjects in the regime of anticipation: Software studies and subjectivity. *Subjectivity, 6*(4), 391–405. https://doi.org/10.1057/sub.2013.12

Mackenzie, A., & Vurdubakis, T. (2011). Codes and codings in crisis. *Theory, Culture & Society, 28*(6), 3–23. https://doi.org/10.1177/0263276411424761

Mayer-Schönberger, V., & Cukier, K. (2013). *Big data: A revolution that will transform how we live, work and think.* John Murray.

Mazzucato, M. (2013). *The entrepreneurial state.* Anthem Press.

Milan, S. (2017). Data activism as the new frontier of media activism. In V. Pickard & G. Yang (Eds.), *Media activism in the digital age* (pp. 151–163). Routledge.

Milan, S., & Treré, E. (2019). Big data from the South(s): Beyond data universalism. *Television & New Media, 20*(4), 319–335. https://doi.org/10.1177/1527476419837739

Mitchell, S. S. D. (2019). Warning! You're entering a sick zone. *Online Information Review, 43*(6), 1046–1062. https://doi.org/10.1108/oir-03-2018-0075

Mosco, V. (2014). *To the cloud: Big data in a turbulent world.* Paradigm Publishers.

Neff, G., & Nafus, D. (2016). *Self-tracking* (1st ed.). MIT Press.

Neff, G., Tanweer, A., Fiore-Gartland, B., & Osburn, L. (2017). Critique and contribute: A practice-based framework for improving critical data studies and data science. *Big Data, 5*(2), 85–97. https://doi.org/10.1089/big.2016.0050

Noble, S. U. (2018). *Algorithms of oppression: How search engines reinforce racism.* NYU Press.

Parks, L., & Starosielski, N. (Eds.). (2015). *Signal traffic: Critical studies of media infrastructures.* University of Illinois Press.

Puschmann, C., & Burgess, J. (2014). Metaphors of big data. *International Journal of Communication, 8*, 20. http://ijoc.org/index.php/ijoc/article/viewFile/2169/1162

Qiu, J. L. (2016). *Goodbye iSlave.* University of Illinois Press.

Rajão, R., & Jarke, J. (2018). The materiality of data transparency and the (re) configuration of environmental activism in the Brazilian Amazon. *Social Movement Studies, 17*(3), 318–332.

Richterich, A. (2018). *The big data agenda: Data ethics and critical data studies.* University of Westminster Press.

Rogers, R. (2013). *Digital methods.* MIT Press.

Ruppert, E. (2011). Population objects: Interpassive subjects. *Sociology, 45*, 218–233.

Sadowski, J. (2019). When data is capital: Datafication, accumulation, and extraction. *Big Data & Society, 6*(1). https://doi.org/10.1177/2053951718820549

Scassa, T. (2019). Ownership and control over publicly accessible platform data. *Online Information Review, 43*(6), 986–1002. https://doi.org/10.1108/oir-02-2018-0053

Turner, F. (2006). *From counterculture to cyberculture: Stewart Brand, the whole earth network, and the rise of digital utopianism*. University of Chicago Press.

van Dijck, J. (2014). Datafication, dataism and dataveillance: Big data between scientific paradigm and ideology. *Surveillance & Society, 12*(2), 197–208.

van Dijck, J., Poell, T., & de Waal, M. (2018). *The platform society*. Oxford University Press.

van Es, K., Schäfer, M. T., & Wieringa, M. (2021). Tool criticism and the computational turn. A 'methodological moment' in media and communication studies. *M & K Medien & Kommunikationswissenschaft, 69*(1), 46–64. https://www.nomos-elibrary.de/10.5771/1615-634X-2021-1-46/tool-criticism-and-the-computational-turn-a-methodological-moment-in-media-and-communication-studies-volume-69-2021-issue-1

Wagemans, A., & Witschge, T. (2019). Examining innovation as process: Action research in journalism studies. *Convergence: The International Journal of Research into New Media Technologies, 135485651983488*. https://doi.org/10.1177/1354856519834880

Webster, J. G. (2017). Three myths of digital media. *Convergence*. http://journals.sagepub.com/doi/abs/10.1177/1354856517700385

Zuboff, S. (2019). *The age of surveillance capitalism: The fight for the future at the new frontier of power*. Profile Books.

Open Access This chapter is licensed under the terms of the Creative Commons Attribution 4.0 International License (http://creativecommons.org/licenses/by/4.0/), which permits use, sharing, adaptation, distribution and reproduction in any medium or format, as long as you give appropriate credit to the original author(s) and the source, provide a link to the Creative Commons licence and indicate if changes were made.

The images or other third party material in this chapter are included in the chapter's Creative Commons licence, unless indicated otherwise in a credit line to the material. If material is not included in the chapter's Creative Commons licence and your intended use is not permitted by statutory regulation or exceeds the permitted use, you will need to obtain permission directly from the copyright holder.

PART I

Global Infrastructures and Local Invisibilities

Data Power and Counter-power with Chinese Characteristics

Jack Linchuan Qiu

Introduction

How to make sense of the People's Republic of China (PRC) as a global data superpower? Conventional wisdom dictates that China is viewed as the mystical Other, so much so that it has become a fetish—much like Japan used to be a while ago as exemplified by the "Japanese school girl watch" column in *Wired* Magazine. I argue, however, that China represents a very different kind of fetish, full of contradictions, caught between the iron fist of the Chinese Communist Party (CCP) on the one hand and the invisible hand of free-wheeling high-tech capitalism on the other. For some, China is fetishised as the ultimate "Black Mirror" writ large with one-fifth of the world's population being subjugated as if they are 1.4 billion guinea pigs being captured in a gigantic panoptic lab (Roberts, 2020; Strittmatter, 2020). For others, it is fetishised as the utopia of a neoliberal data economy, smart cities, artificial intelligence (AI), and miraculous rates of market expansion (Tse, 2015; Nylander, 2017; Lee, 2018).

J. L. Qiu (✉)
Department of Communications and New Media, National University of Singapore, Singapore, Singapore

© The Author(s) 2022

A. Hepp et al. (eds.), *New Perspectives in Critical Data Studies*, Transforming Communications – Studies in Cross-Media Research, https://doi.org/10.1007/978-3-030-96180-0_2

Both these fetishised visions are, however, partial and misleading. My task here is to argue against both of them by sharing an analysis that is more holistic and historicised than conventional approaches, considering China's internal conflicts and its relations with the external world. Such an attempt would bring us closer to the multifaceted reality of data power and counter-power in China (Lindtner, 2020; Wang, 2019, 2020), which is an essential part of the evolving global internet that critical scholars are grappling with (Qiu, 2019). In this chapter, I shall borrow from primary sources and secondary materials in Chinese and in English, in addition to fieldwork and interviews that I have conducted along with colleagues and students from Hong Kong in the past few years.

This chapter will begin by introducing and problematising the popular discourse of China as an "AI superpower". It then argues for a new critical approach that interrogates the complicated reality of Chinese data industries and a holistic framework that is historicised and conflictual, both along geopolitical fault lines surrounding China and within the country along social class cleavages. Providing illustrations from the continual history of Chinese computing in the 1950s to contemporary struggles along datafied picket lines in recent years, I propose that this new holistic approach has four novelties, which are particularly noteworthy.

First, both conventional views on China's data industries, whether utopian or dystopian, are etic observations from external parties, whereas the approach suggested in this chapter emphasises emic perspectives and innate logics from the inside out. This subverts the usual assumption about a unified, global system of data science that prevails over local, national, and regional systems. It also departs from the tendencies of techno-orientalism (Roh et al., 2015) that exoticises and dramatises China as fundamentally different, if not incomprehensible, as do Japan, Korea, and other Asian societies.

Second, a common practice among China specialists is to see the computing and data industries as a recent development that belongs exclusively to the post-Mao era since 1978. Similarly scholars examining data structures of the twenty-first century tend to conceptualise their subject matter as confined to the digital era. This chapter, however, argues otherwise: scholars today ignore the Maoist era before 1978 at our peril; ditto for pre-digital, analogue, even vacuum tube-based computing. While there is change and transformation over time, our holistic approach highlights historical continuity in understanding data power formations.

Third, contemporary analysts often construe Chinese data industries in the shadow of Silicon Valley, although now the former is emerging to challenge the latter. This view fails to see other possibilities of collaboration and symbiosis between China and the US while ignoring other regional dynamics, for example, between Japan and China since the 1960s, or the new development of Chinese IT companies going overseas, for example, and becoming major players in Africa. This chapter situates China in a network of global and geopolitical relationships that is dynamic and multifaceted. Most importantly, I do not presume any predetermined trajectory. The path of development is context-contingent, shaped by institutional inertia, while the major turning points tend to be moments of precarity, when Beijing perceives existential threats and would use data industries, among other instruments, to ensure survival.

What can threaten Beijing? Or more precisely, what can influence the CCP's perceptions of existential threats? Externally, there are geopolitical competition and regional conflicts that can be traced back to the Korean War. Internally, there are social class antagonism and struggles between elite-led and grassroots-driven models of developing the IT industry. Both constitute national security concerns that are central to China's top-level policy decision-making. To fully understand it, we have no choice but to confront key statist forces such as the Chinese military and the formation of counter-power in Chinese factories, IT companies, online and offline. This entails a conflict-oriented framework that differs greatly from neoliberal analysts who insist on viewing data industries as nothing but corporate, private entities, as showcased in Kai-Fu Lee's bestseller *AI Superpowers* (2018).

AI Superpower?

Kai-Fu Lee, former Google Vice President, now Chairman and CEO of Sinovation Ventures based in Beijing, is known for his *AI Superpowers: China, Silicon Valley, and the New World Order* (2018). As I write in September 2019, this book leads Amazon listings in the US: #3 in AI & Semantics, #2 in Robotics & Automation, and #1 in Automation Engineering. Brought up in Taiwan and educated in the US as a top AI researcher, Lee held executive positions at Apple and Microsoft before becoming the President of Google China. After Google left China in

2010, Lee started his own tech venture capital business in Beijing, and he remains upbeat about the future of data industries in the country. Although the genre of fetishising China as a dreamland for AI technology and business was already established (e.g., Tse, 2015; Nylander, 2017), Lee's 2018 book did more than any other volume in simplifying and romanticising China as an emerging AI superpower that has challenged the global supremacy of Silicon Valley and even started to surpass it.

Lee juxtaposes the Chinese model with the US model of AI development. While the US has Google, Uber, Amazon, and Facebook, China has Alipay and WeChat Pay, the ubiquitous mobile payment systems, TikTok the addictive short-video app, Pinduoduo the Chinese version of Groupon, but more powerful, and the food delivery and sharable bike sectors that Lee celebrates. This is in spite of notable efforts from within the industries, be they labour disputes among food-delivery couriers (Sun, 2019) or environmental concerns for abandoned shareable bikes or even the lack of sustainability for the business model itself (Zheng, 2019).

Despite China's AI underbelly, which Lee should be fully aware of, he presents a rosy picture from the perspective of a data scientist who craves more data and the perspective of a business entrepreneur who dreams about constant market expansion. He also contends that the rise of China as an AI superpower will benefit the world because it shakes up the unipolar world dominated by US tech giants. More competition shall work to the advantage of AI developers in both countries, maintains Lee. Americans should learn from the Chinese, or they risk losing their leading edge. "I've spent decades deeply embedded in both Silicon Valley and China's tech scene", wrote Lee, "I can tell you that Silicon Valley looks downright sluggish compared to its competitor across the Pacific" (2018: 15). According to him, American tech companies need to try harder to get more abundant data, more hungry entrepreneurs, and better AI scientists, while US government agencies need to learn from the CCP to improve its policy environment for AI technology.

Lee categorises AI into four types (ibid.: 136), out of which China is starting to take the lead in "Internet AI" and "perception AI", while becoming equal to the US in "autonomous AI". Silicon Valley will only be able to retain leadership in "business AI". China is catching up, even surpassing the US, so rapidly because, as Lee claims, it has more data. This is due not only to the much larger population size of the PRC, but Chinese entrepreneurs are also more tenacious, and they use a "go heavy" approach that is much more labour-intensive than Silicon Valley's typical "go light" approach to product development (ibid.: 70). Lee's argument centred on the sheer quantity of data that unifies the two China fetishes because its logical inference would be to recognise the panoptic surveillance state due to the permission it grants and/or the encouragement it provides for tech companies to collect even more data. The Big Brother can be the best alliance for the Big Other (Zuboff, 2019: 376). This is a key characteristic of China's fledging data power en route to becoming a superpower.

There is some truth in Lee's assessment. But he is wrong with his fixation on the binary opposition between the US and China while forgetting other players, a common tendency among policy analysts and critical scholars of platform economy. In so doing, he ignores the interplay between Beijing and Washington DC that shapes technology on the ground. Moreover, Lee underestimates the internal diversity of the Chinese model from its historical origins to its present state, both full of ambiguities and self-contradictions. He sees China as a single, coherent, and more-or-less insular system while failing to consider the data power of the Chinese military and its associates, as well as the resisting counter-power of Chinese workers and programmers. This mode of thinking is, again, a fetish, a myth repeated daily in commercial media. It does not, however, hold up to scrutiny.

Complex Reality Through Historical and Conflictive Lenses

Myth conceals. A duty of critical scholarship is to reveal. This section offers a cursory overview of what is missing in conventional thinking on China as the ultimate paradise for state-sponsored surveillance capitalism. The goal is not a detailed analysis, but to introduce facts and findings that would unsettle the established China-as-data-superpower discourse, thus preparing ground for a more structured discussion in the next section that shall introduce the new critical approach of this chapter.

Despite talks of automation, data industries in China, like elsewhere, depend on humans for software development and data processing. A quintessential type of "self-programmable labor" (Castells, 1996), Chinese software developers have resisted the exploitative powers of tech giants, most notably, through the "996.ICU" incident (Li, 2019). The code word "996" refers to working every day from 9 a.m. to 9 p.m., 6 days a week. After working such long shifts in the tech industry for a few years, one would end up in ICU, the intensive care unit, when one's life is endangered. As such, "996.ICU" is a campaign among programmers in China against excessive overtime in IT companies. Launched on 26 March 2019, it first appeared as a lengthy document of legal analysis that was posted to GitHub, calling on IT companies to abide by Chinese Labour Law, which stipulates a 40-hour workweek and a maximum of 36 hours of overtime each month; total work time should not exceed 49 hours per week. The 996 arrangement would, however, require employees to work 60–72 hours per week. While most companies see this as a violation of Chinese labour law as a feature of their organisational culture, hence refusing to remunerate extra work, a few tech firms even tried to formally institutionalise it and penalise employees who hope to stick to 8-hour workdays. The post received more than 200,000 "stars" in a few days, turning GitHub into a site of labour struggle which then had a cascade effect through not only social media but also *Worker's Daily*, the party organ newspaper of China's official trade union (ACFTU), which published an editorial in early April expressing support for programmers to protect their legal rights. Within a week, Jack Ma, the boss of Alibaba, fired back, saying that doing excessive overtime is a blessing for the workforce, thus escalating the controversy into the most significant clash of words regarding Chinese programmers working conditions. Who would anticipate such a clash, were China either the utopian or dystopian myth?

It is erroneous to simplify and fetishise China because the recent history of the PRC, including its computing sector as well as social imaginaries of ICTs, has been extraordinarily rich and full of ambiguities. A recent breakthrough is *Information Fantasies: Precarious Mediation in Postsocialist China* by Xiao Liu (2019), which analyses science-fictions, avant-garde cinema, and *qigong* traditional meditation practices in China during the 1980s, the first decade of PRC's post-Mao marketisation reform. These were cultural and social imaginations about technology that reflected "the advent of postsocialist conditions" (Liu, 2019: 26), characterised by ideological incoherence and an ambivalent situation between capitalism and "actually existing socialism". During this period of transition, Chinese "information fantasies" and their "precarious mediation" were powerful and creative, arguably more so than today in terms of its sociopolitical dimensions. And they were joined and promoted by top scientists such as Qian Xueshen, a key figure in China's nuclear programme, who in the 1980s devoted himself to studying "somatic science" of the body and supernatural forces. Will data science and the computing industry pave the way for a socialist future, or will it lead to de-politicisation, rampant marketisation, and terminal alienation? Such inquiries about technology and society were full of paradoxes prior to China's embrace of the internet and its neoliberal data power formation in the mid-1990s. In retrospect, it is apparent that there was nothing preordained about China's emerging data prowess.

Tracing further back to the roots of PRC's IT industry, we cannot ignore some of the groundbreaking achievements of the Maoist era. Scholarly accounts often trace the beginning of computers in China to a November 1955 article in the *People's Daily*. But internal documents show that as early as 1951, the CCP already began making plans to build its own electronics plant based on Soviet scientific literature, in response to the pressing military needs of the Korean War (Lu, 2016). This was the context when, in the early 1950s, the USSR transferred 1942 MiG fighter jets to China in three batches, along with submarines and radars. China took up the task of maintaining these military tools and manufacturing electronic parts for them domestically. A leading example at the time is Factory 774, a.k.a., Beijing Electronic Tubes Factory, home to Asia's largest electronics production line in the early 1960s (Lu, 2016). The following section delves deeper into the Maoist era. For now, it suffices to highlight the need to historicise China's data power all the way back to the early years of the PRC.

Another counter-example is the large-scale social movement in Hong Kong against the Extradition Bill proposed by the authorities in 2019. If Kai-Fu Lee was correct about China's superpower status, if the "go heavy" approach did help foster an omnipotent "Black Mirror"-like system of surveillance, how did the pro-Beijing forces fail to foresee the incoming avalanche of uprising? Despite all the big data, supercomputers, and AI capacity Beijing possesses, why was the Chinese party-state so ineffective in gauging public discontent in Hong Kong? As political scientist John Burns writes, "[W]ithout a fundamental reform of the way intelligence is collected within the [Communist] party to permit more diversity, the party will continue to repeat the mistakes of the past". It is not just the incapacity of AI-powered Chinese authorities, though. Equally important is the ingenuity of Hong Kong's tech-savvy youngsters using a wide range of digital tools (such as Telegram, Bridgefy, and map jams) to coordinate protests, coordinating among themselves, while evading surveillance and bolstering political messages through humour (Dynel & Poppi, 2020). Although activists generally failed in their attempts to produce change, they continue to defy authoritarian control using data as an instrument of large-scale resistance. The case of Hong Kong directly falsifies the myth of China: state-sponsored surveillance capitalism is not invincible. To fully understand China's data power as the thesis, we have to also take into account counter-power as its antithesis before arriving at a synthetical view of the system as a whole.

To debunk the myth of a single Chinese model, I argue that we need to look at it through at least two lenses: one being historical, and the other conflictive. Conceptually this implies we shall deem data power and counter-power as historical products in their technological materiality, and in their sociopolitical meaning, both at moments of radical change triggered by critical existential threats and at mundane times of banal nationalism and cosmopolitan consumerism. Meanwhile, data power and counter-power constitute a conflictive reality at the global, geopolitical, national, and subnational levels, which extends from the hot wars of Korea, Vietnam, and the Taiwan Strait to Cold War confrontations and the ongoing animosity that exists between the US and China today. It is not just external threats but also internal class struggle between farmers, workers, and the underclass on the one hand and the cadres and the super-rich on the other, through data infrastructures, ownership and political-economic arrangements, and contentious issues of distributive justice. The class struggle over data power is fundamentally conflictive (Qiu, 2016a, 2016b), although it also involves negotiations and compromises between the elite and the grassroots—trading co-optation in exchange for recognition; legitimacy in exchange for social security—as observed in other societies and in earlier periods of Chinese history.

At the very bottom of the evolving and conflict-ridden Chinese puzzle are basic questions about power and counter-power. For what goals are the data technologies designed and developed in the PRC? Through what structural performance? Using what division of labour? Under whose control? And, at whose expense? Not only socially, economically, culturally, and politically, but also in terms of environmental costs? At any given time, there are no predetermined answers to these questions. This includes our current era of the so-called AI Superpowers when the answers are still in a formative stage. They remain to be articulated, to be performed and actuated, to be institutionalised.

Chinese Data Power and Counter-power

What then is power and counter-power? They are a pair of concepts central to Castells book *Communication Power*, where power is "the relational capacity that enables a social actor to influence *asymmetrically* the decisions of other social actor(s) in ways that favour the empowered actor's will, interests, and values" (2007: 10, emphasis added). Defined as such, power is the institutionalised, "structural capacity" of imposition. Castells went on to point out that "media are not the holders of power, but they constitute by and large the space where power is decided". The media institutions here would include the computing and data industries.

By counter-power Castells understands "the capacity by social actors to challenge and eventually change the power relations institutionalized in society" (ibid.: 248). He continues: "[I]n known societies, counter-power exists under different forms and with variable intensity, as one of the few natural laws of society, verified throughout history, asserts that wherever is domination, there is resistance to domination, be it political, cultural, economic, psychological, or otherwise … opposed to what they often define as *global capitalism*" (ibid., emphasis added).

Observing from the level of global capitalism, we gain a more holistic view of Chinese power and counter-power, which can be at the same time more nuanced, reflecting the internal complexities of PRC's power formation that are in constant interplay with external forces, as shown in Fig. 1. First, the CCP-led party-state is at the same time a Leninist hierarchical power and a counter-power to the US since the 1950s (and to the USSR during 1960s–1980s). If traced back further, the establishment of the PRC was in itself a revolutionary, anti-imperialist, and anti-capitalist reaction to global capitalist expansion in China prior to 1949. Since then, both the global powers and the PRC itself have triggered counter-power formations, not only resistance but also creative divergence and alternative formations, which borrow selectively from the powers that be at national, regional, and global levels, as can be observed in Chinese computing and data industries. Meanwhile, both Chinese data power and counter-power draw from China's traditional culture, its collectivism and nationalism, its moral values, and translocal networking based on shared identity. The more China globalises technologically, the more distinct values can be drawn from its cultural traditions, be they Confucianist or state-communist.

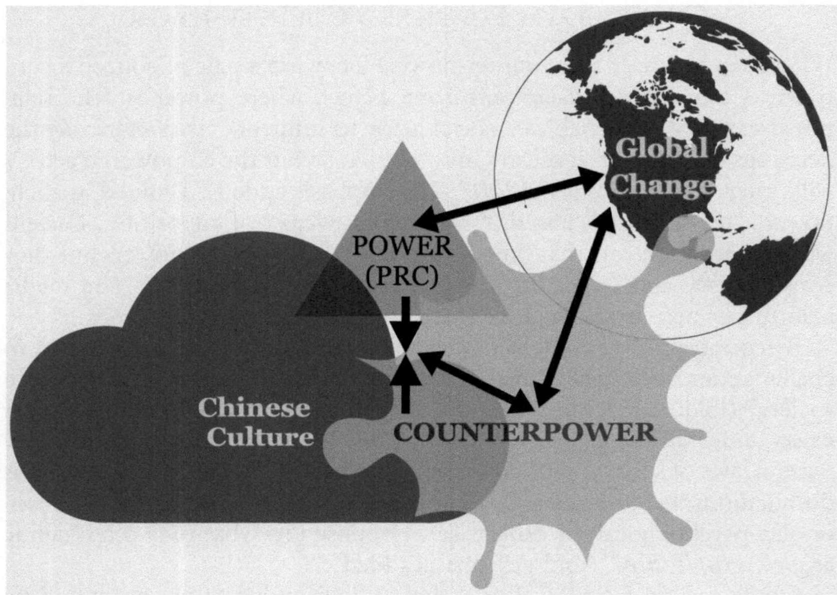

Fig. 1 Chinese data power and counter-power between the national and the global

Seen as such, the Chinese experiences are multi-dimensional and often self-contradictory, leading to the hard question: How did the counter-power end up becoming yet another hegemonic power? This question deserves serious contemplation by all critical data scholars, regarding not only the history of computing in the People's Republic but perhaps the present and future of the various data industries we study as well. After all, distributed computer networks were supposed to decentralise the global economy and serve as a counterweight against mass media empires in the 1990s. Yet, the tech giants of Silicon Valley have further centralised global capitalism into their own hands and Sunstein's dream for "republic.com" (2001) has become more distant than ever in the age of disinformation. Is Silicon Valley's trajectory, from its countercultural origins to its dominant power position today, analogous to that of the PRC? How to make sense of, and even prevent, such regressive movements of a counter-power growing into a dominant power, which then suppresses other counter-powers? This is likely to remain a thorny, yet essential, question for critical data studies in the future.

The case of China, if understood holistically through a historicised perspective that is sensitive to internal and external conflicts, would offer some insights into the aforementioned question. In the following, I illustrate the dynamic model in Fig. 1 with a few selected examples from the PRC's history of computing and the data industries. Together they would inform a more comprehensive and more systematic understanding that traverses both the Maoist and the post-Mao periods while offering an opportunity for us to observe the dialectics of data power and counter-power, within China and beyond.

A good volume to begin with Edward Feigenbaum's *China's Techno-Warriors: National Security and Strategic Competition from the Nuclear to the Information Age*. It documents how, from the horrors of the Korea War, the global superpowers of the US and the Soviet Union were instrumental in breeding Chinese counter-power, institutionalised in the Mao-era's strategic weaponry R&D before the 1980s, whose legacies were influential in the post-Mao era as well. Counterintuitively, Feigenbaum points out: "[T]his structure [of China's hi-tech weapon programs] included comparatively flat hierarchies; extensive horizontal coordination across bureaucratic boundaries; competition; networking; the open exchange of information; peer review; standards-based performance metrics; encouragement of risk-taking behaviour; and the political acceptance of failure" (Feigenbaum, 2003: 6). He also explains how "[o]rganizationally, this national security approach to technology depended on innovative management institutions that coupled top-down Stalinist-style mobilization to structures and incentives more akin to those in contemporary Silicon Valley, based on initiative, personal incentives, risk-taking, and networks of cooperation among experts" (ibid.: 3). In other words, Mao-era's high-tech weaponry research (including computing and telecommunications) was more akin to Silicon Valley in its organisation; yet, this emerging global counter-power was situated at the domestic power centre of the military, avoiding the failure of the civilian-led Soviet computer networking experiments (Peters, 2016).

Established power engenders fledging counter-power, as could be seen in the case of BOE, one of the world's leading display makers for smartphones, laptops, tablets, and televisions, and probably the most well-studied IT company in Chinese-language literature due to the groundbreaking work of Lu Feng (Lu, 2016). Although international media focuses on the likes of Huawei and ZTE, the value of BOE lies in its straddling of the Maoist and post-Mao eras, in both military and civilian sectors. The company's roots lie in the early 1950s when China learned from the Soviets in building vacuum-tube computers. But around the turn of the 1960s, they entered the semiconductor business while emulating the US as well as Japan. In 1963, the first Japanese semiconductor exhibition in Beijing attracted huge crowds and China began importing Japanese semiconductor fabrication machinery in 1968. This was followed by China's "electronic Great Leap Forward" (Wang, 2015) in the early 1970s when about 6000 electronic factories sprung up all over the country. Most of these were civilian and organised along the Maoist principle of the "mass line (*qunzhong luxian*)" stressing the involvement of ordinary workers, farmers, and soldiers in technology development and deployment. Employing more than half a million workers, most of these were grassroots-level computing and data-processing units utilising semiconductor parts from BOE, which by now had changed gear to support both the military and civilian sectors (Lu, 2016). The most important Maoist principle is "autonomy and self-reliance (*dulizizhu ziligengsheng*)", which united the military-led high-tech R&D and grassroots-driven electronic "leap forward", and its influence lasts to this day. According to oral histories from BOE, this Maoist spirit was crucial to the company's difficult transformation during the 1980s and 1990s, when it almost went bankrupt, but survived and made a dramatic comeback since the turn of the century to become a dominant global player thanks to the spirit of "autonomy and self-reliance" (ibid.).

Unlike the mainstream discourse that China only "opened up" to external influence after 1978, the PRC was embedded in transnational exchange regarding science, technology, and society during the Maoist era. For instance, Dallas Smythe, the prominent Canadian political economist and critical communication scholar, visited China during 1972–1973. After the trip, he wrote the legendary essay "After Bicycles, What?" (1994: 230–244) to introduce his observations in China and proposals for socialist media—such as a two-way television system operating much like an "electronic *tatzupao*" to ensure horizontal and interactive communication to meet collective social needs—that would be fundamentally different from capitalist media, especially commercial television. Arguing along the line of "autonomy and self-reliance", Smythe maintained that we need radical alternative imaginations of socialist technology and its own development criteria while discarding the capitalist yardsticks of individualism, consumerism, and "planned obsolescence". This example illustrates an important aspect of international counter-power solidarity, in this case, between Cultural Revolution China and Canada trying to exit from the shadow of Americana.

Smythe's proposals were grounded in the "electronic great leap forward" at the time nearing the tail end of the Maoist era when Chinese workers (e.g., those working in Shanghai's garment factories) established their own "barefoot electricians" (Wang, 2015). The expression came from the "barefoot doctors" in the era of the socialist countryside, where self-educated villagers, with some basic medical training, lived with farmers and innovated to meet patients' local needs. Similarly, in Shanghai, 450 "barefoot electricians" emerged from ordinary workers to help maintain, improve, programme, and de-mythify automated looms, to "control electronics without knowing the ABC" as the saying went, following the Maoist "mass line" principle for electronic technology, also known as the "Shanghai model" at the time. The large-scale grassroots movement influenced Smythe as well as other critical media scholars such as Armand Mattelart and Seth Siegelaub, whose edited volume *Communication and Class Struggle Vol.2: Liberation, Socialism* (1983) includes the minutes of a worker-engineer meeting from Shanghai. This suggests that China and the world have always been connected—that the PRC was not only at the receiving end of technology transfer but it was also an exporter and source of inspiration for Western critical scholars to envision alternative models of development all the way back to the Maoist years.

Ironically, a few years later in the early 1980s, "mass line" technopolitics was abandoned in post-Mao China (Wang, 2015). In its place was imposed the power dominance of imported IBM computers that Chinese workers saw as tools of disempowerment. Chinese-style Luddite resistance followed suit, as did subnational conflicts at both city and organisational level. Counter-power formations at subnational levels became more salient than overall national policy. As would be seen later from instances of worker resistance along the assembly line (Qiu, 2016b) to those in the data mine (such as the 996.ICU), the spectre of Maoism, its "mass line" politics and "autonomy" principles, has continued to haunt China's burgeoning data power projects.

A turning point in the labour-capital relationship within China's IT industries was the tragedies at Foxconn, where 14 workers committed suicide one after another within a few months in 2010, because they could not bear the inhumane exploitation and alienation at the world's largest electronics factory. As Chan and Pun (2010) argue, suicides are an extreme form of protest, and we can consider them an extreme mode of counter-power. The desperate resistance by Foxconn workers spurred a tidal wave of nationwide strikes in 2010 as well as transnational counter-power solidarity such as the "anti-iSlave" campaign (Qiu, 2016a). According to media reports (Motherboard, 2019; Reuters, 2019), many migrant workers have returned from the sweatshops to their home villages in recent years, only to become another type of labour, "tagging labour" as the occupation is now called, for China's rapidly growing data and AI industries. These are, more precisely, "AAI (artificial artificial intelligence)" (Aytes, 2012: 80), when workers perform repetitive, tedious tasks of tagging online content, training machine learning algorithms, while receiving low pay and working long hours under poor conditions, in ways that are similar to way Amazon's Mechanical Turk Human Intelligence Tasks system operates, although in the Chinese case this situation emerged as a consequence of the CCP's infrastructural investments into high-speed internet provision for remote, rural parts of the country.

China's new data infrastructures also afford new forms of activism by China's "network labor", which has become a counterweight to the establishment (Qiu, 2016b). Digital and social media have been used to not only reinforce and extend the picket line but also initiate unexpected campaigns in cyberspace as seen in the 2009 Jinjiang 360-degree sports apparel factory strike when garment workers formed an alliance with hackers to launch Search Engine Optimisation attacks against exploitation and managerial suppression. Digital picket-line struggles have become indispensable and organic to labour movements in the PRC in recent years, partly due to the increasing popularity of short-video sites such as TikTok and Kuaishou among the working classes and partly due to the prevalence of capitalist platforms (e.g., Didi and Meituan) that have become essential parts of the urban infrastructure (Chen & Qiu, 2019; Sun, 2019). Similar to the use of GitHub, during the 996.ICU movement, Chinese activists, gig workers, and factory workers (most notably in 2018 at Jasic, an industrial robot manufacturer) have used novel means to combat censorship by the party-state or their company management using, for instance, innovative data visualisations or zero-value cryptocurrency transactions (so that the censored information will remain accessible on the global blockchain). When top-down power attempts to deactivate alternative networks, grassroots counter-power from different lineages (re)activate new connections, creating new convergences of resistance forces.

Our final example is Transsion, a Chinese company that now presides over a large share of the African smartphone market. With origins in the *shanzhai* informal-economy innovation system (Lindtner, 2020), Transsion became famous in 2016 due to its facial recognition algorithm developed for the detection and beautification of dark-skin faces, a market need from African consumers that was for long ignored by other phone manufacturers, such as Apple, Samsung, or Huawei (Jiemian, 2017; Lu, 2020). It's not just pretty selfies. Transsion also targets Africa's low-end markets, for instance, through its large batteries designed for rural users. The rise of China's data prowess, in this case, may indeed present prospects for a new form of decolonial technology design. The Chinese counter-power, becoming a dominant player in the developing world, may indeed trigger indigenous development on the African continent and throughout the Global South. It would be premature to dismiss this future possibility of a global counter-power movement, inspired and enabled by the likes of Transsion, against the hegemony of Silicon Valley and new forms of data colonialism (Couldry & Mejias, 2019).

Conclusion

This chapter first outlined and debunked the China fetish that either celebrates Beijing's stance supporting surveillance capitalism or demonises it as the worst of Big Brother-type practices. Such conventional thinking fixates either on the CCP party-state and Xi Jinping's "Central Network Security and Informatization Leading Group" or on entrepreneurial success stories and the sheer quantity of data and size of the market as Kai-Fu Lee did (2018). These are important aspects of China's data power but they are oversimplified and can be misleading because they perceive today's reality as natural or predetermined, because they only conceptualise power in the political and economic establishment while forgetting the essential role played by counter-power.

From the historicised and conflictual perspective proposed in this chapter, we may summarise the Chinese data industry's historical journey as having taken place across four phases: It started with (1) a Soviet birth in the 1950s during the Korean War, followed by (2) the Maoist "Electronic Great Leap Forward", around the time of the Sino-Soviet border conflict in 1969. This was a formative phase supplying the "organisational gene" for China's strategic enterprises such as BOE. Then, there was (3) the PRC's neoliberal turn in the 1980s, which brought with it new internal conflicts, ideological ambiguities, and external isolation in the aftermath of Tiananmen, when the internet started to become popular in the mid-1990s. Finally, since the mid-2010s, China entered (4) the "New Normal" era, characterised by lower economic growth, heightened social control, the emergence of alternative social movements, and the convergence of contentious factors—including geopolitical frictions between China and the US—in ways that are diverse, dynamic, often unforeseen or unpredictable, within the PRC and globally.

Unique as each phase is, the four periods are also similar in that they are characterised by the interplay between power and counter-power, especially around issues of national security—in terms of geopolitics (e.g., Korean War) or internal stability (such as Tiananmen). The dialectics between power and counter-power is the *yin* and *yang* of the Chinese model introduced in this chapter. Their interplay is not only antithetical to each other, but they also necessitate, reproduce, co-create, and strengthen each other, although in different historical contexts the specific constitution of that interplay would vary.

During the Maoist era from the 1950s to mid-1980s, the dominant power in China was the military-political complex, which, at a global level, worked as counterweight against the bipolar powers of the Cold War: the US and the USSR. Since the late 1980s, the structure has metamorphosised into a political-industrial-military complex, where the goals of the military still matter, but not as much as the IT industry giants such as Huawei; and ultimately it is the CCP political elite who remain at the fore. While data power operates, almost invincibly, in imposing corporeal control over Chinese bodies, the counter-power forces—diverse as they are—are also breaking loose, creating alternative networks and switching on new, unforeseen connections of resistance. The spectre of revolutionary Maoist "mass line" principles remains an important repertoire for activists and startups to grasp, engendering counter-power formations that have become increasingly large scale, multi-sectoral, and trans-border. Meanwhile, they remain collective endeavours, especially among the lower classes that are united by common existential threats, brought about, for instance, by the Chinese gig economy and platform capitalism.

So, what is the final assessment of the rise of the Chinese data industry's model? Is it the worst dystopia or the most perfect utopia? My conclusion is that neither is the case. It is still too early to tell: Will China represent anti-capitalism or hyper-capitalism at the very end? Is Beijing's top-down approach going to completely control bottom-up formations and horizontal networking? Will Chinese programmers stage more effective resistance, or will robots and AI dominate—even the power of CCP?

It is important to reiterate that ours is not a bipolar world of the US vis-à-vis China. It is a multi-layered, conflictive reality that produces power and counter-power with Chinese characteristics, by which I mean historical products that are collective, contingent, and cross-border, involving not only the US and the USSR, but also Japan, Canada, Europe, Africa, and the world altogether. While the impact of Silicon Valley is not to be dismissed, it is erroneous to neglect the multilateral influences upon China by the likes of Japanese IT companies (Steinberg, 2019) as well as China's influence overseas through cases such as Transsion. It is also increasingly common that mutual influence emerges through joint projects, such as the EU-China Co-Funding Mechanism that operates under the Horizon 2020 framework.[1]

The new perspective proposed in this chapter is emic, dynamic, conflict-sensitive, and historically holistic. Despite dramatic changes in the PRC since the 1950s, we continue to see continuity from the Maoist through to the post-Mao era, from the time of vacuum tubes to today's big data era. There is no preordained trajectory, for good or for bad. Rather, the development path is contingent, forming precariously at critical moments of national security concerns and depending both externally upon geopolitics and internally upon class struggle. History, in this sense, remains a decisive factor in shaping and explaining the Chinese model of the computing and data industries. And history can only be fully understood when we pay attention to the conflicts therein.

References

Aytes, A. (2012). Return of the crowds: Mechanical Turk and neoliberal states of exception. In T. Scholz (Ed.), *Digital Labor: The Internet as Playground and Factory* (pp. 79–97). Routledge.

Burns, J. (2019, September 6), *A failure of intelligence: Why Beijing is liable to repeat its mistakes in Hong Kong*. Hong Kong Free Press. https://bit.ly/2KuM7eb

Castells, M. (1996). *The rise of the network society*. Blackwell.

Castells, M. (2007). *Communication power*. Oxford University Press.

[1] http://chinainnovationfunding.eu/

Chan, J., & Pun, N. (2010). Suicide as protest for the new generation of Chinese migrant workers: Foxconn, global capital, and the state. *The Asia-Pacific Journal*. http://japanfocus.org/-Jenny-Chan/3408

Chen, J. Y., & Qiu, J. L. (2019). Digital utility: Datafication, regulation, labor, and Didi's platformization of urban transport in China. *Chinese Journal of Communication, 12*(3), 274–289.

Couldry, N., & Mejias, U. A. (2019). Data colonialism: Rethinking big data's relation to the contemporary subject. *Television & New Media, 20*(4), 336–349.

Dynel, M., & Poppi, F. I. M. (2020-online first). Caveat emptor: Boycott through digital humor on the wave of the 2019 Hong Kong protests. *Information, Communication & Society*. https://doi.org/10.1080/1369118X.2020.1757134

Feigenbaum, E. A. (2003). *China's Techno-Warriors: National security and strategic competition from the nuclear to the information age*. Stanford University Press.

Jiemian. (2017, October 19). Transsion, the king of mobile phone in Africa makes a fortune in low-profile. *Pandaily*. https://bit.ly/3jyXRZE

Lee, K.-F. (2018). *AI superpowers: China Silicon valley, and the new world order*. Houghton Mifflin Harcourt.

Li, X. (2019). The 996.ICU movement in China: Changing employment relations and labour agency in the tech industry. *Made in China Journal*. https://bit.ly/2KnkfZf

Lindtner, S. (2020). *Prototype nation: China and the contested promise of innovation*. Princeton University Press.

Liu, X. (2019). *Information fantasies: Precarious mediation in post-socialist China*. University of Minnesota Press.

Lu, F. (2016). *The transformation of light: An enterprise's industrial history*. Contemporary China Press. (in Chinese).

Lu, M. (2020). *'Deep ploughing': A mobile phone company's technology translations between China and Ghana*. Unpublished doctoral dissertation. The Chinese University of Hong Kong.

Mattelart, A., & Siegelaub, S. (Eds.). (1983). *Communication and class struggle vol. 2: Liberation, socialism*. International General.

Motherboard. (2019, January 19). *Field visit to the "AI village" in Henam*. https://www.lieyunwang.com/archives/451120 (in Chinese)

Nylander, J. (2017). *Shenzhen superstars: How China's smartest city is challenging Silicon valley*. http://www.shenzhensuperstars.com/

Peters, B. (2016). *How not to network a nation: The uneasy history of the soviet internet*. MIT Press.

Qiu, J. L. (2016a). *Goodbye iSlave: A manifesto for digital abolition*. University of Illinois Press.

Qiu, J. L. (2016b). Social media on the picket line. *Media, Culture & Society, 38*(4), 619–633.

Qiu, J. L. (2019). The global Internet. In D. Hesmondhalgh & J. Curran (Eds.), *Media and society* (6th ed., pp. 3–20). Bloomsbury.

Reuters. (2019, June 28). Data and cheap labor fuel China's AI dreams. Available: https://www.reuters.com/video/watch/data-and-cheap-labor-fuelchinas-ai-drea-id567776675?chan=8gwsyvzx

Roberts, M. (2020). *Censored: Distraction and diversion inside China's great firewall*. Princeton University Press.

Roh, D., Huang, B., & NIiu G. (Eds.). (2015). *Techno-orientalism: Imagining Asia in speculative fiction, history, and media*. Rutgers University Press.

Smythe, D. (1994). *Counterclockwise*. Westview Press.

Steinberg, M. (2019). The Platform Economy: How Japan transformed the consumer Internet. University of Minnesota Press.

Strittmatter, K. (2020). *We have been harmonized: Life in China's surveillance state*. HarperCollins.

Sun, P. (2019). Your order, their labor: An exploration of algorithms and laboring on food delivery platforms in China. *Chinese Journal of Communication, 12*(3), 308–323.

Sunstein, C. (2001). *Republic.com*. Princeton University Press.

Tse, E. (2015). *China's disruptors: How Alibaba, Xiaomi, Tencent, and other companies are changing the rules of business*. Portfolio / Penguin.

Wang, H. (2015). From "Barefoot Electrician" to "Electronic Supervisor": Technology and labor politics in the information industry of China. *Open Times, 3*, 34–48. (in Chinese).

Wang, J. (2019). *The other digital China: Nonconfrontational activism on the social web*. Harvard University Press.

Wang, X. (2020). *Blockchain chicken farm: And other stories of tech in China's countryside*. FSG Originals.

Zheng, A. (2019). *How to burn $2 Billion in 15 months for dummies*. https://medium.com/swlh/how-to-burn-2-billion-in-15-months-for-dummies-bc325d3b06f6

Zuboff, S. (2019). *The age of surveillance capitalism*. Public Affairs.

Open Access This chapter is licensed under the terms of the Creative Commons Attribution 4.0 International License (http://creativecommons.org/licenses/by/4.0/), which permits use, sharing, adaptation, distribution and reproduction in any medium or format, as long as you give appropriate credit to the original author(s) and the source, provide a link to the Creative Commons licence and indicate if changes were made.

The images or other third party material in this chapter are included in the chapter's Creative Commons licence, unless indicated otherwise in a credit line to the material. If material is not included in the chapter's Creative Commons licence and your intended use is not permitted by statutory regulation or exceeds the permitted use, you will need to obtain permission directly from the copyright holder.

Transnational Networks of Influence: The Twitter Presence of the Quantified Self and Maker Movements' Organizational Elites

Anne Schmitz, Heiko Kirschner, and Andreas Hepp

Introduction

Our imaginations of data and data power are not only driven by state agencies and outstanding 'corporate actors' such as media companies like Apple, Google, and Facebook. Just as vital in the process are 'collective actors' characterised by the ways in which they 'act on media' (Kannengießer & Kubitschko, 2017, p. 1): These collectivities make media and the infrastructures that undergird them the focus of their engagement, precisely because the latter have become so relevant to society. Social movements such as the Open Source or Indymedia movement that 'act on media' have a long legacy of scholarly attention; however, pioneer communities such as the Quantified Self (QS) and Maker movements have not yet been examined more closely in this respect: While the QS movement is

A. Schmitz • H. Kirschner • A. Hepp (✉)
ZeMKI, Centre for Media, Communication and Information Research, University of Bremen, Bremen, Germany
e-mail: a.schmitz@uni-bremen.de; andreas.hepp@uni-bremen.de

© The Author(s) 2022

A. Hepp et al. (eds.), *New Perspectives in Critical Data Studies*, Transforming Communications – Studies in Cross-Media Research,
https://doi.org/10.1007/978-3-030-96180-0_3

concerned with data practices of self-measurement and self-optimisation through the intensive use of wearables and other tracking technologies, the Maker movement is characterised by the collaborative development of (digital) manufacturing processes based on 3D printers or micro-computers such as the Arduino and the RaspberryPi.

Both pioneer communities share with social movements their open structure and their aim to stimulate societal change. Likewise, they share with them a sense of mission, on the basis of which they are committed to spreading their ideas as globally as possible. The difference is that pioneer communities—even if they refer to themselves as a movement—share many more traits with the commercial world and established politics and are curated by an organisational elite that can often be traced back to the San Francisco Bay Area (Hepp, 2020a, pp. 33–40). In the case of the QS movement, their main curatorial instruments are publications such as *Wired* magazine and the QS Lab Berkeley website quantifiedself.com, events such as QS conferences, meetups, and prototyping institutions (e.g. the Quantified Self Institute in Groningen, the Netherlands). In the case of the Maker movement, these instruments include publications such as *Make:* magazine or events like Maker Faires and meetings in local makerspace.

However, besides these particular curatorial forms, online networking operates as a vital function for keeping these pioneer communities together transnationally. In both communities, their organisational elite, that is, their principal organisers and decision makers, connect on Twitter to share the latest information and to make announcements. It is this online networking that will be the focus of this chapter to deepen our understanding of pioneer communities and their transnational spread. Through these online networks, the organisational elite of pioneer communities is kept together—but is also able to disseminate its own ideas, such as those regarding the significance of digital technology and data. With reference to these preliminary thoughts, we will address the following three research questions in this chapter: How are the organisational elites of both pioneer communities connected transnationally? What patterns and peculiarities can be identified in terms of account types as well as thematic orientation? And, what similarities and differences exist for individual countries and between each community?

To answer these questions, we will reconstruct followee-networks of the organisational elite of both movements as they play out on Twitter.[1] This approach is closely tied to our media ethnography through which we have already been able to determine the central members of the transnational organisational elites (Hepp, 2018, 2020b), and based on prior knowledge from which, we define the seed accounts to reconstruct the Twitter networks analysed here. Starting from these prominent members of the organisational elites, we aim to (a) trace the *connections* of these accounts to each other and the patterns within their common followee-networks, (b) identify *further transnational key accounts* of the organisational elite on Twitter, and (c) describe their *similarities and differences* on a transnational and national level as well as between both pioneer communities.

While the basis of this chapter is, first of all, a well-defined empirical study, we are interested in something else as well: We want to make apparent the extent to which the two pioneer communities discussed here are transnationally networked and how they act invisibly in a way that aims to spread their own ideas and imaginaries globally, at least according to the wishes of the organisational elite. In many local understandings of self-tracking and making, we find 'translations' (Fredriksson & Pallas, 2017, p. 119) of these imaginaries. Through such an investigation, we want to draw attention to the fact that critical data studies, especially when they make the ambivalences of the local and the global their subject, should also consider the somewhat invisible engagement of pioneer communities.

In sum, we want to substantiate the thesis that the Twitter network of the QS movement can best be described as a 'network of opinion leaders' and that of the Maker movement as a 'network of heterogeneous organisations'. However, in both cases the Twitter analysis underlines the significance of the members of the organisational elite from the San Francisco Bay Area in regard to the pioneer community's transnational figuration. In this sense, Twitter is an instrument used by the organisational elite to establish their ideas and ideologies across each community and spread them globally.

[1] We would like to thank our colleagues Cornelius Puschmann, Yannis Theocharis, and especially Stefanie Walter for their valuable support and remarks on preparing and visualising the Twitter data, as well as Marc Kushin for his careful proofreading.

In what follows we will briefly summarise the current state of research on the QS and Maker movements' organisational elite and the challenges inherent in an investigation of such figurations based on Twitter data. We will then introduce our methodology and present a comparative analysis of each community's elite network. To conclude, we will reflect on how our findings offer an insight into the overall figuration of these pioneer communities and their importance for critical data studies and its future perspectives.

State of Research: The QS and Maker Movements' Organisational Elites

In many respects the QS and Maker movements are intimately related: Both date back to the mid-2000s, both were formed in the San Francisco Bay Area, both were 'founded' by former editors and journalists (Gary Wolf and Kevin Kelly from *Wired* in the case of QS, and Dale Dougherty from O'Reilly Media in the case of the Maker movement), and both managed to orchestrate the extension of their influence and notoriety from the US to Europe and other parts of the world. However, there are also clear differences between both movements that can be identified through the orientation of their practices (the self vs. manufacturing), their visions of media-related collectivity and societal transformation, their events, and the reach of their published works (e.g. websites, journals, and reports) (see Hepp, 2016).

Previous research has shown interest in both pioneer communities, particularly from the point of view of their ordinary members. In the case of the QS movement, this concerns the everyday practices of tracking and self-measurement (Crawford et al., 2015; Didžiokaitė et al., 2018; Lomborg & Frandsen, 2015; Pantzar & Ruckenstein, 2014), the movement's proximity to emerging approaches to personal health (Ajana, 2017; Nafus, 2016; Lupton, 2015; Sharon, 2017; Williamson, 2015), and data security and surveillance issues related to the practice of self-tracking (Abend & Fuchs, 2016; Esmonde, 2019; Fotopoulou, 2018; Lupton, 2014; Sharon & Zandbergen, 2016; Swan, 2013). A number of studies have also investigated the public discourse surrounding the QS movement, be it in the technology magazine *Wired* (Ruckenstein & Pantzar, 2017) or general media coverage (Hepp et al., 2021a, 2021b). While this research is rich in nature and can only be touched upon here in rudimentary form, a study of the QS movement and its organisational elite's engagement does not exist. Apart from obligatory references to the 'founders' of QS in the introduction to various articles and chapters, Gina Neff and Dawn Nafus' book *Self-Tracking* (2016) deserves special mention here. However, they neglect to discuss the pioneer community's elite in any real detail.

Perhaps tangential, but still comparable in regard to the lack of emphasis on the organisational elite, is research into the Maker movement. Researched topics include makerspaces as localities of innovation and learning (Barniskis, 2013; Davies, 2017; Lange, 2015; Peppler et al., 2016; Toombs et al., 2014), the relation of the Maker movement to the do-it-yourself and hacker movements (Hunsinger & Schrock, 2016; Ratto & Boler, 2014), new forms of civic participation through the Maker movement and its events (Kostakis et al., 2015; Nascimento & Pólvora, 2016; Richterich, 2017), the engagement of the Maker movement in (industrial) development (Irani, 2015; Ramsauer & Firessnig, 2016), and a general reflection on 'making' as a (countercultural and pedagogic) practice (Gauntlett, 2018). Likewise, studies into reporting on the Maker movement can also be found which focuses either on the pioneer community's publications in *Make:* magazine (Nguyen, 2016; Sivek, 2011) or on the general discourse on the Maker movement (Hepp, Benz, & Simon, 2021b). Once again, the role of the organisational elite in the Maker movement is discussed only marginally, even in publications that focus on its historical roots (Turner, 2018).

In our own research, we have compared the organisational elite of the QS and Maker movements in Germany, Great Britain, and the US in regard to how they curate their transnational reach (Hepp, 2018, 2020b). With this chapter we want to delve deeper into the network of both on Twitter. There has, in fact, been little research into the Twitter activities of both movements. One rare exception in the case of the Maker movement is a study by Menichinelli (2016), who looked into the Twitter connections of fablabs, makerspaces, and hackerspaces. Menichinelli's analysis provides evidence of a globally spread community on Twitter loosely organised into several sub-communities born out of geographically divergent hubs, each with differences in their organisational structures. However, his analysis focuses solely on different kinds of spaces and their networking and not on a reconstruction of the various accounts of the organisational elite.

Twitter as a platform enables three different 'communication modes' (Bruns & Moe, 2014, p. 16). First is interpersonal communication through the @mention and the @reply functions, and direct messages. Second is communication within networks of followers and followees through original tweets. And, third are potentially large communication streams that are structured and topic-related using #hashtags. These different modes of communication also structure the research of Twitter data to some degree. We find conversation analysis at the level of tweets and retweets (boyd et al., 2010; Paßmann et al., 2014) while another branch of research focuses on the investigation of connections from the networks of single or multiple accounts and their communication flows (González-Bailón & Wang, 2016; Gruzd et al., 2001; Xu et al., 2014). Furthermore, research has been carried out that looks into the dynamics of events based on hashtags (Highfield et al., 2013; Gaffney, 2010; Leavitt, 2013; Lotan et al., 2011).

By reconstructing the Twitter networks of both movements' organisational elites, we consider this study to reside in the tradition of the second line of research mentioned above. Our approach does not focus on the 'volatile networks' (Maireder & Schlögl, 2015, p. 120) of particular events but the analysis of networks on the basis of followee lists, and seeks to identify more stable networks that can influence and structure observation and patterns of practice. Twitter network analyses based on follower and followee lists often concentrate on the structural patterns of these networks as well as particular key accounts as bridges or brokers within them (e.g. González-Bailón & Wang, 2016; Sajuria et al., 2015; Theocharis et al., 2017).

While this kind of research reflects the way Twitter structures data, it also poses considerable challenges. First, there is the challenge of its changing architecture and the growing restrictions applied to the Twitter API which results in limitations in research design (Bruns, 2019; Puschmann, 2019). Second, researching social phenomena solely through Twitter data risks falling short in adequately grasping the subject in question, as these kinds of data cannot be understood isolated from wider social and cultural contexts. Networks on social media platforms and other kinds of 'digital traces' are always socially and culturally embedded and the challenge is how to link this contextual information with online data (Hepp et al., 2018; Tilson et al., 2010). This means that we not only follow arguments that Twitter data are themselves a product of 'interpretative work' (Bowker, 2013, p. 170) that should be understood 'by the means used to handle and process it' (Puschmann & Burgess, 2014, p. 1702), we also want to explore the online activities of the organisational elite in close relation to other ethnographic data we have on their activities.

Methodological Approach: Contextualised Twitter Network Analysis

As already mentioned, our Twitter network analysis is part of a larger project that investigates both movements using a media ethnographic approach. Based on 234 qualitative interviews with members of both pioneer communities, participant observations at twelve major events and twenty-five local meetups and spaces, we were able to identify the organisational elite. Referring to research on organisational leaders in social movements and scenes (Cammaerts, 2005; Hitzler & Niederbacher, 2010; Nepstad & Bob, 2006), we define such members as those who (a) take responsibility for important events, (b) publish widely on and in the name of the movement, or (c) speak publicly on behalf of the pioneer community (Hepp, 2020b). Typically, the key actors of the organisational elite combine all three criteria.

The entry point for our Twitter analysis is, therefore, twenty-one manually selected seed accounts of the QS movement from the US, the Netherlands, the UK, and Germany and nineteen seed accounts of the Maker movement from the US, Italy, the UK, and Germany.[2] Our selection includes two account types: personal accounts and organisational accounts. The personal accounts of the QS community include, for example, the 'founders', Gary Wolf (@agaricus) and Kevin Kelly (@Kevin2Kelly), and organisers of local QS meetups. The organisational accounts of QS are mainly divided into the Twitter appearances of the official website from QS Labs Berkeley (@quantifiedself) and its national and regional offshoots. The corpus for the Maker movement consists of a pool of personal accounts including the founder Dale Dougherty (@dalepd) as well as committed organisers of local makerspaces, a few entrepreneurs, and

[2] The composition of these countries is based on the fact that the two pioneer communities originated in the US. For the European context, the Netherlands also set the course for the Quantified Self movement, as the first European QS Conference took place in Amsterdam and the QS Institute was established at Hanze University in Groningen. Furthermore, the fact that the largest European Maker Faire took place in Rome, Italy also plays a central role in the Maker movement. The comparison of Germany and the UK is particularly interesting, as these two countries differ significantly in their value orientation towards technology.

journalists related to the Maker movement. The organisational accounts belong mainly to Maker Media, the American company that published the community-focused *Make:* magazine (@make) and also organises and licenses the community's events such as the Maker Faire (@makerfaire). However, it is important to note that Maker Media went bankrupt in the summer of 2019.

In order to generate the Twitter network, we captured our seed accounts by using the Twitter API via R and the rtweet package and crawled their followee lists afterwards. The data were collected in February 2020, processed, and then visualised using the open-source software Gephi and the Force Atlas 2 algorithm. In total, we identified 15,162 followee accounts for QS and 18,667 for the Maker movement. However, since the focus of our analysis is trained on connections within the communities' network of our seed accounts, we only labelled accounts within the transnational networks with an in-degree of ≥5. Consequently, a followee is only shown by name within the network if the account is followed by at least five seed accounts. The more seed accounts following the newly identified account, the bigger the node and label size in the network visualisation. Accordingly, the more seed accounts following a user, the more importance we attribute to the account for the debate on Twitter. This procedure resulted in a transnational followee-network of eighty-two labelled accounts for QS and 218 labelled accounts for the Maker movement, including the seed accounts (see Table 1). However, it must be

Table 1 Data set of both pioneer communities

Pioneer community	*Country*	*Seed accounts*	*Total followees*	*Labelled accounts (in-degree)*
Quantified Self	US	6	2396	**67** (≥3)
	The Netherlands	5	8930	**81** (≥3)
	UK	5	3317	**22** (≥3)
	Germany	5	1409	**21** (≥3)
	US/NL/UK/DE	21	15,162	**88** (≥5)
Maker	US	4	5834	**259** (≥3)
	Italy	4	8544	**270** (≥3)
	UK	5	2788	**35** (≥3)
	Germany	6	3035	**59** (≥3)
	US/IT/UK/DE	19	18,667	**218** (≥5)

stressed that this filtering process for the visualisation is manually determined by us in order to focus on the organisational elite and new accounts with a connection to as many seed accounts as possible.

We repeated this sampling procedure for each country within our study. Due to a smaller amount of entry points we labelled accounts at an in-degree of ≥3 within the national networks.

This research design enables us to come to conclusions about how the seed accounts are transnationally connected with one another on Twitter and allows us to discover further accounts with overlaps to the seed accounts and thematic groups within the networks. Moreover, it shows links between each country and points towards potential information brokers. However, it must be clearly stated that the study says nothing about the actual interaction of the organisational elites on Twitter.

Ultimately, the results are then interpreted and contextualised against the background of the transnational and national scale to uncover the dynamics within the overall figuration of the organisational elite.

QS Movement: A Network of Opinion Leaders

The QS organisational elite's Twitter network can be described as a network of opinion leaders. In adopting this classical term, seminally coined by Elihu Katz and Paul Lazarsfeld (1955), we want to point out the following: In the QS movement, it is primarily the accounts of prominent, individual 'influential' (Katz & Lazarsfeld, 1955, p. 150) members of the organisational elite that make up the Twitter network. At this point, it is already important to see our Twitter analysis as a contextualising study: If one were to look purely at the tweets of the QS opinion leaders, their influence would hardly be measurable because it is not necessarily developed in Twitter interactions. In the original sense of Katz and Lazarsfeld's reflections, the role of the 'opinion leader' mainly unfolds through personal communication, for example, at meetups and conferences. Nevertheless, the accounts of these opinion leaders—even if they rarely tweet—are networked with other accounts in a very specific way.

The Transnational Network

The twenty-one seed accounts for QS are divided into fifteen personal and six organisational accounts. However, the resulting followee-network with eighty-eight labelled nodes shows the same pattern as it is also made up of mostly personal accounts. Only eight labelled organisational accounts can be found within the network, which are mostly companies for self-tracking wearables like Fitbit or Withings, or news platforms for digital healthcare such as MobiHealthNews and Rock Health. The network is, therefore, very homogenous.

Looking more closely at the network (see Fig. 1), the nodes with the highest in-degree from our seed accounts belong to US accounts: Gary Wolf (in-degree of 13), followed by the official QS account (in-degree of

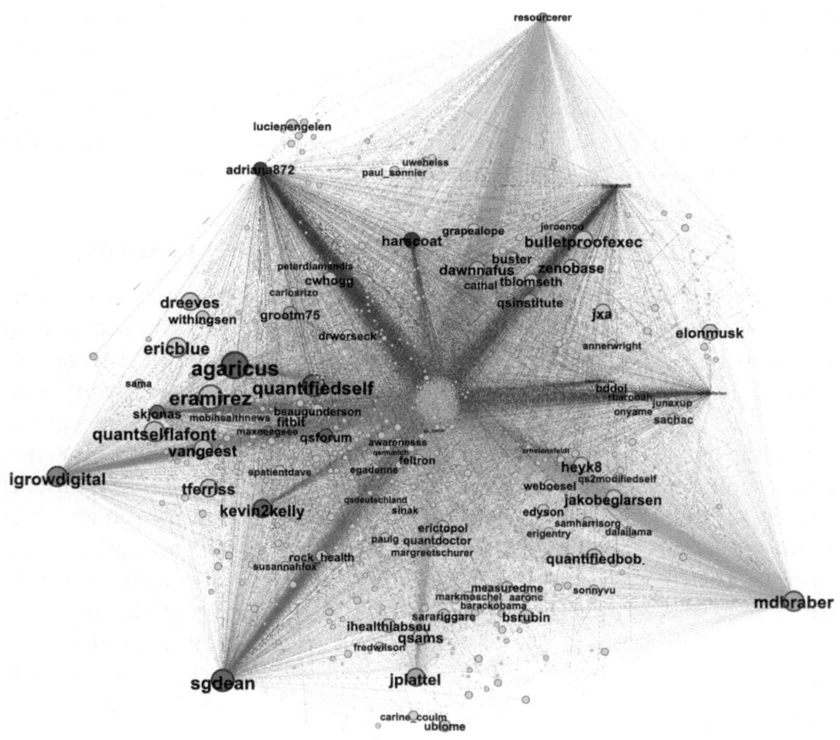

Fig. 1 The transnational followee-network of the QS organisational elite based on twenty-one seed accounts and eighty-eight labelled nodes filtered through an in-degree of ≥5. Created via Gephi, Force Atlas 2. Legend for seed accounts: red—the US; orange—the Netherlands; purple—the UK; green—Germany

11) and Steven Dean (@sgdean, in-degree of 11), a QS meetup organiser in the US. The Dutch seed accounts of Maarten den Braber (@mdbraber, in-degree of 10) and Joost Plattel (@jplattel, in-degree of 9), both Dutch meetup organisers, also have noteworthy amounts of followers among the seed accounts.

Moreover, the network uncovers further key actors within the followee-network who have not previously appeared in our media ethnography but are followed by the majority of our seed accounts and seem to be relevant connections on Twitter. In this regard, we identify Ernesto Ramirez (@eramirez, in-degree of 12), who leads the San Diego QS meetup, as well as Paul LaFontaine (@quantselflafont, in-degree of 10) who is the QS meetup organiser in Denver.

Other 'new' accounts, which stand out in the network due to their overlaps within the organisational elite, can be clustered into a thematic group of eleven male (tech and/or health) entrepreneurs who are located in the US. Of these, Eric Blue (@ericblue), founder and CEO of activeOS, has the highest in-degree of 10 and David Asprey (@bulletproofexec), founder of Bulletproof, and David Reeves (@dreeves), co-founder of LegUp, have nine followers among the seed accounts.

Another, albeit smaller, thematic group of eight accounts can be identified around Dawn Nafus (@dawnnafus, in-degree of 8), Thomas Blomseth (@tblomseth, in-degree of 7), Jakob Eg Larsen (@Jakobeglarsen, in-degree of 8), Sara Riggare (@sarariggare, in-degree of 6), and four other accounts, who research topics related to QS or health. Blomseth, Larsen, and Riggare are even official collaboration partners with QS Labs Berkeley.

Authors of books or journalists related to QS can be grouped into a third group, including Tim Ferriss (@Tferris, in-degree of 9), Kate Farnady (@heyk8, in-degree of 8), Buster Benson (@Buster, in-degree of 7), and six other accounts.

If we look at the labelled followee-network of QS as a whole, it becomes clear that it is heavily influenced by personal accounts in the US—in terms of high in-degrees of the seed accounts, as well as the large number of overlapping followees which can be traced back to the US. Dutch accounts seem to demonstrate moderate importance, represented through the overlaps of the seed accounts to Joost Plattel and Maarten den Braber as well as a number of other accounts such as Yuri van Geest (@vangeest) and Lucien Engelen. By contrast, neither the seed accounts from Germany and the UK seem to play a major role in the transnational Twitter network, nor are there any significant new accounts from both countries with an in-degree of ≥5. Rare exceptions are Florian Schumacher's seed account (@igrowdigital, in-degree of 10), a German meetup organiser, and Denis Harscoat's (@harscoat, in-degree of 8) and Adriana Lukas' (@adriana872, in-degree of 7), both QS London meetup organisers.

The National Context

This resonates with the national followee-networks of the QS movements' organisational elite, as demonstrated by the QS Twitter networks in their respective national contexts.

The network of the organisational elite in the US (Fig. 2a) is based on six seed accounts—four personal and two organisational accounts—resulting in a followee-network of sixty-seven labelled nodes with an in-degree of ≥3. Once more, Gary Wolf's dominant position (in-degree of 5) becomes clear from a national perspective with even more followers (among the seed accounts) than the official QS account (in-degree of 4). Furthermore, the presence of four Dutch seed accounts in the US network, especially Joost Plattel's with an in-degree of 4, is particularly striking. This indicates that these accounts might serve as bridges and potential information brokers between the two countries. In an attenuated form this can also be said for Adriana Lukas and Denis Harscoat, both with an in-degree of 3.

The national network of the Netherlands (see Fig. 2b) is built upon five Dutch seed accounts and spans over eighty-one labelled nodes with an in-degree of ≥3. That means that the number of users with a higher in-degree in the network is comparatively high and that many accounts are followed by at least three seed accounts. Within the network Lucien Engelen, author of the book *Augmented Health Care*, represents the account with the highest in-degree of 5. The seed account with the greatest number of followers belongs to Martijn Aslander (@resourcerer, QS Amsterdam) with an in-degree of 4. The impact of the US seed accounts—especially through Gary Wolf with an in-degree of 4—can also be seen in the Dutch network. Besides the accounts, which already appeared in the transnational network, two entrepreneurs from the Netherlands, Rutger van Zuidam (@rutgervz, founder of odysseyhack) and Daan Dohmen (@daandohmen, founder of FocusCura), are of relative importance in this national network. The presence of researchers in this network is also particularly strong (@sarariggare, @jakobeglarsen, @tblomseth, @drworseck, and

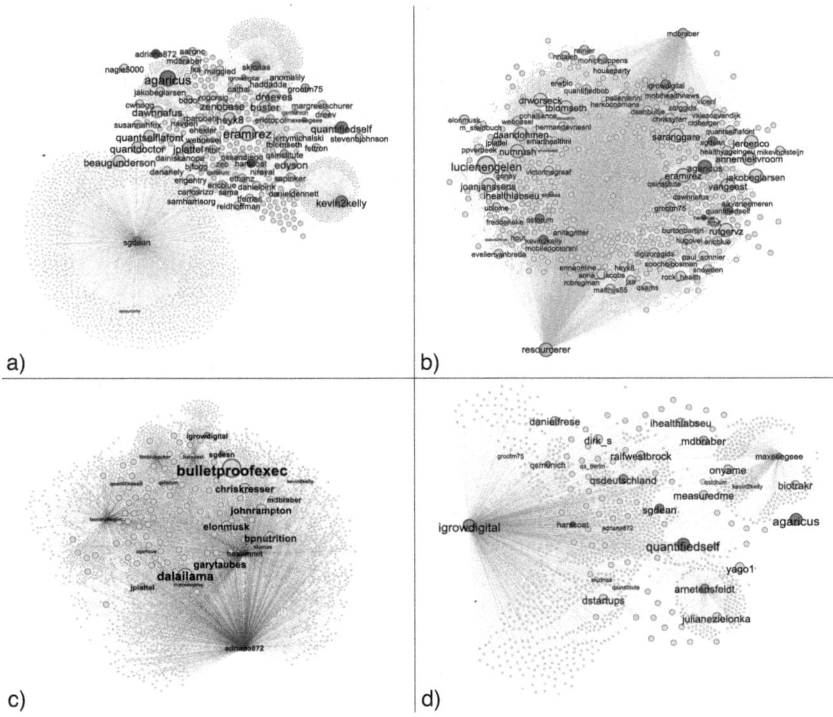

Fig. 2 The national followee-networks of the QS organisational elite from the US, the Netherlands, the UK, and Germany. (a) The US network is based on six seed accounts, resulting in sixty-seven labelled nodes (in-degree of ≥3); (b) the Dutch network is based on five seed accounts resulting in eighty-one labelled nodes (in-degree ≥3); (c) the UK network is based on five seed accounts resulting in twenty-two labelled nodes (in-degree of ≥3); and (d) the German network is based on six seed accounts resulting in twenty-one labelled nodes (in-degree of ≥3). The network is created via Gephi, Force Atlas 2

@AnneMiekVroom). This might be explained by the fact that the first (now closed) QS Institute for research was located in the Netherlands.

The UK network (see Fig. 2c) is different in some ways. It comprises only twenty-two labelled nodes, based on five seed accounts. So, in contrast to the networks described above, we see a relatively small number of accounts followed by more than three seed accounts. Instead, the network is structured around only a few accounts with large overlaps, most notably from the already mentioned David Asprey, founder of Bulletproof, a company selling nutrition supplements, as well as the account of the current Dalai Lama (@dalailama). Slightly less prominent are the organisational account of Bulletproof (@bpnutrition) itself, as well as the personal accounts of Chris Kresser (@chriskresser), a health blogger and Paleo diet enthusiast; John Rampton (@johnrampton), founder of NatureBox, a health-focused foodbox; and Gary Taubes (@garytaubes), author of *The Case Against Sugar*—all with an in-degree of three. This suggests that the UK network tends to have a thematic focus on topics around nutrition, well-being, and biohacking rather than self-tracking. By contrast, the classic QS accounts play a rather subordinate role within this network. Accordingly, the connections to foreign QS seed accounts are also weak.

The German network (see Fig. 2d) is based on five seed accounts and includes twenty-one labelled nodes, with an in-degree of ≥3 and is, therefore, also quite small. Within the network the German seed account of Florian Schumacher, who has led all QS meetup activities in Germany in recent years stands out with an in-degree of 4. Parallel to the national networks of the Netherlands, the accounts of Gary Wolf (in-degree of 4) and QS (in-degree of 4) prominently appear. This allows us to assume that the German QS accounts are also co-oriented towards the American QS Twitter community. Maarten den Braber seems to be another broker connecting the German and Dutch communities. The network also reveals other key actors in the German context. These are mostly German-based accounts from (tech) business-related actors such as Daniel Frese (@ Danielfrese, entrepreneur), Dirk Spannaus (@dirk_s, founder of TwentyZen), Ralf Westbrock (@RalfWestbrock, consultant and coach for innovation), Andreas Schreiber (@onyame, former QS Cologne meetup organiser, co-founder of Medando), and, in a rare exception, a female entrepreneur and speaker on digital transformation and health, Juliane Zielonka (@JulianeZielonka)—all have an in-degree of 3. Consequently, what we can see is, again, a connection of the organisational elite to the startup and entrepreneur scene, represented here by mostly middle-aged, German businessmen.

Ultimately, the comparison of national contexts does well to demonstrate that the QS organisational elite is also mainly connected to individual 'opinion leaders' with different thematic foci, but still maintains a strong connection to founders or researchers. The national UK network distinguishes itself through a slightly shifted focus to nutrition and bio-hacking but is still represented by individual accounts.

Maker: A Network of Heterogeneous Organisations

The Twitter networks of the QS movement described so far clearly contrasts with that of the Maker movement. To put it more emphatically, the latter can be described as a network of 'heterogeneous organisations'. In defining them as such, we mean to say that while individual opinion leaders also appear in this network, it is dominated by a variety of organisational accounts.

The Transnational Network

The nineteen seed accounts of the organisational elite are divided into eleven personal accounts and eight organisational accounts and are, therefore, quite balanced. However, within the transnational followee-network of 218 labelled accounts (in-degree of ≥5), we uncovered a variety of organisational account types next to our seed accounts which were mainly related to Maker Media. These additional organisational accounts belong mainly to technology companies, community platforms, or journalistic outlets. Moreover, with almost the same number of entry points, the resulting followee-network is more than twice as large as the transnational QS followee-network. Consequently, we see a much greater overlap of the seed account in regard to their followees (Fig. 3).

Looking at the transnational Maker network in more detail it becomes clear that the US-based seed accounts are again the most prominent in the transnational network: The seed accounts with the highest in-degree belong to *Make:* magazine (in-degree of 14), Maker Faire (in-degree of 13), and Dale Dougherty (in-degree of 11). The Italian founder of the micro-controller Arduino, Massimo Banzi (@mbanzi), is another seed account which reaches an in-degree of 11. The most prominent seed accounts for the UK and Germany show a thematic continuity within this network. For the UK these are accounts for the UK Maker Faire (@makerfaire_uk) and *Hackspace* magazine (@hackspacemag). Similarly, for Germany it is the accounts for the German Maker Faire (@makerfairede) and the German edition of *Make:* magazine (@makemagazinde) that represent the most prominent seed accounts.

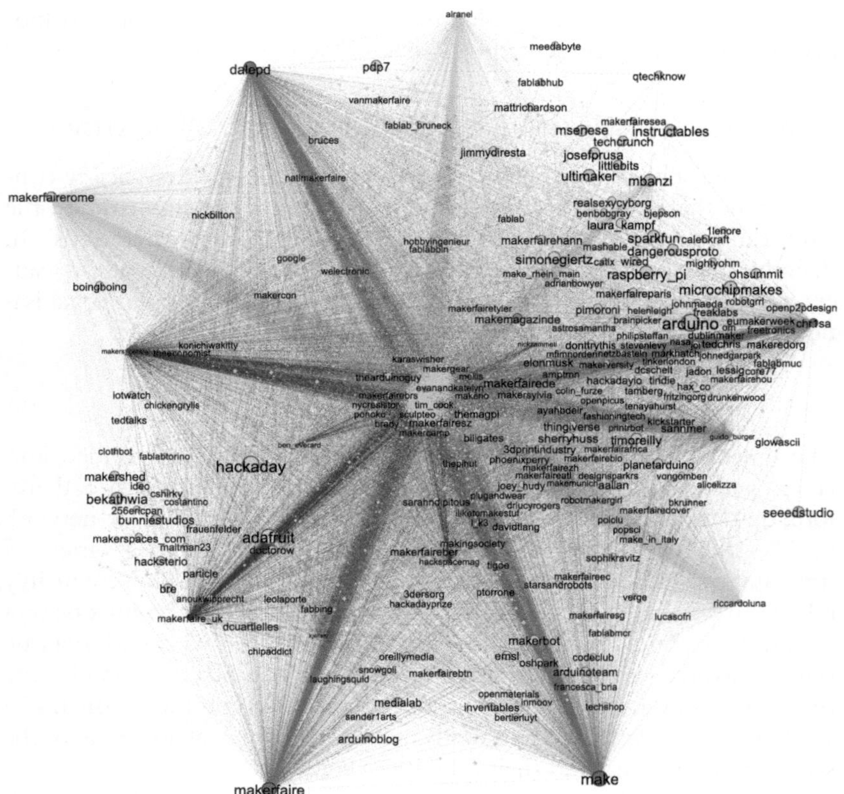

Fig. 3 The transnational followee-network of the Maker organisational elite is based on 19 seed accounts and 218 labelled nodes filtered through an in-degree of ≥5. Created via Gephi, Force Atlas 2. Legend for seed accounts: red—the US; blue—Italy; purple—the UK; green—Germany

The highest in-degree, however, is not held by a seed account, but Arduino's organisational account (@arduino, in-degree of 16), followed by Hackaday (@hackaday, in-degree of 15), a community platform, who have the highest number of overlapping followers among the seed accounts.

In strong contrast to the transnational QS network, in this Maker network the ten accounts which are followed most by the seed accounts are organisational accounts and appear to be of higher importance for the Maker debate than it is for QS on Twitter. Besides the organisational accounts related to Maker Media (eighteen Maker Faire accounts), the organisational accounts can be clustered into thematic groups around technology companies that manufacture, for example, microcontrollers (e.g. @arduino and @raspberry_pi), 3D-printers (@Ultimakers), or other soft- and hardware (e.g. @adafruit, @microship_makes, and @sparkfun); community platforms such as Hackaday, Thingiverse (@thingiverse), or Instructables (@instructables); journalistic outlets (e.g. @Techcrunch, @ oreillymedia, and @wired); and regular Maker festivals and projects (e.g. @ littlebits, @dangerousprototype, and @makershed), as well as a few local spaces (e.g. @fablabbln, @fablabmcr, and @gablab_bruneck).

Among the few personal accounts which exist within this network, another thematic group can be identified around popular YouTubers and influencers such as Simone Giertz (@SimoneGiertz, in-degree of 11), Becky Stern (@bekathwia, in-degree of 11), Laura Kampf (@laura_kampf, in-degree of 9), Jimmy Diresta (@JimmyDiResta, in-degree of 8), Naomi Wu (@RealSexyCyborg, in-degree 8), and Colin Furze (@colin_furze). Connections to entrepreneurs can only be seen sporadically in this network (@elonmusk, @josefprusa).

Altogether, the transnational followee-network of the Maker's organisational elite appears as a heterogeneous structure, with several thematic groups within the pool of organisational accounts and just a few influential personal accounts. Interestingly, Elon Musk's account is the only labelled account, which appears in both the transnational Maker network and the QS network.

The National Context

Once again, when we look at each country separately we can observe that the national networks are also mostly dominated by organisational accounts.

The US network (Fig. 4a) is the second largest network in our sample in terms of labelled accounts. It stretches from our 4 seed accounts over 259 labelled nodes with an in-degree of ≥3. All four seed accounts from the US have the same in-degree of 3 and thus follow one another. Interestingly, Massimo Banzi's account has an in-degree of 4 within the American network, which means that all seed accounts follow him. This indicates that he might be an influential information broker for the

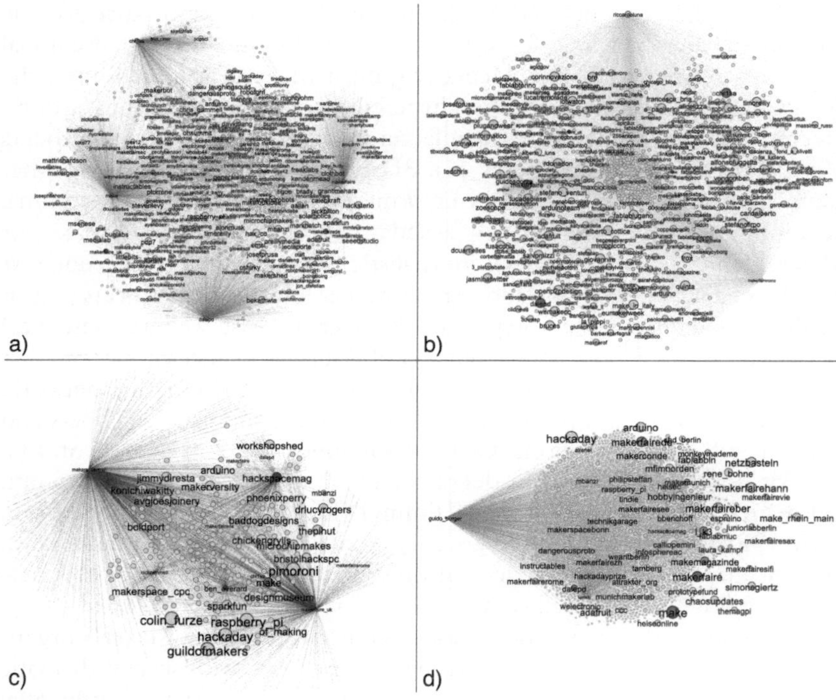

Fig. 4 The national followee-networks of the Maker organisational elite from the US, Italy, the UK, and Germany. (a) The US network is based on 4 seed accounts, resulting in 259 labelled nodes (in-degree of ≥3); (b) the Italian network is based on 4 seed accounts resulting in 270 labelled nodes (in-degree of ≥3); (c) the UK network is based on 5 seed accounts resulting in 35 labelled nodes (in-degree of ≥3); and (d) the German network is based on 6 seed accounts resulting in 59 labelled nodes (in-degree ≥3). The network is created via Gephi, Force Atlas 2

American context on Twitter. The accounts from the UK and Germany have nearly no visibility within the US network. Congruent with the transnational network, there are more organisational account types than personal account types. Taking all accounts with an in-degree of 4 together, we find thirty organisational accounts which can be grouped to more or less the same thematic groups as in the previous network.

The network of the organisational elite of the Italian Maker movement (Fig. 4b) is the largest network in our sample. It spans from 4 seed accounts to over 270 labelled nodes with an in-degree of ≥3. The users with the highest in-degree are the American seed accounts of Chris Anderson (@chr1sa, CEO of 3Drobotis, journalist, and author), Dale Dougherty, and the organisational account of the Maker Faire. On this basis, the Italian network also seems to be oriented towards the Maker movement's US origins. Massimo Banzi and Riccardo Luna (@riccardoluna), former *Wired* author and curator of the Maker Faire Rome, share an in-degree of 3, whereas the other seed accounts are almost invisible within the network. Nonetheless, we see various 'new' accounts. In contrast to the two previous networks the Italian network is dominated by personal account types. Only six organisational accounts with an in-degree of 4 can be identified among the labelled accounts. In regard to the personal accounts with an in-degree of 4, two thematic groups stand out: (mostly male) Italian authors and journalists (e.g. @lucadebiase, @lucatremolada, and @alicelizza) and (mostly male) Italian entrepreneurs related to (maker-) technology (e.g. @giovannire, @Rdonadon, and @Maxciociola).

The UK network (Fig. 4c) is the smallest network within our sample of the Maker movement regarding the labelled networks: Here the network of our five seed accounts stretches over thirty-five nodes with an in-degree of ≥3. The most followed British seed account belongs to the *Hackspace* magazine, whereas the German personal seed account of Guido Burger (@ guidofmakers), an influential maker and speaker, even reaches an in-degree of 4 and can be considered a bridge to the German Maker community. Another seed account with at least an in-degree of 3 is the American *Make:* magazine. Remarkable in this network is that there is no user in the network that is followed by all five seed accounts and just four accounts with an in-degree of 4: Colin Furze, a British YouTuber and -Maker with nine-million subscribers, is the only personal account type while there are three organisational account types—Pimoroni (@pimoroni), a re-seller of maker hardware (particularly in relation to RaspberryPi), Hackaday, and the RaspberryPi account itself. The greater prominence of RaspberryPi compared to the Italian Arduino can be explained through RaspberryPi's origins in Cambridge. Among the accounts with an in-degree of 3 there are mostly organisational accounts related to tech companies and only three personal accounts, two of them are also YouTuber-makers (@jimmydiresta and @avgjoesjoinery).

The network of the organisational elite of the German Maker movement (Fig. 4d) spans from the six seed accounts over fifty-six labelled nodes with an in-degree of ≥3. The seed user Guido Burger stands out in the network as he follows more users that any of the other seed account, but he is only followed by one other seed account (in-degree of 1). Also isolated is Kjell Otto's seed account (@kjellski) with an in-degree of 0, which seems to indicate that he is less relevant and connected within the German Twitter debate than the other seed accounts. The highest in-degree among the German seed accounts is held by the Maker Faire Germany, but *Make:* magazine US (in-degree of 6) reaches the highest value among all seed accounts, followed by Maker Faire US. This suggests that the German maker scene also orientates itself towards the origins of the US Maker movement.

If we look at the followees, Hackaday shows the highest overlap of all six seed accounts. An overlap of five accounts can be achieved by the organisational accounts of the biggest maker events and projects in Germany: Maker Faire Berlin (@makerfaireber) and Maker Faire Hannover (@makerfairehann), as well as Netzbasteln (@netzbasteln), a German DIY-radio show, and, once again, the Arduino account. From this overlap it can be concluded that organisational accounts predominate here as well. Furthermore, the accounts with an in-degree of 3 confirm this: This includes other German maker events (e.g. @mfimnorden and @maker-conde), a makerspace (@fablabbln), the Chaos Computer Club's account (@chaosupdates), and Adafruit, in contrast to only three personal accounts belonging to makers (@hobbyingenier, @rene_bohne, and @ simonegiertz).

In summary, we gain a deeper insight into the Twitter network of the Maker movement if we consider it as being made up of mostly heterogeneous organisations. However, those accounts bridging the national contexts are accounts that belong either to Maker Media or to individuals and organisations that cooperate closely with them. We argue, therefore, that accounts related to Maker Media comprise an organisational core in the pioneer community and its representation on Twitter. However, around this core, further accounts are grouped together such as technologies, projects, and local spaces, which are seen in national contexts as well as in the transnational networks.

Conclusion

The transnational followee-networks of the organisational elite and their national contexts on Twitter show similarities, yet, the composition of the networks of both movements is fundamentally different. With reference to our research questions, we can summarise that the transnational networks of both communities differ in size, as the Maker network is more than twice the size of the QS network (218 vs. 88 labelled nodes). However, they are quite similar in that the seed accounts as well as newly identified accounts in the US dominate in both transnational networks, which speak for an orientation by both pioneer communities towards the origins of the movement in the San Francisco Bay Area. The presence and impact of the German and UK-based accounts seem to be rather low. Specific accounts from the Netherlands for QS and Italy for the Maker movement are relatively present in the transnational Twitter networks, which indicate their potential functionality as bridges.

In regard to our second research question we can state that the QS organisational elite on Twitter is represented as a network of opinion leaders as indicated by the vast majority of personal accounts which belong, on the one hand, to QS meetup organisers who are heavily involved in the core activities of arranging and organising QS events, and, on the other, through the connections of QS seed accounts to influential entrepreneurs and founders. The many Twitter connections to the founders of tech startups found here could serve as an indication of financing and stabilisation attempts by community leaders. By contrast, the organisational elite of the Maker movement is represented on Twitter as a network of heterogeneous organisations which range thematically from Maker Media-related accounts to tech companies, community platforms, journalistic outlets, and specific maker events. This indicates, on the one hand, already established collaboration partners and ways to spread the Maker ideology through various channels (platforms and journalistic outlets) and underlines the importance of digital maker technologies for the movement on the other. In both Twitter communities, significantly more male accounts are visible when looking at the personal account types.

For our third research question we can summarise that the QS organisational elite's Twitter network is not only smaller in regard to the number of accounts but also less intensely overlapped compared to the Maker network. This also applies to the national contexts, as the transnational thematic groups are also reflected at the national level. The national networks of the QS movement are also more individualised and seem to be more fragile in their overall institutional structure. The national Twitter networks of the Maker movement's organisational elites are larger in size, more tightly meshed, and consist of various account types (with Italy being a notable exception) in comparison to QS. Moreover, the connections to established companies reflect a potentially more solid structure of the movement even in the national contexts. In regard to national particularities, the QS community has a thematic proximity to dietary and biohacking concerns as represented by the British network. In the Maker community, the Italian network stands out for its many personal accounts instead of the dominance of organisational accounts found in other Maker networks. In both pioneer communities it is noticeable that the German and British national networks have significantly less overlaps.

The results can be related to our previous media ethnography. As our research shows, both differ in the way they are curated by their respective organisational elite (Hepp, 2020b): The QS movement is based on the model of an 'unenforced trademark', that is, the legally incomplete trademark protection of the term 'quantified self' which prevents others from securing the copyright. Such a strategy allows the movement's founders to facilitate a discourse of belonging and exclusion around the movement's principal ideas. In the case of the Maker movement, curating is carried out through a 'franchise model', specifically the development of the concept and the corporate identity of *Make:* magazine or the Maker Faires by Maker Media (and since 2019, the *Make:* Community), which are both licensed in different forms. The Twitter networks of each community correspond in structure and character to these two models: The 'unenforced trademark' model relies much more on looser forms of organisation and the commitment of individuals who hold the pioneer community together through conferences and meetings. In the franchise model, established organisations have a much higher priority; alongside Maker Media and its successor, the *Make:* Community, other publishing organisations, companies, and local spaces.

What can be concluded from our research for the perspective of critical data studies? First of all, it becomes apparent how intensively pioneer communities—albeit in different ways—are networked transnationally through their organisational elites. Pioneer communities are important collective actors for technology-related change in that their organisational elites want to spread globally certain imaginaries of societal transformation and associated social practices and that they are—while partly invisible in their engagement—astonishingly successful in doing so. Typically, it is first and foremost their ideas and imaginaries that spread, on the basis of which particular institutions such as local spaces are created, experimental practices are locally established, or previously existing institutions are overhauled, for example, when community workshops become makerspaces or the ideas of the Quantified Self movement spread in local sports groups. Certainly, comprehensive 'translations' (Fredriksson & Pallas, 2017, p. 119) of the original ideas and imaginations take place in these processes.[3] However, the curatorial status of the pioneer communities' organisational elite remains in place.

Pioneer communities themselves are highly ambivalent phenomena. As is apparent, not least from the self-description of the 'movement' as emerging from the ideas of the American counterculture—as is generally the case in the San Francisco Bay Area and the Silicon Valley tech industry (Castells, 2001; Turner, 2006). However, specific to pioneer communities is a particular way of referring to countercultural ideas of self-empowerment as can be seen in the self-measurement of the QS movement or in the tinkering of the Maker movement. This always happens simultaneously with the infiltration of the 'Californian ideology' (Barbrook & Cameron, 1996, p. 44), which media-related core can be identified in the assumption of the direct formability of society through digital technologies. Perhaps such internal ambivalence explains the principal connectivity of the ideas and imaginations of pioneer communities in different local contexts. Accordingly, it seems fundamental to us to consider the role of pioneer communities in critical data studies if the latter want to understand how the 'thinking' (Daub, 2020) of Silicon Valley and its tech industry spreads globally. If critical data studies wants to consider this, they will find it difficult to avoid an examination of pioneer communities.

[3] With reference to a different cultural context than the one to which our own empirical studies are based, this is vividly illustrated by the example of the adaptation of the Maker Movement in China, where makerspaces are implemented by the state as places of technical innovation (Lindtner, 2020).

References

Abend, P., & Fuchs, M. (Eds.). (2016). *Quantified selves and statistical bodies transcript*. Bielefeld.

Ajana, B. (2017). Digital health and the biopolitics of the quantified self. *Digital Health, 3*, 1–18.

Barbrook, R., & Cameron, A. (1996). The Californian ideology. *Science as Culture, 6*(1), 44–72.

Barniskis, S. C. (2013). Makerspaces and teaching artists. *The Teaching Artist Journal, 12*, 6–14.

Bowker, G. C. (2013). Data flakes. In L. Gitelman (Ed.), *'Raw data' is an oxymoron* (pp. 167–171). MIT Press.

boyd, d., Golder, S., & Lotan, D. (2010). *Conversational aspects of retweeting on Twitter*. Presented at the HICSS-43. IEEE, Kauai, HI, January 6.

Bruns, A. (2019). Social media platforms and their fight against critical scholarly research. *Information, Communication & Society, 22*(11), 1544–1566.

Bruns, A., & Moe, H. (2014). Structural layers of communication on Twitter. In K. Weller, A. Bruns, J. Burgess, M. Mahrt, & C. Puschmann (Eds.), *Twitter and society* (pp. 15–28). Peter Lang.

Cammaerts, B. (2005). ICT-usage among transnational social movements in the networked society. In E. Silverstone (Ed.), *Media, technology and everyday life in Europe*. Ashgate.

Castells, M. (2001). *The Internet galaxy. Reflections on the Internet, business, and society*. Oxford University Press.

Crawford, K., Lingel, J., & Karppi, T. (2015). Our metrics, ourselves. *European Journal of Cultural Studies, 18*, 479–496.

Daub, A. (2020). *What tech calls thinking*. Macmillan.

Davies, S. R. (2017). *Hackerspaces*. Polity Press.

Didžiokaitė, G., Saukko, P., & Greiffenhagen, C. (2018). The mundane experience of everyday calorie trackers. *New Media & Society, 20*, 1470–1487.

Esmonde, K. (2019). Accommodating and resisting self-surveillance in women's running and fitness tracking practices. *Qualitative Research in Sport, Exercise and Health, 12*(1), 1–15.

Fotopoulou, A. (2018). From networked to quantified self. In Z. Papacharissi (Ed.), *A networked self* (pp. 144–159). Routledge.

Fredriksson, M., & Pallas, J. (2017). The localities of mediatization: How organizations translate mediatization into everyday practices. In O. Driessens, G. Bolin, A. Hepp, & S. Hjarvard (Eds.), *Dynamics of mediatization* (pp. 119–136). Palgrave Macmillan.

Gaffney, D. (2010). *Quantifying online activism. Extending the frontiers of society on-line*. Presented at the WebSci10, Raleigh, North Carolina, 26–27 April 2010.

Gauntlett, D. (2018). *Making is connecting*. Polity Press.

González-Bailón, & Wang, N. (2016). Networked discontent: The anatomy of protest campaigns in social media. *Social Networks, 44*, 95–104.

Gruzd, A., Wellman, B., & Takhteyev, Y. (2001). Imagining Twitter as an imagined community *american behavioural. Scientist, 55*(10), 1294–1318.

Hepp, A. (2016). Pioneer communities. *Media, Culture and Society, 38*(6), 918–933.

Hepp, A. (2018). What makes a maker? *Nordisk Tidsskrift for Informationsvidenskab Og Kulturformidling, 7*(2), 3–18.

Hepp, A. (2020a). *Deep mediatization*. Polity Press.

Hepp, A. (2020b). The fragility of curating a pioneer community: Deep mediatization and the spread of the quantified self and maker movements. *International Journal of Cultural Studies, 23*(6), 932–950.

Hepp, A., Breiter, A., & Friemel, T. (2018). Digital traces in context. *International Journal of Communication, 12*, 439–449.

Hepp, A., Alpen, S., & Simon, P. (2021a). Beyond empowerment, experimentation and reasoning: The public discourse around the Quantified Self movement. *Communications, 46*(1), 27–51.

Hepp, A., Benz, S., & Simon, P. (2021b). Zwischen Utopie und Dystopie: Wie der öffentliche Diskurs über die Maker- und Quantified-Self-Bewegung in Deutschland und Großbritannien die Pioniergemeinschaften zu Treibern tiefgreifender Mediatisierung macht. *M & K Medien & Kommunikationswissenschaft, 69*(2), S. 270–298. https://doi.org/10.5771/1615-634X-2021-2-211

Highfield, T., Harrington, S., & Bruns, A. (2013). Twitter as a technology for audiencing and fandom. *Information, Communication & Society, 16*(3), 315–339.

Hitzler, R., & Niederbacher, A. (2010). *Leben in Szenen: Formen jugendlicher Vergemeinschaftung heute*. VS.

Hunsinger, J., & Schrock, A. (2016). The democratization of hacking and making. *New Media & Society, 18*, 535–538.

Irani, L. (2015). Hackathons and the making of entrepreneurial citizenship. *Science, Technology & Human Values, 40*(5), 799–824.

Kannengießer, S., & Kubitschko, S. (2017). Acting on media. *Media and Communication, 5*, 1–4.

Katz, E., & Lazarsfeld, P. F. (1955). *Personal influence. The part played by people in mass communication*. Free Press.

Kostakis, V., Niaros, V., & Giotitsas, C. (2015). Production and governance in hackerspaces. *International Journal of Cultural Studies, 18*, 555–573.

Lange, B. (2015). Fablabs und hackerspaces. *Ökologisches Wirtschaften, 30*, 8–9.

Leavitt, A. (2013). Exploring the cultural salience of Twitter memes. In K. Weller, A. Bruns, J. Burgess, M. Mahrt, & C. Puschmann (Eds.), *Twitter and society* (pp. 137–154). Peter Lang.

Lindtner, S. M. (2020). *Prototype nation*. Princeton University Press.

Lomborg, S., & Frandsen, K. (2015). Self-tracking as communication. *Information, Communication & Society, 19*, 1015–1027.

Lotan, G., Graeff, E., Ananny, M., Gaffney, D., Pearce, I., & boyd, d. (2011). Information flows during the 2011 Tunisian and Egyptian revolutions. *International Journal of Communication, 5*, 1375–1405.

Lupton, D. (2014). Self-tracking modes. Available at SSRN 2483549.

Lupton, D. (2015). Quantified sex. *Culture, Health & Sexuality, 17*, 440–453.

Maireder, A., & Schlögl, S. (2015). Twitter-Öffentlichkeiten. In A. Maireder, J. Ausserhofer, C. Schuhmann, & M. Taddicken (Eds.), *Digitale Methoden* (pp. 115–139). Kommunikationswissenschaft.

Menichinelli, M. (2016). Mapping the structure of the global maker laboratories community through Twitter connections. In C. Levallois, M. Marchand, T. Mata, & A. Panisson (Eds.), *Twitter for research handbook 2015–2016* (pp. 47–62). EMLYON Press.

Nafus, D. (Ed.). (2016). *Quantified*. MIT Press.

Nascimento, S., & Pólvora, A. (2016). Maker cultures and the prospects for technological action. *Science and Engineering Ethics*, Online First, 1–20.

Neff, G., & Nafus, D. (2016). *Self-tracking*. MIT Press.

Nepstad, S., & Bob, C. (2006). When do leaders matter? *Mobilization: An International Quarterly, 11*, 1–22.

Nguyen, J. (2016). Make magazine and the social reproduction of DIY science and technology. *Cultural Politics, 12*, 233–252.

Pantzar, M., & Ruckenstein, M. (2014). The heart of everyday analytics. *Consumption Markets & Culture, 18*, 92–109.

Paßmann, J., Boeschoten, T., & Schäfer, M. T. (2014). The gift of the gab. In K. Weller, A. Bruns, J. Burgess, M. Mahrt, & C. Puschmann (Eds.), *Twitter and society* (pp. 331–344). Peter Lang.

Peppler, K., Halverson, E., & Kafai, Y. B. (Eds.). (2016). *Makeology*. Routledge.

Puschmann, C. (2019). An end to the wild west of social media research. *Information, Communication & Society, 22*(11), 1582–1589.

Puschmann, C., & Burgess, J. (2014). Metaphors of big data. *International Journal of Communication, 8*, 16900–11709.

Ramsauer, C., & Firessnig, M. (2016). Einfluss der Maker Movement auf die Forschung und Entwicklung. In Techno-Ökonomie-Forum (Ed.), *Industrial engineering und management* (pp. 43–61). Springer Gabler.

Ratto, M., & Boler, M. (Eds.). (2014). *DIY citizenship*. MIT Press.

Richterich, A. (2017). Hacking events. *Convergence*, Online First. https://doi.org/10.1177/1354856517709405

Ruckenstein, M., & Pantzar, M. (2017). Beyond the quantified self. *New Media & Society, 19*, 401–418.

Sajuria, J., Van Heerde-Hudson, J., Hudson, D., Dasandi, N., & Theocharis, Y. (2015). Tweeting alone? *American Politics Research, 43*(4), 708–738.

Sharon, T. (2017). Self-tracking for health and the quantified self. *Philosophy & Technology, 30*, 93–121.

Sharon, T., & Zandbergen, D. (2016). From data fetishism to quantifying selves. *New Media & Society, 19*, 1695–1709.

Sivek, S. C. (2011). We need a showing of all hands. *Journal of Communication Inquiry, 35*, 187–209.

Swan, M. (2013). The quantified self. *Big Data, 1*, 85–99.

Theocharis, Y., Vitoratou, S., & Sajuria, J. (2017). Civil society in times of crisis. *Journal of Computer-Mediated Communication, 22*(5), 248–265.

Tilson, D., Lyytinen, K., & Sørensen, C. (2010). Research commentary-digital infrastructures. *Information Systems Research, 21*(4), 748–759.

Toombs, A., Bardzell, S., & Bardzell, J. (2014). Becoming makers. *Journal of Peer Production, 5*, 1–8.

Turner, F. (2018). Millenarian tinkering. *Technology and Culture, 9*, 160–S182.

Turner, F. (2006). *From Counterculture to Cyberculture: Stewart Brand, the Whole Earth Network, and the Rise of Digital Utopianism.* The University of Chicago Press.

Williamson, B. (2015). Algorithmic skin. *Sport, Education and Society, 20*, 133–151.

Xu, W. W., Sang, Y., Blasiola, S., & Park, H. W. (2014). Predicting opinion leaders in Twitter activism networks. *American Behavioral Scientist, 58*(10), 1278–1293.

Open Access This chapter is licensed under the terms of the Creative Commons Attribution 4.0 International License (http://creativecommons.org/licenses/by/4.0/), which permits use, sharing, adaptation, distribution and reproduction in any medium or format, as long as you give appropriate credit to the original author(s) and the source, provide a link to the Creative Commons licence and indicate if changes were made.

The images or other third party material in this chapter are included in the chapter's Creative Commons licence, unless indicated otherwise in a credit line to the material. If material is not included in the chapter's Creative Commons licence and your intended use is not permitted by statutory regulation or exceeds the permitted use, you will need to obtain permission directly from the copyright holder.

The Power of Data Science Ontogeny: Thick Data Studies on the Indian IT Skill Tutoring Microcosm

Nimmi Rangaswamy and Haripriya Narasimhan

Introduction

> Now you can pursue world class courses on Indian soil … we are grooming a new generation of global Indians. … Join now and be the best. (A 'computer institute' brochure in small town India)

India's large under-twenty-five adult population has stoked huge demand for technical, IT job-ready education. India is home to the largest under-twenty-five demographic profile in the world, 604,394,787 people and 49.91 per cent of the total population (Census, 2011), requiring a

N. Rangaswamy (✉)
Kohli Centre on Intelligent Systems, International Institute of Information Technology, IIIT, Hyderabad, India
e-mail: nimmir@iith.ac.in

H. Narasimhan
Department of Social Anthropology and Sociology, The International Institute of Technology, Hyderabad, India
e-mail: haripriya@iith.ac.in

© The Author(s) 2022

A. Hepp et al. (eds.), *New Perspectives in Critical Data Studies*, Transforming Communications – Studies in Cross-Media Research,
https://doi.org/10.1007/978-3-030-96180-0_4

widespread, skill-oriented educational model, equipping youth to thrive in highly dynamic job markets. Strewn across India are private skill tutoring training centres and this chapter draws from ethnographic research conducted in Ameerpet, arguably India's largest IT skilling hub, a suburb in the Hyderabad Metropolis in the state of Telangana, South India, and Kumbakonam, a mid-sized town in the state of Tamil Nadu, South India. We undertook this research to explore the notion of information technology's (IT's) pervasiveness existing in a physical tutoring model of classroom teaching. Students marginalised in the more formal and competitive education system flock to Ameerpet-like IT skill hubs in preparation for the cut-throat job market. Small towns like Kumbakonam lack quality educational institutions like the Indian Institutes of Technology (IITs) or state-run public educational institutes. The IT skill tutorial institutes function like undergraduate colleges imparting engineering education, some of them are even like 'finishing schools' claiming to teach 'spoken English' and soft skills like 'social interaction and presentation' accompanying 'world class' computing skills. In the course of time, "there is a mystique of 'excellence' with commercial computer-training centres peddling a rhapsodic invocation of excellence which expect it to perform miracles in the absence of basic systemic changes" (National Employability Report, 2016). Both Ameerpet and Kumbakonam are about the aspiring college-going student population engaged in transforming opportunities within their lifetime. Aspirations are no longer about employment alone; it is about getting a toehold in a technology-driven, globalising world. Kumbakonam and Hyderabad are local contexts transforming to embrace global aspirations through hyper-local ecosystems such as the IT tutorial hub, a stepping stone to IT employment.

Fifty years ago, a call for a reformation of academic statistics pointed to the existence of an as-yet unrecognised *science* that learnt from data (Davenport & Patil, 2012). The catchy name 'data science' urged academic statistics to expand beyond theoretical statistics and statistical modelling to data preparation and presentation and on prediction rather than inference. Data sciences is the generalisable extraction of knowledge from data and is also an information science focused on the collection, analysis, visualisation, and management of large amounts of data (ibid.). For young adults in India, the power of data science ontogeny influences the future of work and a career in a new digital age with all its accompanying challenges. The digital environment and future educational and industrial human resources need data science programs. In the past five years, the data sciences evolved techniques for the (automated) analysis of data and expected data scientists to be high-ranking professionals "making discoveries while swimming in data", which implies implications for new business directions (ibid.). Data scientists are envisioned to work with technologies so as to put big data to use, sometimes in the service of new solutions, services, and applications.

Formal education in the data sciences, broad in scope and dynamic in content, remains a challenge to traditional educational departments. Data scientists, even today in India, acquire big data skills outside of the formal education system. Additionally, pedagogic concerns in the creation of knowledge from big data, evolving and expanding a formal curriculum to keep pace with cutting-edge data science technologies create further challenges. Around 1.5 million students in India graduate each year with an engineering undergraduate degree (Varshney, 2006). The lack of a job-ready education system and a burgeoning demand for IT skills have rendered the Indian youth ill-suited to gaining employment in technical professions. The need of the hour is a comprehensive, across the board, scalable, skill-oriented educational system equipping young people to survive and thrive in highly dynamic job markets (LaDousa, 2007). Gaps in skill education and suitability for the IT job market are currently being addressed for job-oriented skill tutoring. The advent of online education and blended learning in the Global North points to a possible solution for India too, but online learning for the Indian consumer has thus far consisted primarily of importing courses designed in and for the Global North posing many challenges in deriving value out of the online courses in their current form. Massive Open Online Courses (MOOCs) continue to remain contextual in nature lacking student retention from sign-up to course-end (LaDousa, 2005; Rosé et al., 2014). The geographical scale of the Indian sub-continent, variation in language and cultural contexts, and diversity in the levels of student capacities to learn and participate in the education system present additional challenges to scale quality educational systems (LaDousa, 2007).

A groundswell of the informal skilling industry in India has been actively responding to this skill gap since the 1980s—a response that the current education system is unable to deliver (Patibandla & Petersen, 2002). IT tutoring institutes dot the skyline of Indian metropolises, cities, and small towns imparting IT skills that cater directly to the job market. Students join these institutes in large numbers, attracted by the short-term time commitment and the clear match between their goal of attaining work and what the institutes offer (Joshi et al., 2018). Over three decades of commercially driven IT, skill hubs have evolved into tutoring models that youth segments in India are not only flocking to but finding educational value in persisting with their training. This chapter probes the Indian commercial skill tutoring market that continues to proliferate as our research field. We investigate the research site to understand the social profiles of learners and the perceived efficacy of the Ameerpet-like ecosystem and small-town micro hubs such as Kumbakonam over the formal education system in India for IT industry-ready skills.

An Overview of Technical Education, Higher Education, and Unemployability in India

This section will elaborate on the context of higher education in India, addressing gaps in the education system, especially in the areas of employability and job readiness. India is home to the largest under-twenty-five demographic profile in the world—604,394,787 people (49.91 per cent of the total population) (Census, 2011; Statistics Times, 2016; Joshua, 2014). The combined forces of an ill-qualified education system and a burgeoning international IT job sector has stoked the youth in India to seek job-ready data science education for gainful employment. Around 1.5 million students graduate each year with an undergraduate engineering degree (NIC, 2017). The combined strengths of the Indian government and corporate bodies, well aware of the widespread demand for job-oriented data science education, are yet to successfully address the situation. In 2017, there were a total of 6447 approved technical institutes, which enrolled 2,871,007 students (All India Survey of Higher Education, 2012). Out of these institutions, the top government-backed institutions that are recognised in the Global North, such as the Indian Institute of Technology (IIT), make up only a small fraction of the total intake (the total number of seats offered in all Government Funded Technical Institutes [GFTIs] in 2017 was a mere 36,200) (Mohammad et al., 2014). There is no denying the acute shortage of technical schools in the country to cater for the 1.5 million young Indians leaving high school every year. More importantly, competition for admittance is extremely high. The acceptance rates at these institutions are the lowest in the world by a wide margin, with the IIT acceptance rate of 0.7 per cent in 2014 being eight times less than that of Ivy League institutions such as

Yale and Harvard (Quality Council of India, 2016; Toppr, 2015). Given the extraordinary level of competition in quality conscious, 'first tier' institutions, many industrious students turn to a variety of other lower-quality technical institutions. These 'second-tier' institutions vary widely in terms of regulation, the number of students, examination patterns, syllabus, the quality of their faculty, and fee structures. State universities are run by the governments of each of these states and territories and cater to anywhere between 67,000 and 120,000 undergraduate students, distributed among their affiliate colleges. The best teachers opt for top ranking colleges or private institutions (which can match corporate pay packages) leaving many other schools thirsty for quality teaching staff (Kapur, 2010). Most low-quality institutions are unable to address the diverse backgrounds of their student body in terms of linguistic variation, varying levels of pre-existing knowledge, and previous training. Students we spoke to in Ameerpet, all of whom were from 'second-tier' institutions, alluded to the lack of out-of-class tutoring, mentorship, or guidance at their institutes. The uneven quality of teaching, a limited focus on practical knowledge, and the lack of participative classroom culture create an exam-focused atmosphere, with students focusing on memorising material rather than developing practical knowledge of the subject. A technical report places 18.4 per cent of the total number of engineering graduates generally employable and only 3.2 per cent suitable for jobs in the IT industry (National Employability Report, 2016)—emphatic evidence of the huge skill gap arising out of the Indian science and technology education system.

Jiazhi and Steinmüller (2021) plot and analyse the rise and growth of leadership and business coaching tutorship in China as a response to a changing sociopolitical environment. The movement of people from rural areas into the urban labour markets striving for upward social mobility views education and self-improvement as the primary means of realising that desire. The leadership and business coaching programme, apart from offering skill sets and job aspirations, signals the hopes and anxieties of the rising Chinese middle class in a new political context of entrepreneurship and emerging capitalist business culture. Our fieldwork in Ameerpet and Kumbakonam, apart from resonating with the above study, speaks to the educational gaps in the teaching of science and engineering in the context of a rising Indian off-shoring job sector.

Methodology and Field Sites

This chapter is informed by primary data collected through a variety of qualitative methods. Between June 2017 and March 2018, we conducted fieldwork in Ameerpet, a dense bustling commercial suburb of Hyderabad city and the largest IT skill tutoring hub in India, largely run by the private sector, offering a plethora of courses ranging from robotic process automation to manual testing courses. Our ethnographic methods afforded us a first-hand experience of Ameerpet's social geography and enrolling in a course for six weeks helped us engage in an immersive classroom experience to get a contextual feel for the quality of teaching, tutors, class infrastructure, and student interactions. Unstructured interviews with students, tutors, and managers of IT skill training institutes helped us gather first-person data about the hub. These face-to-face, in-depth interviews helped us understand the importance and value of the Ameerpet institutes for students, the quality of tutoring, and the relevance of the syllabi. We developed social profiles of students and tutors to comprehend the motivations to study or teach at Ameerpet. Our depth of immersion and our interactions with key stakeholders in the Ameerpet tutoring system offered points of view to evaluate the offerings and implications of its ecosystem. In Kumbakonam, interviews were conducted among students at an IT training institute and staff at two schools and a few other IT training institutes in the town. Visits to engineering colleges on the outskirts of town and focus group meetings with students yielded interesting observations. Importantly, we produced an index of the technology infrastructure, computers, software, and peripheral devices servicing the IT skill training in classrooms. All our interviews were recorded, transcribed, and coded manually to analyse our research questions. More importantly, handwritten notes from field and classroom observations in regard to tutoring and pedagogic styles, student response, and classroom participation contributed significantly to the coding and analysis of data. We employed the inductive approach, also known as inductive reasoning, beginning with observations and the development of arguments based on these observations. We observed recurring patterns, resemblances, and regularities in the data to arrive at conclusions. These recurring patterns were developed into themes and aided in the manual coding of transcripts from which we were able to draw insights.

A response to the demand for basic and advanced IT skills in India is not only opportunistic but also pedagogic and industry-ready. This response has been realised in the form of skill tutoring classes, driven and managed by the private sector offering short, condensed material promising job readiness in less turnover time than allegedly provided for in the formal Indian education system. Classroom pedagogic tools cater to and are customised for the effective imparting of IT skills to compete in the job market. The Ameerpet and Kumbakonam marketing and the course structure of classes identify and target employability as the key goal of students enrolled in their classes, offering a clear link between the skills they learn and the job market they are preparing to enter (Joshi et al., 2018). Estimates vary about the number of students taking these courses in Ameerpet—anywhere between 60,000 and 100,000 students per month (The Economist, 2017). Each institute in the hub offers multiple courses ranging from introductions to MS Office to Robotic Process Automation, with a fee structure varying across institutes and courses, from INR 2000 to as much as INR 35,000 for a single course (i.e. US $30–$550). The fees charged depend on the reputation of the tutorial centre and the quality of teaching and the teachers they employ. The turnover time for these institutions is rapid—courses last from one to six months and multiple batches for the same course are held in quick succession. The managers of the Ameerpet institutes update courses to tally with skills in the job market, mining information from online job portals and their own industry contacts to keep track of the latest demand for jobs. Kumbakonam town makes an interesting contrast to the metropolis of Hyderabad to showcase the ubiquity of computers and the internet. A town located 300 kilometres away from the metropolitan city of Chennai in South India is trying to negotiate its image as a *mofussil*, an "inferior" space, as a place "marked by slowness, by absence of the new and recent" (Fox, 1969). Kumbakonam is actively resisting the above definition by bringing the 'computer' into its 'everyday'. A visitor to Kumbakonam will be struck by the presence of a large number of 'net cafes', or 'browsing centres. Ethnographic studies in Kumbakonam town were conducted in the summer of 2009 when India had consolidated the IT wave.[1] Kumbakonam is a 'temple town', literally, a pilgrim town full of temples, situated on the banks of the river Kaveri with good transport links to the nearby cities of Chennai and Trichy. The Mahamaham festival, held every twelve years, during the Tamil month of Maasi (February–March) brings

[1] This research was funded by ESRC early career fellowship awarded to HN (Grant number—**RES-063-27-0089**).

devotees to the town to take a 'holy' dip in the Kaveri waters. Kumbakonam is also known as a trading centre for textiles, brass utensils, and gold jewellery. It may be a dusty town, but certainly not a sleepy one. The streets are constantly full of people. Long-time residents of Kumbakonam recall the town as being one of the best in the state for its schools and colleges. There are three schools that are over a century old. The 'Silver tongued' singer Srinivasa Sastri, the mathematical genius Srinivasa Ramanujam, and the well-known agricultural scientist M.S. Swaminathan are just some of the famous 'sons of the soil'. It is also a town that has recently seen a spate of 'modernising' infrastructural development comprising apartment complexes, a new university, a three-star hotel, a 'cutting edge' hospital, and at least one television channel dedicated to 24-hour news about Kumbakonam. The past 20 years has infused the city with a new 'life blood' in the form of aspirational behaviours to acquire modern devices and to adapt to the computing age.

The Ameerpet IT Skill Hub: There Is a Skill Just Around the Corner

There are 66 types of SAP courses on offer in the hub.—Ameerpet Instructor

Ameerpet draws students from all corners of India, who move to Hyderabad city specifically to enrol in a variety of courses. Their primary motivation is securing a job or upgrading their IT skills to advance an existing job profile. The hub attracts learners from neighbouring districts in Telangana to nearby states such as Andhra Pradesh and Orissa to the far-flung northeast states such as Manipur. International students, particularly from the Middle East and South-East Asian countries, are present in small numbers. The students usually have graduate degrees in different streams, most without work experience; there is also a considerable section of employed learners who join courses to upgrade their IT skills. Despite varied educational backgrounds, none of these learners has studied in India's top-tier colleges—most have degrees from unranked private colleges and state universities located closer to their homes. They choose to come to Ameerpet to learn the same skills which are primarily oriented towards teaching a focused set of content, one that is relevant for a specific job profile (Joshi et al., 2018). The language of jobs and job titles captivate prospective students and help them seek out courses that might align with a potentially active job profile. Students aim and aspire for specific job profiles and seek to enrol in courses offering specific skills related to those profiles. While students gain skills as they progress in a course and are aware of the need for multi-skilling, they prefer to take courses that are clearly tailored for a job profile instead of courses geared towards a generic skill set.

A prospective IT skill learner 'discovers' Ameerpet, seeks out specific institutions, and tutors with the help of a robust information network. Anthropological literature alerts us to the significance of informal networks in low-resource, developing economies. In these spaces, it becomes crucial to understand and tap into informal networks in order to disseminate information and pitch products (Espinoza, 1999). As Anant, 23, a student from the state of Orissa in Central-West India doing a Quality Testing course, told us, "I came here because my school mate who has now completed his engineering program and worked for a year told me about this place. I am in my final year and have dropped a semester to learn at Ameerpet. I am taking multiple courses on Java, Python and web design too." A good number of students we spoke to had similar stories to share about friends, colleagues, and family members (who have themselves taken courses in the hub and found work in the IT industry) directing them towards Ameerpet. Deepak, male, 21, another student in the same course from the relatively nearby city of Visakhapatnam, told us, "My elder brother got a job from here only. Everyone knows about this place; it is the best coaching centre." This reputation of Ameerpet being the 'best coaching centre' is fortified in the way information circulates through students who have benefitted and found jobs—many do not even attempt to probe alternative options such as exploring local institutes or online classes before settling for Ameerpet (Joshi et al., 2018). References from immediate social networks rely on existing trust in the social relations that recommend Ameerpet institutes. Further, social relations among students from similar socio-economic standing and with similar demands bolster faith in the selection criteria offered by these networks. Tutors at the Ameerpet institutes are ex-employees or moonlighting in the hub while holding down a day job in the IT industry, which assures students of the authenticity of their learning experience. Ameerpet success stories hover around institutions' placement records, as does anecdotal evidence from those who have found jobs after their time in Ameerpet.

Job attainment is a central goal among learners who enrol in Ameerpet and landing a job in the IT industry is their top priority when enrolling at an Ameerpet institute. In a discussion amongst several students regarding placements and prospective salaries, Vishali, a young woman in her 20s, from the far away North-eastern state of Tripura, said, "What's the problem in starting from zero? We can start from the base, get experience then move ahead." In another discussion, Prakash, male, early 20s, a fresh graduate taking an SAP course, echoed this statement, "I want to get a job in the IT sector. … I don't care about the salary. Growth and money will come later, once the job is there." Students hear and believe that stable jobs in the IT industry, along with opportunities to go abroad, are definitional as career prospects (Joshi et al., 2018). Many were content with a non-technical work profile in the IT sector—a toehold in the IT industry promised a 'future of possibilities'. Tutoring institutions are aware of this demand and adopt several strategies to ensure course offerings align with the skills required for the IT market and constantly upgrade skill requirements. Managers of tutoring centres browse job portals and scour their networks in the IT industry to gather 'know how'. As, Sankar, a co-director at a fifteen-year-old institute, said, he spent several hours a week trying to figure out the current in-demand courses by looking at job portals. As Raman, a manager at a recently opened institute, explained to us, "If a student comes and says they have seen a course on a job website, and I don't have that course, then students will not come to me. Even if that course is not right for them and later I make them take another course, first I need to have the latest course on offer."

Instructors are a key industry link—a majority of them are employed IT professionals who moonlight as tutors in Ameerpet. They have insider knowledge of job profiles and what skills are in demand, the kind of skills and knowledge needed to learn tools, and modes of classroom delivery of course material so that students may fare well in job interviews. For instance, instructors stress the difference between 'course knowledge' and 'job interview proficiency'—tutors highlight sections of coursework that are important for interviews and train students to answer basic yet critical interview questions. Tutors decide course syllabi and could choose to not teach redundant software tools in favour of new industry standards. In the testing course undertaken by the first author, the instructor explained that he wouldn't be teaching a tool called QTP, QuickTest Professional, since it was no longer valuable in the job market (despite being listed in the syllabus), and would instead focus on 'Selenium', an open-source testing software platform, which was, in the instructor's words, "all the rage in learning testing tools".

The Ameerpet hub addresses job readiness and employability, key concerns for students looking for an employable job profile in India. Learners are alert to the alignment of a course towards a job profile for which they clearly see value in a job market. Institutions in Ameerpet manually scan job portals and use industry connections in order to identify the latest tools and IT industry trends. Online platforms can significantly refine this process with data mining tools to match what technical and soft skills recruiters are looking for at any given time and place. Project-oriented courses focus on teaching the entire skill set required for entry-level work in a particular job profile or stream. For example, the Testing Tools course in Ameerpet begins with an overview of software engineering and development; moves on to teaching Java programming skills that testers will require; students then move on to the use of various automation applications; and finally, students create and execute test cases as they would in a workplace. Similarly, an Android development course that lasts for two months begins with software installation, moves to coding in Java, and then to an overview of various APIs and their implementation for software applications. It is the packaging of an end-to-end learning and tutoring course structure that proves successful for the Ameerpet tutoring institute. The Ameerpet tutoring ecosystem, despite a narrow focus on the off-shored IT skill requirements, offered a spectrum of sometimes mind-boggling courses to meet the differentiated demands of a globalised IT job market.

The Coaching Micro Hubs of Kumbakonam

Now it's all 'e' ... 'e-publishing', e-education, e-commerce....—Ramanan, Owner of a Computer Coaching Centre

A visitor to Kumbakonam will be struck by the presence of a large number of 'net cafes', or 'browsing centres'. It would appear that Kumbakonam is perennially online. One such 'cafe' is located in a non-descript building on a main road, right next to a stall making fried fast food. A very small room is divided into six cubicles on either side, each fitted with one computer, connected to the internet. All the government contract tenders are applied for online. High school exam results are published and train tickets are booked online. Even passport forms are now filled out online, as are applications for bank jobs. Most of the clients in internet cafes are, not surprisingly, college students who come to check their email, download music, voice chat with friends and family, fill out application forms, and search for jobs. Naseer jokingly said that, in addition to name and address, one is now required to provide a mobile number and an email ID for most job and university applications; information technology is omnipresent in this *mofussil* town. The effect of IT, in the form of the mushrooming of tutoring centres designed exclusively for imparting skills and training in computer applications, is referred to as 'computer coaching centres', 'computer class', or 'computer institute'. Students in small towns like Kumbakonam feel marginalised, because of the lack of a global/cosmopolitan 'atmosphere'. They wish to participate in the growth they see in cities like Chennai and consequently abandon any interest in traditional occupations (Gupta, 2005). This lack of 'atmosphere' in the town works as a strong impediment to their perceived ability of achieving this goal. This results in a state of "educated unemployment" as more of a reality than the dream of participating in the so-called IT boom.

Being literate has increasingly come to mean being 'computer literate'. The perception is that "India has changed, the state of Tamilnadu has changed and, Kumbakonam town has changed as well". "Now it's all 'e' … 'e-publishing', e-education, e-commerce", says Ramanan, the owner of a 'computer coaching' centre. Anandi, an eighteen-year-old girl, studying at an arts college outside the town, and a student at Ramanan's institute, said that computers are omnipresent (*engum* computer) and "computers are everywhere and in everything", because "from small things to satellite launches everything is done through computers". Another student, Venkat, thought that he "didn't know anything" before he joined Ramanan's class because he did not know how to use a computer. Without a degree of computer literacy, any other kind of knowledge is considered irrelevant. During our informal interactions with high school students, at the cusp of choosing an undergraduate education, it became apparent that engineering was the subject of choice and Infosys, one of the first home-grown IT giants in India, was considered as one of the most desirable companies to work for. Computers are seen as a gateway to the outside world. Not one participant expressed a desire to study medicine or law.

Schools in Kumbakonam hold hopes that in ten years' time, "there will be a computer in each classroom". Coming from the principal of a school where blackboards on wooden stands separate one classroom from another, this statement might seem premature and overly optimistic. But the fact that all government administrative work is now computer-based has allowed for a discourse on "the omnipresence" of computers in everyday life. The principal acknowledged that they hoped to create only "awareness" and not necessarily equip students with the required knowledge on how to use computers. Educational institutions in small towns are engaged in attempts to change their image as one of lacking resources and infrastructure facilities by investing in IT in a bid to regain a certain status and operationalise 'modern' education content (Kumar, 2006). At the undergraduate level, the resource and infrastructural gap between private and state-funded institutions is glaringly apparent, as are the gaps between students coming from Tamil and English language schools.

Private colleges have, to some degree, exploded in Kumbakonam over the last ten years including two engineering schools. The engineering courses' curricula are relatively demanding. Students write eleven papers per semester and twenty-two papers per year. In four years of study, they write eighty-eight papers in total. Students taking computer science as one of the four subjects in high school learn C and C++ in school but because they are not taught particularly well, they do not possess a full understanding of these languages. For those students who are first-generation college-goers, or who studied in a school where Tamil was the medium of instruction, an engineering education is full of obstacles that must be overcome, the least of which is the college's location (Fuller & Narasimhan, 2006). We argue that, yet again, the challenges of the low-quality formal education system are mitigated by the rise and evolution of commercial IT skill tutoring centres that bridge education and job readiness.

Computer Coaching Centres

Computer is like Laadam (rein) for a horse ... everything is held by computers.—
Saravanan, E-Cube coaching centre

One can see fluorescent pink and green wall posters across Kumbakonam announcing classes specifically set up to teach specific computer applications in anything from ten days to six months. It is striking to see that the IT industry is seen as the only field where progress can be achieved. As Ramanan said, "Students don't know the famous manufacturing firm Simpsons, but they all know about Wipro and Infosys". It is companies like these that students would prefer to join after their training, to partake, in some sense, in the globalisation experience. Computer coaching arrived in Kumbakonam in the early 2000s. Ramanan's 'Srinivasan computers' is one such ‘coaching centre’, functioning out of the living room of an ancestral house in Kumbakonam. Ramanan is an unmarried, forty-year-old native of a smaller town nearby who had worked in Chennai teaching computing at a local institute. He decided to settle down in Kumbakonam and works in a local school during the day as manager of their computer lab. Ramanan’s institute started in 1998 with one computer. Now he has ten to twelve “systems” and fifty batches of ten to twelve students per batch, each lasting for six months. A mix of both engineering and non-engineering, students, male and female, living in Kumbakonam or small towns nearby attend these classes. Students felt that their teachers in college were unable to teach them the skills they required to find work. For instance, in regard to C++, only fifteen software applications are taught on the college syllabus. They feel compelled to come to institutes (like Ramanan’s) in order to learn the whole gamut of applications related to C++. Even if the current engineering syllabus at mainstream colleges is considered ‘good’, students are quick to criticise the staff for their inability to teach complicated languages and programs and their inability to speak English fluently (the running joke among students is that those who fail to get jobs become faculty in engineering colleges).

Branches of popular institutes (such as NIIT and Aptech that have a national presence) apparently had to close down branches in Kumbakonam due to a lack of trained faculty. The fee structure varies from one institute to another. Ramanan’s institute proudly claims that “no course will cost more than INR 1500 or 25 $US”. Computer Sciences Corporation (CSC) is a tutoring franchise and opened a branch in the late 1990s. The owner is a well-off farmer’s son and has registered his disdain for agricultural occupations by founding CSC. CSC remains open from 7 a.m. to 9 p.m. Students receive about an hour’s tuition per day. With 16 staff members and 330 students, CSC doubles its student attendance in the summer months. CSC teaches ‘fundamentals’, a word commonly used to denote a package of basic computer applications such as MS Word, Excel, PowerPoint, and Access, charging INR 2000 or US $70 for a two-month course. Most of the teaching is done in Tamil since many CSC learners are from rural areas. Much like the tutorial hubs of Hyderabad, students from out of town live in hostels or rent apartments in Kumbakonam and attend evening classes such as those run by Ramanan. Those from surrounding villages leave their homes at 6 a.m. to commute to class. The Professor said, “[W]e don’t push them hard … we let them learn at their pace. … Students in Chennai are assumed to be able to ‘tackle’ the outside world, while students from *mofussils* lag behind. In Chennai, students 'know the atmosphere', and know how to find work. Students in Kumbakonam, on the other hand, lack communication skills, motivation and understanding. Tests are, therefore, a struggle."

E-Cube, another computer coaching institute, presented itself very attractively, located in a fluorescent green building on the first floor above a tea kiosk selling local beverages. E-Cube has a small veranda at the front of the building leading to a large hall in which several plastic curtains separate 'class' sections. One end of the wall is lined with computers, again covered by curtains. The person running this institute, Saravanan, who works as faculty in the computer science department of a local college, said, "[C]omputers are like *laadam* (rein) for a horse ... everything is held together by computers".

The institute is also involved in 'body shopping', web design, and various consultancy projects. Sujatha, the receptionist, has a master's in IT at the college where Saravanan was her teacher and commutes from a village about ten miles from Kumbakonam. Discussing the problems students face when studying programming languages, she said, "[M]oney problem [affordability], language problem [inability to understand English] ... problems are obvious and we try to bring solutions". The commonly seen 'flexi boards' in Tamil Nadu require a fair degree of computer work. According to E-Cube's owners, most parents assume that their children can survive just by making flexi boards, and prefer that they study 'computers' instead of more traditional occupations. Institutes like E-Cube 'take students through the "first step" of gaining some knowledge in computers', offering a 'stepping stone' towards advancement and opportunities to 'study further' in Chennai. Tutors like Saravanan begin by making students "learn how to switch on a computer and familiarise themselves with its functions". Institutes set up in the hope that the density of Kumbakonam and the nearby areas would attract large numbers of young adult learners. Posters on the walls at this institute announce to visitors, "Now you are in the world of success". The front office is supposedly 'posh', with nice sofas and well-dressed female staff on the front desk. Glass doors lead to classrooms labelled 'computer lab', which look very sophisticated in the context of a small town. The idea is to provide students with a visual imagery of a modern setting which they may not have access to, even in the colleges they go to for their main studies.

Another critical requirement for towns like Kumbakonam is the prospect of a job, preferably in the IT industry. Learners felt the acute lack of privileges that cities and metropolises in India afford; the possibility of employment that begins with the 'campus interview' or the opportunity a good quality formal educational institution can offer to attract IT companies to recruit their graduates. A college gains its reputation partly through the firms that recruit from each cohort, which in turn depends on the reputation the college already enjoys as a consequence of students achieving high grades. Not many industries visit Kumbakonam colleges for interviews but that has not constrained the young who hope and aspire to land an IT job in the cities of Tamil Nadu.

Beyond Developing IT Skills to Employment

Ravi comes from the Thevar community, traditionally a small agricultural community in Kumbakonam. His father is a farmer, but more importantly, he is also a traditional village chief (naataamai). But Ravi is not interested in agriculture or getting involved in the numerous conflicts that take place both within and between villages. Like most young adults and students in the town, he is the first member of his family to go to college. As his parents and his extended kin group do not speak English, Ravi is really "frustrated, finding it difficult to practice speaking the English language at home". In the company of friends and college mates, Ravi is further handicapped by another problem because, he said, "if a person wishes to speak too often in English that person is teased, ridiculed, mocked, and almost ostracised. Such social control acts as a barrier to my desires to 'develop' (munnetram)." Ravi feels he will have to leave Kumbakonam and go to Chennai in order to develop his English language skills.

The tutoring hub's learning model develops a robust understanding of the cultural and contextual aspects that may affect an individual student's education; the why and what of students flocking to the hub; the professional goal of these learners in seeking courses offered by the tutoring hub. In tweaking course content and pedagogic style to match student learning capacities and in revising the syllabus to reflect the current market and job readiness, the hubs cater for a wide range of students coming from diverse socio-economic backgrounds. The tutoring hub in Hyderabad and the coaching centres in Kumbakonam not only teach IT skills but also offer additional services to guide students in the enhancement of their job prospects. Students are able to download and fill sample résumé templates; staff share links and forward job announcements by email; certificates are provided upon the completion of coursework; and work experience is provided off-site where the opportunity arises when a member of staff also work in the wider IT sector. A range of soft skills are imparted in the course of software training. Part of developing a 'personality' suited for a global India is in the attainment of fluent English skills and most 'personality development' classes include 'spoken English' classes. In Kumbakonam, the most popular has been operational for almost thirty years. It is run by a retired English professor and his wife. The course runs from April to June, during summer vacations, and costs an individual learner INR 1800 or US $25. The class recruits several hundred students each year. Workbooks and files given to students are mostly the type of English grammar found in high school textbooks. The manager of the institute stated that students are taught to write grammatically correct English and then to speak in English. The standard of English among the learners was so poor that "they needed to learn to write properly before they spoke properly". Courses that go beyond IT skill building offer learners the opportunity to develop their 'creative writing', as well as their 'public speaking' skills and conduct mock interviews and group discussions. As one student explained, "we are taught how to get on to the stage and [these institutions] are similar to finishing schools … the idea to develop soft skills required by industry".

Job attainment is a central goal among students who enrol at tutorial hubs. Many of them were not concerned with their future job profile, or salary even, as Prakash, male, early 20s, a fresh graduate taking an SAP course, said, "I want to get a job in the IT sector. … I don't even mind if I end up with a non-technical work profile. Growth and money will come later, once the job is there" (Joshi et al., 2018). Tutoring centres adopt strategies to ensure course offerings align with the skills required. For example, a quality testing tool called QTP, no longer valuable in the job market, was replaced by 'Selenium', an open-source testing software platform. Tutoring centres offered job-enhancing services, such as résumé/CV templates, links to job portals and email forwards of possible opportunities and, more perhaps of more value, lab environments complete with desktops, employee ID cards, and biometric entry systems, where students learn the ropes of in simulated offshore workplaces. Some of the 'live labs' in the Ameerpet hub provide hands-on training in IT projects and offshore assignments "to reduce the gestation period" of a future employee in the IT industry. Employment and employability are not only student-recruiting mantras for hubs in Ameerpet and Kumbakonam, they function as gateways to IT-enabled employment for a sizable section of young

Indians while addressing the severe crunch in resources and manpower arising from the established Indian institutional educational sector.

The vignettes we offer in this chapter make a case for looking at 'data studies' from an ethnographic perspective. Our research uncovers a 'program' of upward mobility that has been scarcely investigated in India, though myriad studies exist on the desire for English language education (see LaDousa, 2005). Extensive studies of the IT industry's growth in India tend not to tap into the large, informal, and semi-formal training infrastructure which is key to the industry's 'success' (e.g., see Patibandla & Petersen, 2002). Computers have captured the imagination of the upwardly mobile middle classes since the 1990s but had largely escaped critical examination (as Krishna Kumar points out in LaDousa (2007)). Our aim here is to not just 'fill this gap' in literature but push forward a theoretical perspective on data from an ethnographic lens. What directions might a perusal of 'data' in the Indian context look like? The excerpts presented here articulate the varied expectations of those undergoing training in these institutes. It is largely about livelihood and yet it is also about aspirations towards a different way of life. Students at these institutes keep themselves abreast of the latest developments in the field. The objective is not merely to get a job, but to be in a position to use one's training as a vantage point from which to achieve upward mobility. In Kumbakonam, that meant moving to Chennai for jobs. In Chennai, it meant taking up an on-site project in the USA. But the aim is always beyond the immediate city, town, or one's home town.

Conclusion

> *Indian higher education seems like an enigma wrapped in a contradiction. Pockets of excellent teaching and research are surrounded by a sea of substandard colleges. The best graduates compete successfully in the world job market, but unemployment at home is the reality for many. Scholarship is often superseded by politics and, in many institutions, crisis is the norm. A system which was at one time highly selective has opened its doors to large numbers, yet at the same time there is conflict and sometimes violence over what remains a scarce commodity.* (Altbach, 1993, p. 4)

IT tutoring curricula operates and is built with the express purpose of providing youth with skills that eventually lead to gainful employment. Globally, people are reskilling and upskilling themselves in the hopes of becoming more competitive in the labour market, but how will these skills translate into employment opportunities? What are the most effective ways for people to learn and apply ICT skills across diverse population types and socio-economic contexts (Garrido et al., 2009)? Employability encompasses a combination of factors that demand an exploration of the role IT skills and skilling play in educational contexts and their relevance for IT-enabled employment in particular. The challenge for researchers is to speak of employability, drawing on specific educational ecosystems and their evolution against the broader context of globalisation and the intense competition for knowledge labour. Private, market-driven IT tutoring non-governmental organisations training young Indians to improve employment opportunities act as *liaisons* between the desirous IT sector and the means to acquire capabilities to get a toehold on the ladder of skill upgradation.

Globalising countries like India are transforming into information technology service hubs but are paying scant attention to the creation of new education models to suit employment in the IT industries. Government expenditure on education in India is only 3.8 per cent of GDP, compared to the world average of 4.8 per cent (All India Survey of Higher Education Report, 2012). Despite the large numbers of engineering graduates (Mahajan, 2014), only 18 per cent of graduating engineers in India were employable in roles for the ICT industry. The poor quality of privately provided engineering education in India is attributed to the high levels of competition of getting a place in one of a small pool of quality institutes, poor government investment in technical education, and what Akerlof (1970) calls "informational asymmetry" between institutions and students. The latter has pushed quality engineering education out of the market to allow "lemons" (or low-quality education) to dominate the field.

IT skills and coaching hubs in Ameerpet and Kumbakonam have functioned to fill gaps in job readiness and employability, which are key concerns for students. In places like Kumbakonam, the 'pain' of living and being educated in the 'provinces' point to the Indian state's inability to deliver "progressive, successful" educational paths for its young population (Kambhampati, 2002). In spite of an ineffectual education system, the young learners in Ameerpet and Kumbakonam remained extremely optimistic about their place in a global India. They do not subscribe to Gohain's view of "drugged incuriosity and intellectual paralysis" characteristic of many marginalised spaces and their peoples (Gohain, 1997). The hubs do not offer Ivy League education, nor are they necessarily institutes of excellence. However, what they do well through a widespread, scalable, skill-oriented educational system is to channel the young population in India through routes where they might find ways to survive and thrive in highly dynamic job markets. As global IT sectors become knowledge-centric, the skills related to information-intensive employability make visible the growing gap in the ability of existing educational systems and IT job readiness. This chapter has sought to offer a specific response to bridge gaps in data science education and skill development, thereby addressing a key condition for narrowing the employment and skill gap, one negative symptom of a burgeoning IT sector.

References

Akerlof, G. A. (1970). The market for lemons: Quality uncertainty and the market mechanism. *Quarterly Journal of Economics, 84*(3), 488–500.

All India Survey of Higher Education Report. (2012). https://dhte.py.gov.in/sites/default/files/aishe-final-report-2011-12.pdf

Altbach, P. G. (1993). The dilemma of change in Indian higher education. *Higher Education, 26*(1), 3–20.

Ameerpet. (2017). India's unofficial IT training hub. *The Economist*, March 30, 2017. https://www.economist.com/business/2017/03/30/ameerpet-indias-unofficial-it-training-hub

Census of India. (2011). https://censusindia.gov.in/2011-common/censusdata2011.html

Davenport, T. H., & Patil, D. J. (2012). Data scientist: The sexiest job of the 21st century. *Harvard Business Review, 90*(10), 70–76.

Espinoza, V. (1999). Social networks among the urban poor: Inequality and integration in a Latin American city. In B. Wellman (Ed.), *Networks in the global village* (pp. 147–184). Westview Press.

Fox, R. (1969). *From zamindar to ballot box: Community change in a north Indian market town.* Cornell University Press.

Fuller, C. J., & Narasimhan, H. (2006). Engineering colleges, 'exposure' and information technology professionals in Tamilnadu. *Economic and Political Weekly, 41*(3), 258–262.

Garrido, M., Badshah, A., & Coward, C. (2009). *Deconstructing ICT skills and employability*, Vol 5, Issue 2 | Summer 2009 (Special Issue: ICT Skills & Employability), Information Technologies and International Development.

Gohain, H. (1997). In search of scholarship. *Economic and Political Weekly, 32*(1–2), 26–27.

Gupta, D. (2005). Whither the Indian village: culture and agriculture in 'rural' India. *Economic and Political Weekly, 40*(8), 751–758.

Jiazhi, F., & Steinmüller, H. (2021). Leadership programmes: Success, self-improvement, and relationship management among new middle-class Chinese. *Ethnos*. https://doi.org/10.1080/00141844.2020.1867605

Joshi, M., Joshi, T., & Rangaswamy, N. (2018). Scaling classroom IT skill tutoring: A case study from India. *Proceedings of the 2018 ACM Conference on Human Factors in Computing Systems* Paper No.: 630. 1–12.

Joshua, A. (2014). Over a quarter of enrolments in rural India are in private schools. *The Hindu*, January 16, 2014. http://www.thehindu.com/features/education/school/over-a-quarter-of-enrolments-in-rural-india-are-in-privateschools/article5580441

Kambhampati, U. S. (2002). The software industry and development: The case of India. *Progress in Development Studies, 2*(1), 23–45.

Kapur, D. (2010). Indian higher education. In C. Clotfelter (Ed.), *American universities in a global market*. University of Chicago Press.

Kumar, N. (2006). Provincialism in modern India: The multiple narratives of education and their pain. *Modern Asian Studies, 40*(2), 397–423.

LaDousa, C. (2005). Disparate markets: Language, nation, and education in north India. *American Ethnologist, 32*(3), 460–478.

LaDousa, C. (2007). Liberalisation, privatisation, modernisation, and schooling in India: An interview with Krishna Kumar. *Globalisation, Societies and Education, 5*(2), 137–152.

Mahajan, A. (2014). 1.5 million engineering pass outs in India every year, fewer getting hired. *Dazeinfo*, October 28, 2014. https://dazeinfo.com/2014/10/28/1-5-million-engineering-pass-outs-india-every-year-fewer-getting-hired-trends/

Mohammad, M., Alsaig, A., & Alsaig, A. (2014). Introducing context-awareness to MOOC systems. *Proceedings of the 3rd International Conference on Context-Aware Systems and Applications*. ICST, Brussels, Belgium. 48–56.

National employability report: Engineers. (2016). *Annual report 2016*. Aspiring Mind. https://www.scribd.com/document/350743380/National-Employability-Report-Engineers-Annual-Report-2016

NIC. (2017). JOSSA seat information. http://josaa.nic.in/SeatInfo/root/SeatMatrix.aspx

Patibandla, M., & Petersen, B. (2002). Role of transnational corporations in the evolution of a high-tech industry: The case of India's software industry. *World Development, 30*(9), 1561–1577.

Population of India. (2016). Statistics Times. http://mospi.nic.in/statistical-year-book-india/2016/171

Quality in School Education Report. Quality Council of India, New Delhi. (2016). https://pdf4pro.com/view/quality-in-school-education-quality-council-of-india-4b246a.html

Rosé, C. P., Carlson, R., Yang, D., Wen, M., Resnick, L., Goldman, P., & Sherer, J. (2014). Social factors that contribute to attrition in MOOCs. *Proceedings of the first ACM conference on Learning @ scale conference (L@S '14)* ACM, New York, NY, USA, 197–198.

Toppr. (2015). Why IIT JEE is the toughest exam in the world? https://www.toppr.com/bytes/iit-jee-toughest-exam/

Varshney, L. R. (2006). *Private engineering education in India: Market failures and regulatory solutions.* Unpublished manuscript, Department of Science, Technology, and Public Policy, Massachusetts Institute of Technology, Cambridge, MA.

Open Access This chapter is licensed under the terms of the Creative Commons Attribution 4.0 International License (http://creativecommons.org/licenses/by/4.0/), which permits use, sharing, adaptation, distribution and reproduction in any medium or format, as long as you give appropriate credit to the original author(s) and the source, provide a link to the Creative Commons licence and indicate if changes were made.

The images or other third party material in this chapter are included in the chapter's Creative Commons licence, unless indicated otherwise in a credit line to the material. If material is not included in the chapter's Creative Commons licence and your intended use is not permitted by statutory regulation or exceeds the permitted use, you will need to obtain permission directly from the copyright holder.

Fighting the "System": A Pilot Project on the Opacity of Algorithms in Political Communication

Jonathan Bonneau, Laurence Grondin-Robillard, Marc Ménard, and André Mondoux

Too much freedom can lead to the soul's decay.
—Prince.

Introduction

Adopting critical perspectives in digital technology research faces several challenges. From the outset, the first consists, if we want to open up thinking about their economic, political, and social issues and consequences, of the question of the so-called neutrality of technology. Whether it is technics as an "ontological role" (Heidegger, 1958), collective memory (Stiegler, 1994), individuation dynamic (Simondon, 1989), or the

J. Bonneau • L. Grondin-Robillard • M. Ménard • A. Mondoux (✉)
UQAM Montreal, Montreal, Canada
e-mail: bonneau.jonathan@uqam.ca; grondin_robillard.laurence@uqam.ca; menard.marc@uqam.ca; mondoux.andre@uqam.ca

© The Author(s) 2022
A. Hepp et al. (eds.), *New Perspectives in Critical Data Studies*, Transforming Communications – Studies in Cross-Media Research, https://doi.org/10.1007/978-3-030-96180-0_5

capitalism phenomenon (Lefevbre, 1971), several contributions have marked the will to assign to technics attributes which go beyond simple, neutral instrumentalisation to recognise a role of co-instituting social dynamics. In addition to this continuing challenge, contemporary studies of algorithms and artificial intelligence face additional obstacles. The algorithms fundamentally lack transparency (Castets-Renard, 2018), thus inducing the need to audit them (Mittelstadt et al., 2016). This opacity is made all the greater since it takes place in a context of social acceleration (Rosa, 2010) which tends to make their presence fleeting—merchant circulation of personal data (Mondoux & Ménard, 2018) and commercial property which make their accountability uncertain (Watson & Nations, 2019) at best. Add to this that they are heterogeneous in nature and often integrated into larger systems (Kitchin, 2014) and the reluctance of social media to open up their services to research, it is understandable that it is tempting for critical studies to abandon the empirical dimension to focus on "theoretical" contributions. The aim of this project is to open up "theoretical" reflections on algorithms to the contribution of their empirical study. To do this, we have had to adopt several strategies that we share with you in this chapter, as well as their anchoring within an analytical framework inspired by critical perspectives.

Political Communication in the Age of Algorithms

The use of algorithmic processes (automatisation of the production, circulation, and consumption of data by the use of computational procedures) in political communication is increasing. Assessing the impact of the automatic production, circulation, and delivery of political messages and advertising is challenging because the work carried out by algorithms is still largely hidden. Our current research project is intended to shed light on the contribution made by artificial intelligence, more specifically **recommendation algorithms**, to political advertising and messages in digital social media.

> The essential function of recommender systems is mathematically predicting personal preference. [...] Thematically, recommenders aid users along four key dimensions (which, may or may not overlap): they help users decide what they could or should do next: they help users explore a variety of contextually relevant options: they help users compare those relevant options;

> and, perhaps most critically, they help users discover options and opportunities they might not themselves have imagined. (Schrage, 2020: 5)

We will use a methodology designed to meet the challenges currently faced by research on algorithms (they are not neutral and difficult to study because of their opacity: Kitchin, 2017; Diakopoulos, 2014; Bucher, 2012) and demonstrate that social media only have as much targeting power as their users' contributions as expressed by their actions.

Studies of political communication in industrial societies have traditionally started from the concept of propaganda and its effects on public opinion (Lasswell, 1927; Lippmann, 1922; Maarek, 2008). Whether their perspective was functionalist or critical, classical studies in political communication took as their premise the need to establish a dynamic system ensuring the mass production and circulation of messages that would convince citizens, and inform their political choices, in a context in which they lacked the ability to understand the complexity of social and political dynamics (Ellul, 1962; Herman & Chomsky, 1988; Lippmann, 1922). The concept of propaganda indicates a structural transformation of the modern democratic public sphere (Habermas, 1962), defined by citizens' ability to rationally discuss the ends that are the basis of society. The media play a key role in this type of instrumental communication, since they provide a way of reaching "the masses" (Herman & Chomsky, 1988; Turner, 2018).

The internet has unfolded around a prophetic discourse announcing the concrete realisation of the ideal of the Habermasian public sphere. Digital social media appeared in the aftermath of postmodernity, which is characterised by two powerful tendencies: a crisis of legitimacy for political institutions and hyperindividualism.

With the collapse of grand narratives (Giddens, 1994; Lyotard, 1979), the Habermasian ideal of rational discussion based on common standards has become a mechanism legitimising a new social dynamic based on the primacy of circulation over content (Dean, 2005, 2009). Arguments based on reason are now relativised as personal opinion, and debates on means—rather than ends—now predominate in the political public sphere. The ideal of political communication based on reason becomes a circular communication process in which deliberation takes second place to "an organizational and systemic logic, centered on efficiency, effectiveness, control over the environment, launching operations with a purely utilitarian or strategic basis" (Freitag, 2002: 43; our translation). This phenomenon has been analysed in critical studies of the digital world (Andrejevic, 2013; Morozov & Haas, 2015; Stiegler, 2015) as a new form of social control described through the concept of algorithmic governmentality:

> a form of government essentially fed by raw data (signals that are intrapersonal and a-significant, but quantifiable), operating through the anticipatory configuration of possible events rather than the regulation of behavior, and solely addressing individuals through notifications that trigger reflexes rather than relying on their understanding and will. Thus, the constant reconfiguration, in real time, of individuals' information and physical environments on the basis of "data intelligence"—whether this is called "personalization" or "security metabolism"—is a new form of government. (Rouvroy, 2012, n.p.; our translation)

"Algorithmic governmentality" (De Filippi, 2016) may be said to embody a break with traditional political communication to the extent that it no longer seeks to persuade through rational discourse, but attempts to provoke responses through signals and stimuli. Processes of political communication are seen as legitimate less in relation to "great aims" than because of their pragmatic, technical, quantifiable, and verifiable effectiveness (Nickerson & Rogers, 2014).

Hyperindividualism (Mondoux, 2011) is part of the same dynamic. Now freed from the "yoke" of ideology and all that is political, individuals have become subjects for whom, ultimately, free will in itself is sufficient to justify their values, express themselves, or build their identity: this leads to processes of personalisation. Digital social media have thus been seen as tools of self-expression and the search for identity (Mondoux, 2011; Papacharissi, 2010), as new, more "democratic" information media and, especially, as sources of digital traces through the production of personal and behavioural data (Ménard, Mondoux, Ouellet & Bonenfant, 2016; Berthier & Teboul, 2018). As part of this new dynamic, political communication has also shifted, with the help of digital tools and traces, towards personalisation and microtargeting (hypersegmentation of a large target audience—Barbu, 2013) through the use of data that is produced by individuals (Barocas, 2012; Woolley & Howard, 2018) and processed by recommendation algorithms (Boyd & Reed, 2016; Shorey & Howard, 2016).

While some may see in this a sign that democracy is being restored or enhanced, one thinks about the promises of an "E-Government" trend (Lee et al., 2011), major problems and challenges undeniably exist. One of them is that algorithms contribute to a dynamic characterised as a form of totalisation without totality (Freitag, 2002), that is, the totalisation is not inscribed in symbolic politico-institutional representations ("totality") as it is immanently assumed as immanent and "neutral" (technical abstraction). Hence, the algorithmic governmentality tends to conceal ideology and the political realm: if you accumulate "raw" data (Gitelman, 2013) and produce a quantifiable synthesis, you can then claim to have established a direct relationship with the "Real" (Ménard & Mondoux, 2018) giving rise to an equally objective view of society itself. Deprived of the normative and expressive support of ideology and the political realm, collective reflection and praxis lose their meaning. In this context, the issue of political communication becomes all the more crucial in that a twofold challenge must be met: not only to convince people in terms of ideas but also to (re)legitimise the political realm itself (Sfez, 1992). Political communication must deal with these new dynamics.

The dynamic of individualised communication contributed to the decline of journalism as the main source of mediation with citizens (gatekeeping) (Entman & Usher, 2018; Public Policy Forum, 2017). This left the door wide open to the production of personalised messages that help reduce all messages (whether political, personal, commercial, etc.) to the same level of legitimacy as opinion, in a plethoric jumble of fake news, journalistic information, sentiments, propaganda, disinformation, and even interference between states (as in the case of Russia during the 2016 American presidential election) (Boyd-Barrett, 2019; Spicer, 2018).

The dynamic of personalisation is also (re)produced by the use of recommendation algorithms that tend to confine individuals to a "personal cocoon" (Bodo et al., 2017) or "echo chamber" (Boutyline & Willer, 2011; Pariser, 2011), in which they receive only what resembles (is correlated with) their "profile"; this profile is nourished by their personal opinions and behaviour. Not only are individuals confined to a dynamic that excludes other opinions (since personalising algorithms send content that "complies" with the individual's stated values and opinions—Gao et al., 2010; Sha, 2013), but this same dynamic tends to strengthen and radicalise opinions: in fact, this is one of the main challenges facing a number of Western societies today. In our view, the dynamic of personalisation tends to obscure what is political, giving precedence to "facts" (quantitative objectivations) over law (the political realm) and making it all the more difficult to achieve a genuine emancipatory praxis (Rouvroy & Berns, 2013; Ouellet, Ménard, Bonenfant & Mondoux, 2015).

In response to recent Facebook scandals—the integrity of Facebook debates and exchanges in the public sphere is a major issue—we, like others, argue:

> Strong arguments support the position that algorithmic agents that operate without proper, or flawed, human oversight; or absent of well-defined governance and ethical frameworks, may have negative effects on greater societal norms and values such as the holy triumvirate of *liberté*, *égalité*, *fraternité*—or to put it in the language of the existing legal frameworks, fundamental human rights and freedoms, equality, and social cohesion. (Bodo et al., 2017: 137)

This raises the important question of the political in the age of artificial intelligence and the need to "reintroduce" what is human—both "politics" and everything that is political—in these processes of automation. Artificial intelligence can design any computer algorithm or technological method that allows a machine to simulate part of the human intelligence, that it is to learn, predict, make decisions, or perceive its surroundings. Algorithms can therefore be used in simple interactive media, in which case the entirety of the control is left to a human's will to contribute to this interaction through person-machine communication, often in order to facilitate an arduous or complicated task. Once artificial intelligence is implemented into the process, part of this control is left to the machine and some of the thought process required by a more complex task is translated into an artificial communication monologue completed by the machine itself. With the arrival of massive data collections and machine learning capabilities (such as it can be seen with recommendation algorithms), more and more of this control is being delegated to computer and technological systems, which often dialogues between them in others to compartmentalise the information, augmenting the amount of artificial communication required being produced, which in turn leaves out humanity from most of this process with little to no means of contributing, figuring out, or interfering with these processes.

In disclosing the empirical work carried out by recommendation algorithms, this research will raise awareness among members of the public and decision-makers of the issues involved in automating political messages on digital social networks. Such issues extend well beyond the traditional problem of protecting personal data, and our research can contribute to reflections leading to the development of normative and regulatory frameworks. Lastly, access to algorithms in general, and their lack of transparency (Mittelstadt et al., 2016; Pasquale, 2015), is problematic, especially in the context of privatisation and the economic power of GAFAM (Google, Apple, Facebook, Amazon, Microsoft) (Biancotti & Ciocca, 2018). From the interface to their functions, social media platforms show a tendency to promote the image of citizens as independent individuals in control of the technology they are using (Bruneault & Laflamme, 2020), which is a problem when the advertisement shown on their feed is by nature anchored in social and political dynamics that requires the recognition of the influence generated by other individuals and by the medium itself. For these reasons, our research will contribute to emerging reflections on the socio-political contexts of a truly "social" deployment of artificial intelligence, chiefly in that it will provide an innovative empirical corpus showing how recommendation algorithms act on the basis of citizens' personal profiles on digital social media and how political messages and advertising circulate (what profiles receive what messages, where the messages come from, how frequent they are, etc.).

Research Objectives and Methodology

The chief objective of this research project is to analyse the communicational and socio-political consequences of automating through algorithmisation the production, delivery, and consumption of political messages and advertising, in order to problematise issues related to democracy in a digital social context and their impact on election processes. This objective encompasses four sub-objectives:

1. Carry out an empirical analysis of algorithmic systems used as tools to produce, circulate, and consume political messages and advertising in digital social media, in order to understand how they work.
2. Analyse the relationship between user profiles (described in terms of their geographical, sociocultural, and media diversity) and the political advertising and messages they receive, in order to identify processes of microtargeting (personalisation) carried out by algorithms.
3. Analyse the circulation and targeted delivery of political advertising and messages during the next Canadian federal election campaign (2023) in order to understand how algorithmic political communication can affect election processes.

4. Develop recommendations about the effects of the algorithmisation and microtargeting of political communication on digital citizenship in the public sphere in digital social media, in order to support reflections that will eventually lead to the establishment of statutory or regulatory frameworks.

In this project, we intend to use a research method that will enable us to shed light on the hidden contribution of recommendation algorithms to the production and circulation of political advertising and messages in digital social media. One of the characteristics of algorithms is that they are not neutral: “algorithms are created for purposes that are often far from neutral: to create value and capital; to nudge behavior and structure preferences in a certain way; and to identify, sort and classify people” (Kitchin, 2017). This is a position shared by a number of authors (Bozdag, 2013; Fleischmann & Wallace, 2010; Gillespie, 2014; Mager, 2012). Algorithms are also difficult to study because of their opacity (“black box”), and this makes it difficult to see how their power and influence are exerted (Bucher, 2012; Diakopoulos, 2014). One of the more promising methods available is reverse engineering: “the process of articulating the specifications of a system through a rigorous examination drawing on domain knowledge, observation, and deduction to unearth a model of how that system works” (Diakopoulos, 2014: 404). This strategy is recommended (Bodo et al., 2017) and used by a number of authors (Bodo et al., 2017; Diakopoulos, 2014; Gambs, Aïvodji, Arai, et al., 2019b; Gambs, Aïvodji, & Ther, 2019a; Hannak et al., 2013; Lazer et al., 2014; Mikians et al., 2012; Mukherjee et al., 2013).

Since the information openly available on the platform (through the means of options such as “why I am seeing this content”) are either too broad or sometimes even cryptic (compared to the extent to which a company can define its targeting requirements), we have to rely on external methods of finding the answers to our questions. To extract an algorithm from its “black box”, one of the two following variables must be controlled: inputs (the targeted messages defined by producers) or targets (the profile types of those receiving them). Since we cannot control the messages produced by political entities, we need to study their reception by creating a range of possible targets with controlled profiling criteria.

Establishing and Feeding Control Accounts

The digital social network used in this research project is Facebook; this is because Facebook is easy to use and remains the most popular social media platform. Moreover, Facebook has been continually involved in multiple controversies related to electoral advertising. To achieve our objectives, we have chosen not to use the Facebook accounts of actual participants. It would be difficult to recruit hundreds of people who would be willing to provide access to their Facebook account. Their diligence in keeping a diary, and making sure they recorded the right elements, might have been problematic, and there would be ethical problems associated with the circulation of personal data. In addition, this approach would have to deal with the possibility of behaviour changes throughout the participants' observation period and the introduction of uncontrolled biases. Instead, we have chosen to set up control accounts with profiles managed and fed by automatons (bots). This will facilitate and accelerate operations while making the accounts more uniform (thanks to a controlled environment) and reducing the number of resource persons required to feed active accounts on a daily basis. The automated strategy will also provide for the large-scale capture, categorisation, and archiving of all political advertising and messages received, thus making them complementary to the Big Data infrastructure that we are using.

Methodological criteria used to set up control accounts allow for the following:

- Virtual accounts set up in a given region (without actual travel)
- Maximum speed of execution
- A process that is easily reproduced and taught
- Ethical monitoring throughout the process
- Accessibility of tools by automated systems in development.
- The possibility of increasing the amount of control accounts and their regional, social, and cultural diversity (personalities, range of behaviours, number of marginalised, LGBTQ, or disabled persons, etc.).

At first glance, the project may seem to raise several of the ethical issues raised by AI, mainly the collection of personal data from Facebook profiles and the application of automated tools for mining and analysing social media (Hilyard et al., 2015; Taylor & Pagliari, 2018). But, as more and more studies are finding out, surveys need to go where people are: online (Ouchchy et al., 2020). Our research does and will continue to respect ethical guidelines. Human beings indirectly placed in relation to the control accounts will not be subject to any data collection. It will be necessary to animate the control accounts with content and ensure that they are incorporated into networks of friends while limiting interactions with "real" users to exchanges ensuring participation in a common network. Since users are not themselves the subject of the research, it is not necessary to obtain their consent. No information about users will be compiled and no information, therefore, will be disclosed, whether it is direct, indirect, or related to vulnerable persons. Since interactions with users will be minimal and chiefly limited to the transmission of messages, the control accounts will not cause users any undue loss of time. Impact on the platform (Facebook) will also be minimal to non-existent. Findings will not lead to disclosure of any Facebook security breaches or sensitive information. Loss of resources potentially caused by the control accounts will also be minimal, and the impact on advertisers (and investors) negligible, since 200 witness accounts out of more than 2 billion on Facebook will not have any perceptible effect on their data. Also, in order to pursue our research with ethical consideration (Elovici & al., 2013), we have made sure to only view ads and interact with pages that already had a large number of subscribers, diminishing their cost well under the average of $0.01 per view that the Facebook ad centre charges. This research strategy was approved by our institution's ethics committee in January 2019 for a pilot project focusing on the 2019 federal election campaign (August–October 2019), enabling us to fine-tune the methodology through a pretest based on the creation of approximately 100 control accounts.

Setting up the control accounts proved to be a fastidious business that could not be automated. Facebook requires an email address to provide authentication when an account is created. Microsoft Hotmail was used to satisfy this requirement, since it is currently the only popular email system that does not base registration on association with a cell phone number—a piece of data that cannot easily be accessed or falsified in large numbers. A database was created combining the fields used to open Hotmail and Facebook accounts in order to keep a record of all the information required to open the accounts. Randomly generated last names, first names, and dates of birth (based on Québec population statistics) were used to create email addresses that were undetectable, since email systems themselves suggest combining these elements. Finally, a rule was set up to direct messages from all Hotmail accounts to a single address, in order to simplify the process of monitoring and storing communications generated by the Facebook control accounts.

Creating Profiles and Feeding the Control Accounts

Once established, the Facebook control accounts were provided with individual data and information based on categories that had been identified to build a specific profile for each account.

Number assigned to each profile. This was a way of tracking and archiving profiles from creation to elimination.

Control account names. We created random associations of the most popular Québec last names and first names, and then defined email addresses based on these associations and a birthday derived from the age of the profile.

Age. Profiles were randomly distributed between two age groups: 18–35 and 35–60. Since minors cannot be targeted by political ads, we decided to focus on the age groups most likely to receive the desired messages and split them into two, relying on Facebook ad targeting's available options.

Photographs. To personalise control accounts, we used a bank of royalty-free images for Facebook cover photos (unsplash.com), and a website (thispersondoesnotexist.com) able to generate an infinite supply of portraits of non-existent persons that were used as profile pictures. Photographs were algorithmically generated using general criteria of ethnicity and age. To limit the amount of control accounts and data needing to be analysed during this first phase, and given that Facebook requires that accounts be created in the region in which they will be active, our initial accounts were set up in Montreal, Canada.

All activities, posts, indications that a page was liked, sharing or re-sharing of other Facebook posts, and so on took place according to the following parameters.

Open/closed. Control accounts described as "open" had a network of 100 friends (among the control accounts), without regard for profile type and/or political allegiance, and could "like" most of the major Facebook interest categories (see list below). Posts were written in the first person, contained marks of emphasis ("!"), and were more than 140 characters long. Profiles described as "closed" had at most 40 friends and their interactions were restricted to control accounts with a profile similar to theirs.

Posts were less expressive, more neutral (they were not written in the first person), focused on a single Facebook interest category, and fewer than 100 characters long.

Active/passive. "Active" control accounts progressed towards 30 minutes of activity per day, with several different activities every day (liking, posting, sharing, etc.). "Passive" control accounts were restricted to less than 30 minutes of daily activity.

Positive/negative. Control accounts use a majority of words rated "positive" or "negative" in the Harvard IV-4 dictionary of psychology database (www.wjh.harvard.edu/~inquirer/), often used for sentiment analysis (Crossley et al., 2017).

Interests. The control accounts "liked" pages included in the Facebook "interests" that serve as the basis for advertising categories. We used the following categories:

- Business and industry
- Food
- Entertainment
- Families and relationships
- Fitness and wellness
- Shopping and fashion
- Hobbies and activities
- Sports and outdoors
- Technology

Political party affiliation. Control accounts were randomly assigned a "political profile" dictating which political ads and messages they would like, comment on, and (re)share:

- Conservative Party of Canada
- Liberal Party of Canada
- New Democratic Party
- Bloc Québécois
- Green Party
- Neutral

All activities of the Facebook control accounts were preserved and documented as followed:

- Identification number
- Type of post
- Text of post
- Time taken to put up each post and collect associated data
- Status verification for each post (posted, number of characters, list of words related to sentiments)

These operations allowed us to identify ten profile types, similar to the number involved in traditional targeting grids (Beyer et al., 2014; Lau et al., 2018).

Creating control accounts and feeding them on a daily basis in real time would require significant human resources, leading to prohibitive costs. We therefore chose to use Java-scripted interface manipulation bots to automate these fastidious and voluminous tasks. The bots were able to feed the control accounts automatically through activities (messages, shares, subscriptions, likes, keywords, etc.) that were compatible with their target profile. Bots also provided automatic capture (through screenshots) of ads and messages received in newsfeeds and stored them in a database, thus establishing a controlled environment.

List of automated operations

- Variable length of connection and speed of execution
- Verification of expected connection time for the control account
- Opening of the mobile Web version of Facebook
- "Organic" writing of user IDs and passwords (variable and random speed of writing)
- Skipping Facebook friend suggestions and security recommendations
- First run through Facebook newsfeed; screenshot (observation of long-term effects)
- "Organic" writing of Facebook post
- Second run through Facebook newsfeed; screenshot (observation of short-term effects)
- Disconnection
- Clearing trackers and connection history

Maintenance of the control accounts and collection of the messages and content they received were carried out as follows:

- Automatic organisation of screenshots in files for each control account
- Manual downloading of archives and profile information for each control account
- Compilation of emails sent by Facebook to control accounts
- Manual overview and sorting of images and information provided by Facebook
- Incorporation into a database covering the various sources of information and allowing for cross-referencing between a thematic categorisation (the one used to create the control accounts) and personalisation factors for the control accounts, based on the following criteria: interests (Facebook categories), type of post (in own or followed account, sponsored or suggested page), type of source (governmental services, Facebook group or page [sub-category for political parties], business, news).

Preliminary Findings

A first test, the pilot project, was carried out between August 10 and December 10, 2019, with 100 control accounts activated and (gradually) fed automatically with daily activities (posts, shares, and re-shares). Daily data collection was also automated.

Our first analyses showed that there is a time lag (lasting several days or weeks) before ads appear in the right-hand column or on the news wall of the control accounts. It can also be shown that the time lag is associated with browser "activity", both on Facebook and on using a search engine, and that it is associated, therefore, with collecting cookies. As long as the search engine's browsing history and cache memory are empty, there is no possibility that ads will appear in connected Facebook accounts. Visiting a few websites that use cookies (Amazon, Aldo, Dynamite Clothing, etc.) before making a connection with the Facebook account leads to the appearance of ads, initially in the right-hand column (desktop view). The control account then needs to interact with ads in the right-hand column (by clicking on the links) in order to "activate" ads on the news wall (mobile view and desktop).

In short, although Facebook can provide advertisers with various targeting options (personalised website audiences, personalised mobile app audiences, personalised audiences based on a client list, personalised interaction audiences), the option that will most quickly reach a new Facebook account, for either political or advertising messages, is the "personalised website audience" using browsing history with cookies. This kind of targeting associates people who visit a website with Facebook accounts; the Facebook pixel incorporated into the webpage is one of the ways this is done (Trevisan et al., 2019). With this kind of targeting, advertisers may, for instance, launch a campaign to reach people who have visited a product page on their website, in order to encourage them to come back to the website and continue shopping. They can also create an "audience" consisting of every person who has visited their site over the previous months, in order to share similar new products with them.

We need to carry out further analysis of this observation: despite all our efforts to ensure the ongoing existence of control accounts and their receptivity to advertising content, with only a few exceptions, the majority of these accounts, even on the day before the election or on election day itself, did not receive any advertising from any of Canada's five major political parties. It remains difficult to explain why this is, although we can put forward some hypotheses.

A first possible cause is related to advertising targeting options. It is likely that community managers and/or those responsible for digital marketing in political parties such as the Conservative Party of Canada (CPC) or the Liberal Party of Canada (LPC), each of which spent close to a million Canadian dollars on Facebook advertising, decided to target only the following: people who had interacted with their Facebook pages, people who were on their membership lists, or people who had visited their website. This way of activating ads was in fact validated through our research project test accounts. However, if this hypothesis is confirmed, it remains surprising, since it was our assumption that parties would generally try to increase the number of potential voters.

A second, less likely hypothesis is that no control account was targeted by political advertising because the "Montreal" geolocation was not part of the targeting criteria. Given that all of our control accounts were set up in the same region, it is impossible for us to completely eliminate this factor as a potential cause.

A third possibility is that political parties may have chosen the "broad targeting" option. Broad targeting mostly relies on Facebook's delivery system to find the best people (as defined by Facebook) to show ads to. In other words, the parties might have chosen to let the Facebook algorithm define their targeting. Given that this algorithm is known to create echo chambers, it is likely that control profiles without membership in political groups, or friend networks or browser histories displaying clear political convictions, would not be targeted. From a methodological perspective, despite their divergent ideals, political parties use overlapping keywords to discuss their electoral programme, which means that control profiles could not create politicised posts associated with one party rather than another. In addition, in order to comply with ethical rules governing this kind of research, profiles could not join or participate in Facebook groups because of the requirement to avoid establishing relations with Facebook users.

A fourth hypothesis is simply related to the stages that must be gone through on Facebook before an account is included in targeted advertising based on interests or interactions. Probably to avoid the proliferation of fake accounts, a certain amount of time seems to be required to observe the technical parameters involved in the creation, activation, and activity of a new account, but also to observe its connection network, interactions, activities, and so on. To automate the accounts, therefore, it was not sufficient to deal with technical connection variables; other factors had to be taken into account to respond to Facebook's scrutiny. After several months, we also noticed that accounts with a "passive" level of activity never received targeted advertisements, regardless of any other criterion and independently of the targeting option or the account's browsing habits. All of these accounts were therefore eliminated after a certain time, given that it was impossible to collect data from them. Ensuring that a profile was linked to a more active account through friendship (in this case, with the researchers' account) was also identified as a necessary step for the account to be recognised for targeting.

One last point: these initial tests enabled us to identify the conditions enabling a Facebook account to be "activated", that is, to receive messages and content from the service provider. According to what we now know, these conditions do *not* include how many "likes" are given to pages or posts, how much connection time is involved, how many searches are carried out on Facebook, or how many games are played. However, our tests have shown that to receive content through Facebook, an account must have a browsing history with cookies. In the next stages of the research, it will be important to verify, using massive data, whether variables such as posted or shared content, number of posts, or quality of friends (active or passive) affect the reception of messages and ads in general, and in particular political ads and messages.

Next Steps

Now that the pilot project is finished, we can start preparing for large-scale research to be carried out during the next federal election campaign (fall 2023). Our goal is to enrich the parameters established for control accounts by extending: (a) their *geographical* scope (to cover all of Québec); (b) their *social and cultural* scope (by increasing the diversity of control account profiles to include minorities and marginalised or vulnerable groups in terms of ethnicity, sexual orientation, disability, educational level, etc.); and (c) their scope in terms of *media* (each Facebook profile will be matched with a control account in other digital social media such as Twitter, Instagram, and YouTube). The project will involve the creation of approximately 1500 control accounts.

To prevent piracy, Facebook geolocates account activities, which means that accounts must be automated from the region in which they are set up. To carry this out, we will put together kits including modules consisting of five small computers, already configured with control accounts and profiles specific to the newly targeted regions, and automation scripts (bots) to feed the accounts, gather data, and send it to Montreal. We will rely on our contacts and the Université du Québec network to set up modules in five cities: Chicoutimi (UQAC), Gatineau (UQO), Québec (Université Laval), Sherbrooke (Université de Sherbrooke), and Trois-Rivières (UQTR). Each module will be managed remotely by three high-performance computers in Montreal (UQAM) that will provide the interface for the research group's Big Data architecture. This infrastructure already exists and has been operational since 2015. We will be able to store, analyse, and visualise all of the data from the control accounts in real time.

All political advertising and messages that are received will be used to create a database. Ads and messages will be compared with the federal government's database of officially "registered" advertising in order to detect any issues of conformity or potential interference. We also intend to establish a database of "unofficial" messages and ads, identifying their sources, in order to pave the way for an analysis of the circulation of fake news or any other type of interference. In a second phase, we plan to identify and analyse through correlation which profile types receive political advertising and messages, and how often this occurs; we will also identify and analyse, through correlation, if there is any variation/personalisation of a given message according to the targeted profile (microtargeting).

Conclusion

One of the main preliminary results was that Facebook targeting skills are not like the "hypodermic model effect" (Bineham, 1988), unilateral and automatic. To be able to target, Facebook needs the trace generated by online activities such as using a search engine or visiting websites. Facebook thus needs a larger ecosystem where data circulates openly among commercial partners. It is to be noted that we had to outsmart Facebook and its undesirable accounts detecting strategies in order to get a small glimpse of their algorithms at work. Also, to be noted, geolocalisation by Facebook plays a central role in account creation protocols.

This pilot project gave us a glimpse into Facebook's black box and allowed us to formulate observations that are surprising, to say the least, and that go well beyond the scope of our research. Our purpose was to analyse the communicational and socio-political consequences of automating (algorithmising) the production, delivery, and consumption of political advertising and messages, in order to problematise issues related to democracy in a digital social context and their impact on election processes. However, our preliminary findings convincingly demonstrate that Facebook's ability to detect accounts that fail to comply with community standards is still flawed. This raises an economic question: if "sponsored" posts are seen by all accounts, even duplicates or automatised profiles, are advertisers paying a fair price for their targeted ads?

These findings also lead us to formulate observations which, although they are outside the scope of our current project, should be the subject of future work. We believe there is research to be done on Facebook users and the Facebook algorithm. (1) How are suggestions made in regard to other accounts that "you might know", and how do you become "friends" with other accounts? Some of our control accounts received invitations from other accounts within hours or days of being activated. No response was given to these invitations. In addition, (2) gender seems to have an impact on the number of invitations received from strangers. Control accounts associated with "women" aged 18–35 were the ones that received outside invitations. (3) The more we study Facebook's targeted advertising, the more obvious it becomes that this advertising is lacking in transparency for advertisers and community managers. The fragmentation of users' areas of interest appears to be lacking in documentation and clarity, which may make targeting, and even the classification of business pages, less effective. (4) We also deplore the overall lack of transparency and understanding in relation to Facebook's advertising tools.

The preliminary results also show that it is possible to go beyond the empiricism/theory dichotomy, but at the cost of overcoming several obstacles, mainly refuting technic's neutrality without giving in to technical determinism. This allows to open the "black box" of algorithms (Pasquale, 2015) to reveal empirically their presence by their effects. It also bonifies the research's case when appearing before ethical committees that are not always up to par with bleeding-edge approaches involving "new technologies". Nonetheless, several obstacles remain, mainly that algorithms are still private property. This has consequences when it comes to obtaining social media companies' full collaboration. In our project, for instance, we still had to play a game of hide-and-seek in order to maintain the presence of our control accounts, with Facebook trying to expunge them as fake accounts. More importantly, revealing the algorithms is the basis for any meaningful audit, ethically, politically, and socially.

This non-visibility of the algorithms also has major repercussions on the political front: our preliminary results allowed us to observe the effects of the "machinization of politics" where the values/finalities are being concealed by the means (technics): political goals are being measured by success itself. In other words, circulation is the main goal (Dean, 2009) over the message itself, thus creating a void—or loss of symbolic efficiency (ŽiŽek, 2009). We can translate this notion into two main trends: "empowered" individuals are now emancipated of the disciplinarian yoke of ideology, but at the same time they lose the normative contribution of ideology (transcendental symbolic mediations producing "universal" common values), a void being picked up by algorithmic automatisation. A look at the state of America in the 2020 elections already shows us a possible future for political communication: all values are reduced to the expression of a personal opinion and thus the individual prevails over the institutions and their norms, and at best de facto leaving the latter in the hands of the technical automation projected as a neutral and a "natural" means—nullifying the need for visibility—to achieve goals that are primarily defined in terms of pragmatic efficiency. This brings to mind the Heideggerian warning: the more Man sees himself for the "lord of the earth", the more he confuses his destiny for that of modern technic, as the Dasein succumbs to the lures of the power of power itself.

References

Andrejevic, M. (2013). *Infoglut: How too much information is changing the way we think and know*. Routledge.

Barbu, O. (2013). Advertising, microtargeting and social media. *Procedia—Social and Behavioral Sciences, 163*, 44–49.

Barocas, S. (2012). The Price of precision: Voter microtargeting and its potential harms to the democratic process. In *Proceedings of the first edition workshop on politics, elections and data* (pp. 31–36). ACM. https://doi.org/10.1145/2389661.2389671

Berthier, T., & Teboul, B. (2018). *From digital traces to algorithmic projections*. Elsevier.

Beyer, A., Knutsen, C., & Rasch, B. (2014). Election campaigns, issue focus and voting intentions: Survey experiments of Norwegian voters. *Scandinavian Political Studies, 37*(4), 406–427.

Biancotti, C., & Ciocca, P. (2018). *Data Superpowers in the Age of AI: A Research Agenda*. VOX CEPR Policy Portal. Retrieved January 25, 2019, from https://voxeu.org/article/data-superpowers-age-ai-research-agenda.

Bineham, J. L. (1988). A historical account of the hypodermic model in mass communication. *Communication Monographs, 55*(3), 230–246. https://doi.org/10.1080/03637758809376169

Bodo, B., et al. (2017). Tackling the algorithmic control crisis—The technical, legal, and ethical challenges of research into algorithmic agents. *The Yale Journal of Law & Technology, 19*, 133–180.

Boutyline, A. G., & Willer, R. (2011). *The Social Structure of Political Echo Chambers: Ideology Leads to Asymmetries in Online Political Communication Networks*. http://opensiuc.lib.siu.edu/pn_wp/57

Boyd-Barrett, O. (2019). Fake news and 'RussiaGate' discourses: Propaganda in the post-truth era. *Journalism, 20*(1), 87–91. https://doi.org/10.1177/1464884918806735

Boyd, D., & Reed, L. (2016). *Who controls the public sphere in an era of algorithms?* Data & Society Research Institute.

Bozdag, E. (2013). Bias in algorithmic filtering and personalization. *Ethics and Information Technology, 15*(3), 209–227.

Bruneault, F., & Laflamme, A. S. (2020). AI Ethics: how can information ethics provide a framework to avoid usual conceptual pitfalls? An Overview. *Ai & Society*. https://link.springer.com/article/10.1007/s00146-020-01077-w

Bucher, T. (2012). Want to be on the top?' Algorithmic power and the threat of invisibility on Facebook. *New Media and Society, 14*(7), 1164–1180.

Castets-Renard, C. (2018). "Régulation des algorithmes et gouvernance du machine learning: vers une transparence et 'explicabilité' des décisions algorithmiques ?", *Revue Droit & Affaires*. Assas: Revue Paris II, (15th ed.).

Crossley, S. A., Kyle, K., & McNamara, D. S. (2017). Sentiment analysis and social cognition engine (SEANCE): An automatic tool for sentiment, social cognition, and social-order analysis. *Behav Res, 49*, 803–821. https://doi.org/10.3758/s13428-016-0743-z

De Filippi, P. (2016). Gouvernance algorithmique: vie privée et autonomie individuelle à l'ère des Big Data. In P. De Filippi & D. Bourcier (Eds.), *Open Data & Data Protection: Nouveaux Défis pour la Vie Privée* (p. 2016). Mare & Martin.

Dean, J. (2005). Communicative capitalism: Circulation and the foreclosure of politics. *Cultural Politics, 1*(1), 51–74.

Dean, J. (2009). *Democracy and other neoliberal fantasies: Communicative capitalism and left politics*. Duke University Press.

Diakopoulos, N. (2014). Algorithmic accountability. *Digital. Journalism, 3*(3), 398–415. https://doi.org/10.1080/21670811.2014.976411

Ellul, J. (1962). *Propagandes*. Armand Colin.

Elovici, et al. (2013). Ethical considerations when employing fake identities in online social networks for research. *Science and Engineering Ethics, 20*(4), 1027–1043.

Entman, R. M., & Usher, N. (2018). Framing in a fractured democracy: Impacts of digital technology on ideology, power and cascading network activation. *Journal of Communication, 68*(2), 298–308. https://doi.org/10.1093/joc/jqx019

Fleischmann, K., & Wallace, W. (2010). Value conflicts in computational modeling. *Computer, 43*(7), 57–63.

Freitag, M. (2002). *L'Oubli De La Société*. Presses universitaires de Rennes.

Gambs, S., Aïvodji, U., & Ther, T. (2019a). *GAMIN: An Adversarial Approach to Black-Box Model Inversion*. arXiv preprint:1909.11835.

Gambs, S., Aïvodji, U., Arai, H., Fortineau, O., Hara, S., & Tapp, A. (2019b). Fairwashing: the risk of rationalization. *Proceedings of the 36th International Conference on Machine Learning*, Long Beach, California.

Gao, M., Liu, K., & Wu, Z. (2010). Personalization in web computing and informatics: Theories, techniques, applications, and future research. *Information Systems Frontiers, 12*, 607–629.

Giddens, A. (1994). *Beyond left and right: The future of radical politics*. Polity Press.

Gillespie, T. (2014). The relevance of algorithm. In B. Gillespie & Foot (Eds.), *Media technologies*. MIT Press.

Gitelman, L. (2013). *Raw data is an oxymoron*. MIT Press.

Habermas, J. (1962). *The structural transformation of the public sphere*. MIT Press.

Hannak, A., Sapiezynski, P., Kakhki, A.M., Krishnamurthy, B., Lazer, D., Mislove, A., & Wilson, C. (2013). Measuring personalization of web search. *Proceedings of the 22nd International Conference on World Wide Web*. Association for Computing Machinery, 527–538, 22nd international conference on World Wide Web (WWW 2013), Rio de Janeiro, Brazil.

Heidegger, M. (1958 [1953]). La question de la technique, *Essais et Conférences*, Gallimard, 9–48.

Herman, E. S., & Chomsky, N. (1988). *Manufacturing consent: The political economy of the mass media* (p. 2013). Pantheon Books.

Hilyard, K., Broniatowski, D., & Dredze, M. (2015). *How Far can Twitter Reach in Good Survey Research?* http://www.socialsciencespace.com/2015/04/how-far-can-twitter-reach-in-good-survey-research/

Kitchin, R. (2014). *The data revolution* (1st ed.). SAGE Publications Ltd..

Kitchin, R. (2017). Thinking critically about and researching algorithms. *Information, Communication & Society, 20*(1), 14–29.

Lasswell, H. D. (1927). The theory of political propaganda. *American Political Science Review, 21*(3), 627–631. https://doi.org/10.2307/1945515

Lau, R., Kleinberg, M., & Ditonto, S. (2018). Measuring voter decision strategies in political behavior and public opinion research. *Public Opinion Quarterly, 2*(S1), 325–350.

Lazer, D., Kennedy, R., King, G., & Vespignani, A. (2014). The parable of Google flu: Traps in big data analysis. *Science, 343*(6176), 1203–1205.

Lee, C.-P., Chang, K., & Berry Stokes, F. (2011). Testing the development and diffusion of E-government and E-democracy: A global perspective. *Public Administration Review, 71*(3), 444–454.

Lefevbre, H. (1971). *Vers le Cybernanthrope*. Denoël/Gonthier.

Lippmann, W. (1922). *Public opinion*. Harcourt, Brace and Company.

Lyotard, J.-F. (1979). *La condition post-Moderne*. Les Éditions de Minuit.

Maarek, P. J. (2008). Political marketing. *The International Encyclopaedia of Communication*. American Cancer Society. doi: https://doi.org/10.1002/9781405186407.wbiecp062.

Mager, A. (2012). Algorithmic ideology. *Information Communication & Society, 15*(5), 769–787.

Ménard, M., & Mondoux, A. (2018). Big Data, circuits marchands et accélération sociale. In Mondoux & Ménard (Ed.), *Big Data et Société* (pp. 63–83). PUQ.

Ménard, M., Mondoux, A., Ouellet, M., & Bonenfant, M. (2016). Big Data, gouvernementalité et industrialisation des médiations symboliques et politico-institutionnelles. *Interfaces Numériques, 5*(2), 1.

Mikians, J., Gyarmati, L., Erramilli, V., & Laoutaris, N. (2012). *Detecting Price and Search Discrimination on the Internet*, 79–84. Proceedings of the 11th ACM Workshop on Hot Topics in Networks, ACM, New York.

Mittelstadt, B., Allo, P., Taddeo, M., Wachter, S., & Floridi, L. (2016). The ethics of algorithms: Mapping the debate. *Big Data & Society, 3*(2), 2–68.

Mondoux, A. (2011). Identité numérique et surveillance. *Les Cahiers du Numérique, 7*(1), 49–59. https://doi.org/10.3166/lcn.7.1.49-59

Mondoux, A., & Ménard, M. (2018). Big Data, médiation symbolique et accélération sociale. In N. Aubert (Ed.), *@ La Recherche du Temps—Individus Hyperconnectés, Société Accélérée: Tensions et Transformations* (pp. 217–236). ERES.

Morozov, E., & Haas, P. (2015). *Le Mirage Numérique: Pour une Politique du Big Data*. Les Prairies ordinaires.

Mukherjee, A., Kumar, A., Liu, B., Wang, J., Hsu, M., Castellanos, M., & Ghosh, R. (2013). *Spotting Opinion Spammers Using Behavioral Footprints*. Paper presented at the Proceedings of the 19th ACM SIGKDD international conference on Knowledge discovery and data mining, Chicago, Illinois, USA. doi:https://doi.org/10.1145/2487575.2487580.

Nickerson, D. W., & Rogers, T. (2014). Political campaigns and big data. *Journal of Economic Perspectives, 28*(2), 51–74. https://doi.org/10.1257/jep.28.2.51

Ouchchy, L., Coin, A., & Dubljević, V. (2020). AI in the headlines: The portrayal of the ethical issues of artificial intelligence in the media. *AI & SOCIETY, 35*, 927–936. https://doi.org/10.1007/s00146-020-00965-5

Ouellet, M., Ménard, M., Bonenfant, M., & Mondoux, A. (2015). « Big Data et quantification de soi. La gouvernementalité algorithmique dans le monde numériquement administré ». *Canadian Journal of Communication, 40*(4), 597–613.

Papacharissi, Z. (2010). *A networked self: Identity, community, and culture on social network sites*. Routledge.

Pariser, E. (2011). *The filter bubble: What the internet is hiding from you*. Penguin Books Limited.

Pasquale, F. (2015). *The black box society: The secret algorithms that control money and information*. Harvard University Press. https://doi.org/10.4159/harvard.9780674736061

Public Policy Forum. (2017). *The Shattered Mirror: News, Democracy and Trust in the Digital Age*. Ottawa, Canada. https://shatteredmirror.ca/wp-content/uploads/theShatteredMirror.pdf

Rosa, H. (2010). *Accélération: Une Critique Sociale du Temps*. La Découverte.

Rouvroy, A. (2012). Mise en (n)ombres de la vie même: face à la gouvernementalité algorithmique, repenser le sujet comme puissance. *Mediapart*. https://blogs.mediapart.fr/antoinette-rouvroy/blog/270812/mise-en-nombres-de-la-vie-meme-face-la-gouvernementalite-algorithmique-repenser-le-sujet-com

Rouvroy, A., & Berns, T. (2013). Gouvernementalité algorithmique et perspectives d'émancipation. *Réseaux, 177*, 163–196. https://doi.org/10.3917/res.177.0163

Schrage, M. (2020). *Recommendation engines*. The MIT Press. http://mitpress.mit.edu/9780262539074

Sfez, L. (1992). *Critique de la Communication*. Seuil.

Sha, X. (2013). Personalizing trending content in social media. *Sociology*. Télécom ParisTech, 2013. https://pastel.archives-ouvertes.fr/tel-01226534/

Shorey, S., & Howard, P. N. (2016). Automation, algorithms, and politics, automation, big data and politics: A research review. *International Journal of Communication, 10*, 5032–5055.

Simondon, G. (1989). *Du Mode d'Existence des Objets Techniques*. Aubier.

Spicer, R. N. (2018). Lies, damn lies, alternative facts, fake news, propaganda, Pinocchios, pants on fire, disinformation, misinformation, post-truth, data, and statistics. In *Free speech and false speech* (pp. 1–31). Palgrave Macmillan. https://doi.org/10.1007/978-3-319-69820-5_1

Stiegler, B. (1994). *La Technique et le Temps. Tome 1, La faute d'Epiméthée*. Galilée.

Stiegler, B. (2015). *La Société Automatique. Tome 1, L'Avenir du Travail.*

Taylor, J., & Pagliari, C. (2018). Mining social media data: How are research sponsors and researchers addressing the ethical challenges? *Research Ethics, 14*(2), 1–39. https://doi.org/10.1177/1747016117738559

Trevisan, M., Traverso, S., Bassi, E., & Mellia, M. (2019). 4 years of EU cookie law: Results and lessons learned. *Proceedings on Privacy Enhancing Technologies, 2019*(2), 1. https://doi.org/10.2478/popets-2019-0023

Turner, G. (2018). The media and democracy in the digital era: Is this what we had in mind? *Media International Australia, 1–12*, 1. https://doi.org/10.1177/1329878X18782987

Watson, H. J., & Nations, C. (2019). « Addressing the Growing Need for Algorithmic Transparency". Communications of the Association for Information Systems, vol. 45. https://doi.org/10.17705/1CAIS.04526.

Woolley, S. C., & Howard, P. N. (2018). *Computational propaganda: Political parties, politicians, and political manipulation on social media*. Oxford University Press.

ŽiŽek, S. (2009). *The ticklish subject: The absent Centre of Political Ontology*. Verso.

Open Access This chapter is licensed under the terms of the Creative Commons Attribution 4.0 International License (http://creativecommons.org/licenses/by/4.0/), which permits use, sharing, adaptation, distribution and reproduction in any medium or format, as long as you give appropriate credit to the original author(s) and the source, provide a link to the Creative Commons licence and indicate if changes were made.

The images or other third party material in this chapter are included in the chapter's Creative Commons licence, unless indicated otherwise in a credit line to the material. If material is not included in the chapter's Creative Commons licence and your intended use is not permitted by statutory regulation or exceeds the permitted use, you will need to obtain permission directly from the copyright holder.

Indigenous Peoples, Data, and the Coloniality of Surveillance

Donna Cormack and Tahu Kukutai

Introduction

> We tend to associate practices of population surveillance with Western modernity and the intensification of security routines with the last decade defined by the "Global War on Terror". I suggest, however, that proliferation of methods to monitor and control populations are legacies of the practices that were developed in the colonies to manage civilian populations. (Berda, 2013: 627)

Surveillance is an enduring characteristic of colonialism for Indigenous peoples (Smith, 2009; Smith, 2012). In Aotearoa NZ, as in other settler colonial societies, Māori have long been subject to surveillance by state institutions and agents. While the colonial gaze often makes claims to

D. Cormack (✉)
Te Kupenga Hauora Māori, University of Auckland, Auckland, New Zealand
e-mail: d.cormack@auckland.ac.nz

T. Kukutai
Te Ngira: Institute for Population Research, University of Waikato, Hamilton, New Zealand
e-mail: tahuk@waikato.ac.nz

© The Author(s) 2022

A. Hepp et al. (eds.), *New Perspectives in Critical Data Studies*, Transforming Communications – Studies in Cross-Media Research, https://doi.org/10.1007/978-3-030-96180-0_6

neutrality and objectivity (Smith, 2012), state representations have centred on constructions of difference and deviance, on understandings of Indigenous peoples as dangerous, and on the management of Indigenous resistance to colonialism.

This chapter considers how surveillance functions to regulate and manage Māori peoples within the context of the racialised social divisions fundamental to coloniality. In line with the work of decolonial scholars (e.g. Grosfoguel, 2004, 2016; Maldonado-Torres, 2007; Mignolo, 2007; Quijano, 2007), coloniality is understood as the colonial beliefs, systems, practices, hierarchies, and power relations that persist beyond formal structures and institutions of colonialism (Grosfoguel, 2004). Through this lens, we can interrogate the continuities that current surveillance approaches to Indigenous peoples have with the racialised logics and social orders set in place as part of global systems of imperialism and colonialism. We recognise the diverse histories and experiences of Indigenous peoples globally, while also acknowledging shared experiences of colonial oppression, dispossession, and extraction and how these play out through contemporary forms of data colonialism.

Surveillance practices and techniques have shifted with changing data environments and relations such that algorithmic and biometric surveillance are now commonplace (Kak, 2020; Murphy, 2017). In an era of big data and datafication (Couldry & Yu, 2018), there are myriad possibilities for monitoring individuals and groups in ways that deepen power asymmetries and perpetuate harm. Many kinds of data are generated, captured, and used without consent and sometimes without the knowledge of those from whom the data originated. The secondary use of data often extends far beyond its original purpose (Martin-Sanchez et al., 2017), and data are accumulated and stored in massive data warehouses across the world. A key feature of contemporary state surveillance practices in Aotearoa NZ is the extensive use of big data and linked government datasets for varied purposes, including attempts to predict future behaviours and outcomes (Kukutai & Cormack, 2019). As a member of the Digital Nations "network of the world's most digitally advanced nations",[1] Aotearoa NZ is considered at the leading edge of data innovation. It thus makes for an instructive case study in which to consider the potential implications of data practices for Indigenous peoples more broadly.

[1] digital.govt.nz/digital-government/international-partnerships/digitalnations/

Recognising that resistance has always been a part of Indigenous responses to colonialism, this chapter also explores how Māori are asserting rights to Māori Data Sovereignty to counter and disrupt prevailing data relations, as part of broader Indigenous Data Sovereignty movements. Simultaneously an Indigenous social movement, and a burgeoning field of Indigenous-led research[2] (Carroll et al., 2019; Kukutai & Taylor, 2016; Walter et al., 2021; Walter & Suina, 2019), Māori Data Sovereignty is fundamentally about Māori control of Māori data to advance Māori self-determination (Kukutai & Cormack, 2019). The concept of self-determination is closely connected to the articulation of Indigenous Peoples' rights in the United Nations Declaration on the Rights of Indigenous Peoples (UNDRIP) and domestic treaties.[3] A central tenet of Indigenous self-determination is that Indigenous peoples have an inherent right to be in control of their destinies and to create their own political and legal organisations (Toki, 2017). Seen in this light, Indigenous sovereignty over Indigenous data is an extension of Indigenous peoples' fundamental right to self-determination. This (re)orientation allows us to explore contemporary data colonialism from within Indigenous frameworks of collective self-determination and collective rights. It also encourages alternatives grounded within the knowledges and lived experiences of those peoples who are most impacted negatively by ongoing coloniality, and data colonialism more particularly, and to envision data relations and data practices that are anti-colonial, relational, and collective.

Colonialism and the Racialised Surveillance of Indigenous Peoples

Surveillance was critical to establishing control and managing Indigenous peoples as part of colonisation (Sa'di, 2012). Decolonial scholar Ramón Grosfoguel calls us to be attentive to the centrality of race in colonialism,

[2] Māori Data Sovereignty is a recognised field of research in the Australia New Zealand Standard Research Classification that was revised in 2020. Accessed here: https://www.mbie.govt.nz/science-and-technology/science-and-innovation/research-and-data/anzsrc/

[3] In Aotearoa NZ, two crucial treaties are He Whakaputanga (1835 Declaration of Independence) and Te Tiriti o Waitangi (1840 Treaty of Waitangi). Signed between representatives of Queen Victoria and more than 500 rangatira (chiefs), the Treaty of Waitangi is a broad statement of principles on which the British and Māori made a political compact to found a nation-state and build a government in New Zealand. Today the Treaty is widely accepted to be a constitutional document.

noting that within the "imperial/capitalist/colonial world-system, race constitutes the transversal dividing line that cuts across multiple power relations such as class, sexual and gender relations at a global scale" (2016: 11). For Indigenous peoples, race became the central distinction between coloniser and colonised (Quijano, 2005), separating the "zone of being" from the "zone of nonbeing" (Fanon, 1986; Grosfoguel et al., 2015), the *watched* from the *watchers*.

Ideas about racial difference were used to legitimise European dominance—in some instances this equated with outright extermination—and to disqualify the full participation of Indigenous populations in economic and political life (Ittmann et al., 2010). Over time, racial hierarchies became naturalised and embedded through ideological structures, institutional arrangements (including institutional forms of racial discrimination), and state classifying practices. The latter are instructive in so far as they reveal the deeper ways in which racial structures (Omi & Winant, 1986) prescribed differential access to rights, social goods, and opportunities.

In Aotearoa NZ, as in other settler colonial societies, racial classification practices not only created divisions between coloniser and colonised, but also created hierarchies of difference among native peoples based on perceived racial and cultural proximity to Europeans. Until as late as the 1950s, New Zealand census reports included lengthy commentaries about the imagined Aryan and Asiatic origins of Māori (Kukutai, 2012). The number and growth of the 'Maori-European half-caste' population was closely monitored as their proportion relative to Maori 'full-bloods' was seen as an important indicator of the rate of 'racial amalgamation' (Ward, 1974). Unsurprisingly, many tribes viewed census-taking with suspicion, perceiving it to be linked with taxation, conscription, and land alienation. Tribes aligned with the Kingitanga—a Māori political movement established to preserve Māori autonomy over Māori lands—were especially uncooperative (Kukutai, 2012).

The power to define the boundaries of Māori identity was firmly under setter control and pursued largely for the benefit of the nation-state. State categorisations of race were also at odds with complex and nuanced Māori ways of defining and describing collective belonging that emphasise connections and relations with kin—past, present, and future—and with the natural world (Burgess & Painting, 2020; Mahuika, 2019). However, these racial classifications became the primary categories around which the surveillance of Māori was organised as a state activity in order to measure progress with goals of assimilation and to control or disrupt connections to land. In his analysis of race and colonial land law, Meredith argues that "persuading Maori to embrace European habits, customs, and English language was one measure of getting them to accept the law" and through that persuasive action, access to Māori land (Meredith, 2006: 106). The explicit effect of being declared a European under Section 17 of the Native Land Amendment Act was the Europeanisation of the applicant's land—in effect, the removal of protective mechanisms extended to Māori land.

In tandem with the categorisation of Indigenous peoples into racial categories, the coloniser was constructed as a knowing subject and Indigenous peoples as knowable objects (Quijano, 2007; Smith, 2012). As Linda Tuhiwai Smith (2012) notes, much of the early knowledge about Indigenous peoples was through practices of observing, documenting, and collecting. These practices were not neutral or benign (Smith, 2009), but, rather, represented a colonial proclivity to "'listen in' on the subaltern, whether through surveillance, bio-piracy or reified forms of consumption" (Byrd & Rothberg, 2011: 6). They formed the framework for later forms of more systematic and mass surveillance, including those that rely on supposedly neutral datasets. Over time they created hierarchies of 'epistemic credibility' (Alcoff, 1999) in terms of who could know and who could be known. In this way, the white colonial became the reliable knower and, in relation to surveillance, the credible watcher. Surveillance then is linked to *hierarchies of knowledge* in colonial societies and represents another form of the epistemic violence (Smith, 2012) that is enacted as part of colonial projects.

Surveilling and Managing Indigenous Deviance and Threat

In the colonial common sense story about Māori, Māori difference is often constituted as deviance (Barnes et al., 2012), a narrative evident elsewhere as a common trope of colonisers about those whose lands they invade (de Leeuw et al., 2010). Surveillance was integral to this production of the discourse of surrounding Indigenous compliance with, or deviation from, newly imposed Westernised and capitalist norms (Smith, 2009). While particular manifestations of this discourse varied over time and context, a persistently repeated construction was that of deviance as dangerous. At times, the danger was framed as a biological threat from diseased or unsanitary bodies (Dow, 1999; Wanhalla, 2006), at other times as a moral threat. In dominant elite discourses, Māori were also represented as a physical threat to (non-Māori) property and life, thereby justifying the dispossession and violence that occurred.

In this narrative of danger, the threat of violence or "rebellion" was produced as constant (Smith, 2009). Scholars have suggested that 'anxiety, fear and *angst*' amongst colonisers were central elements of colonisation, manifesting as a "colonial panic" (Fischer-Tiné & Whyte, 2016: 2). Moana Jackson notes that Māori have been constructed as a threat "whenever they have questioned their dispossession or whenever the colonisers wanted to keep them in a position of political powerlessness and economic inequality" (Jackson, 2016: 2). Like all settler colonies, land acquisition was paramount in Aotearoa NZ along with the suppression of Indigenous political and economic autonomy. For at least two decades following the Treaty of Waitangi (Orange, 1987), Māori dominated the produce trade, with one newspaper reporting that the Māori market share in some main centres "was so large indeed as to as to nearly monopolize the market and to exclude the Europeans from competition" (Belich, 1996: 215). The level of Māori control over trade and land was viewed as a distinct threat to the economic ambitions of the rapidly growing settler colonial population and was an important factor in the invasion of the Waikato region and confiscation of more than one million acres from tribes. Labelling Māori as "rebels" was a key part of the process of justifying ongoing dispossession, facilitated through the *Suppression of Rebellion Act 1863* that was "passed to enable the 'legal' suppression of actual and often armed Māori resistance to the depredations of the Crown, and led ultimately to the raupatu or confiscation of thousands of acres of our land" (Jackson, 2016: 7).

Observation, measurement, monitoring, and surveillance were thus linked not only to discursive practices of categorisation but to material incursions into the lives of Indigenous communities. As the focus turned from the security of the land to the management of peoples, "colonial regimes developed sophisticated forms of control through documentation and surveillance" (Berda, 2013: 628) that allowed the state to determine where it needed to intervene:

> The underlying impetus of all this observation and intelligence gathering was to provide a portrait of the progress of colonial rule. It identified

individuals and groups that were adhering to state policies, and singled out those who were not for further remedial discipline. (Smith, 2009: 17)

This "remedial discipline" took many forms for Indigenous communities already subjected to the violent dispossession of lands. Colonial state institutions and agencies, such as the Native Schools, child 'welfare', and policing systems, were instrumental in the ongoing surveillance and enforcement of compliance with state goals of assimilation. While changes in technological capabilities have allowed for the surveillance of Indigenous peoples to occur in new and more sophisticated ways, the underpinning racial and colonial beliefs endure. Although often framed in relation to concepts of safety and security, contemporary state surveillance practices have a primary interest in maintaining state power and control. O'Connell, drawing on Genel (2006), notes that

the paradox of biopolitics is that protection for some is fully tied to harm for others; others who must be positioned as intolerable, outside of humanity. Colonial and imperial logics are built on knowledge practices designed to define and manage populations, and establish the right to rule. (O'Connell 2016: 79)

In Aotearoa NZ, this positioning of some groups as worthy of protection, and others as risky or potentially dangerous, continues to be a fundamental part of the mindset of contemporary surveillant practices that often involve states and corporations ignoring, undermining, or explicitly breaching human rights. Since the early 2000s, state surveillance powers in Aotearoa NZ have increased considerably (Keenan, 2016). The 2002 *Terrorism Suppression Act* significantly expanded state powers of surveillance (Wakeham, 2012) and within five years was invoked to justify state paramilitary raids across the country (Wakeham, 2012). Most of those people arrested in the October 2007 raids were Māori (12 out of 17 people). The 'Tuhoe raids' that took place in Te Urewera were the most violent and included barricading off whole communities, which did not happen in other areas where raids took place (Jackson, 2016; Keenan, 2016). The *Terrorism Suppression Act* was used by the police to carry out intrusive covert surveillance of a number of people prior to the raids; however, no charges were eventually brought under the Act. Wakeham (2012) suggests that the subsequent framing of the raids by the government and the police perpetuated fear, even in the absence of evidence to support any terrorism activity having occurred. The raids also highlighted the contingent nature of citizenship for Indigenous peoples in colonial settler states, whereby Indigenous membership is sufficient to justify state interventions or actions that fundamentally violate individual human rights. The same point can be made of the notorious Northern Territory National Emergency Response in Australia, which also raised serious human rights concerns (Australian Human Rights Commission, 2008).

According to Lloyd and Wolfe (2016), the expanded investment in, and use of, surveillance within nation-states as part of the ostensible 'war on terror' has happened alongside reductions to state investment in social services and the increased involvement of the private sector. The 'paramilitarization of the police' is seen as part of a broader response to the domestic threat of "disaffected, unincorporable masses" (Lloyd & Wolfe, 2016: 109). In fact, the close relationship between the police and the military has been a feature of policing in Aotearoa NZ since colonisation (Hill, 2016). Similar parallels were drawn in a letter to the New Zealand Government from the Special Rapporteurs on the promotion and protection of human rights and the Secretary-General's Special Representative on Human Rights Defenders. In it they "expressed concern at the extent of surveillance, the interception of telephone calls, and the monitoring of computer accounts since 2005. Concern was expressed that the arrest of Māori might be connected to their historic struggle for land and political rights" (Keenan, 2016: 26–27).

In this sense the over-surveillance of Māori in relation to the October 2007 raids was not particularly novel, nor was the use of disproportionate and excessive force (Jackson, 2016). Rather, this was a continuation of a grim history of excessive state force including armed raids on the Māori pacifist community at Parihaka in 1881 (Buchanan, 2018) and on Maungapōhatu in 1916 to arrest the Tuhoe prophet Rua Kenana on concocted charges of sedition (Binney et al., 1979). Māori continue to be policed and incarcerated at significantly higher rates than non-Indigenous New Zealanders and to be treated harshly by the legal system (Human Rights Commission, 2012). Like other Indigenous children in colonial settler states (SCRGSP, 2018; de Leeuw et al., 2010), Māori children also continue to be much more likely to be taken into state 'care', that is, removed from their homes by the state (Office of the Children's Commissioner, 2016), with Māori making up over 60 per cent of all children in foster care in 2017 (Keddell & Hyslop, 2019).

Colonial Surveillance in an Era of Big Data in Aotearoa NZ

Surveillance has taken a dramatic turn in the digitally enhanced era of big data with myriad possibilities for monitoring individuals and groups, including algorithmic surveillance (Murphy, 2017), biometric and facial recognition technologies (Gates, 2011), licence plate readers, CCTV (Waiton, 2010), cell phone data (Gellman & Soltani, 2013), and the monitoring of social media use (Owen, 2017). The secondary use of data often extends far beyond its original purpose and without explicit consent. In Aotearoa NZ, successive governments have enthusiastically embraced the use of large, linked datasets to identify social liabilities and risks and to direct social funding to try and realise greater returns on investment (NZIER, 2016). One of the data innovations at the centre of this data-driven investment approach is the Integrated Data Infrastructure or IDI (Stats NZ, 2018a). The IDI links census data for the whole population with a number of key administrative and survey datasets from a range of sectors and government agencies, including tax, education, health, child welfare, justice and corrections, and police data, including the NZ Police 'gang registry'. The data in the IDI are de-identified once linkage has occurred, and a number of technical safety mechanisms are in place for use of the dataset (Stats NZ, 2018a). However, concerns have been raised about the IDI (Jonas, 2018; Kukutai & Cormack, 2019). Although data are de-identified before being made available to researchers, the linking of multiple data sources enables new forms of surveillance that exist outside of ethics and other privacy or consent mechanisms (O'Connell, 2016). Linkage is generally not based on individuals' informed consent for their data to be included in the IDI, shared between agencies, or linked to other datasets, and there are few mechanisms for opting out. Most data are collected as part of other routine or survey collections, so people may be unaware that the data they provide will be able to be linked to multiple other data sources in a way that allows for them to be tracked over time and across social services.

The Young Parent Payment (YPP) is an example of the construction of social liabilities and the use of data sharing to monitor perceived risk. The programme, which provides financial support for 16–19-year-old parents who meet the eligibility requirements, uses "information and technology to monitor outcomes and financial sanctions to enforce new compulsory social obligations for both parent and child" (Ware et al., 2017: 503).

These obligations include attending budgeting and parenting skills courses and enrolling in a Teen Parent Unit after the baby has reached the age of one. Recipients are subjected to levels of monitoring and surveillance that other beneficiaries are largely exempt from (Ware et al., 2017). As Māori women account for more than half of teenage pregnancies (Stats NZ, 2020a, 2020b), these surveillance practices have a disparate impact on Māori communities.

In the era of big data and extensive data linkage, state surveillance now involves predicting future potential risk (Capatosto, 2017), through the use of predictive risk and actuarial modelling approaches. Governments are increasingly using algorithms to supplement or replace human decision-making, motivated by a desire to reduce costs while meeting targets for service delivery. Cognitive and other sorts of bias can penetrate machine learning and algorithms in various ways. As Capatosto notes, "human beings encode our values, beliefs, and biases into these analytic tools by determining what data is used and for what purpose" (Capatosto, 2017: 3).

Innovations in data linking technologies in Aotearoa NZ have made it possible to identify and track individuals and, to a lesser extent, families, over time, and across their interactions with government-funded institutions. Predictive analytical approaches have been developed in a number of sectors, using historical case data to predict the risk of an 'event' occurring in the future. The rationale is often that early detection enables early intervention and prevention. Internationally, researchers and scholars have identified how these approaches target specific social groups, often entrenching already oppressive social hierarchies (e.g. Benjamin, 2019a, 2019b; Eubanks, 2018). In Aotearoa NZ, predictive tools have been used in the area of youth unemployment (NEET), family violence (SAFVR), and reconviction and reimprisonment (RoC*RoI) (Stats NZ, 2018b). There are major issues to this practice for Māori (Blank et al., 2015; Keddell, 2015, 2016). Over-surveillance of Māori historically, in particular by police, corrections, and other punitive and disciplinary institutions, means that data about Māori are more likely to be included in government datasets. In the context of child protection, one of the main issues affecting the accuracy of predictive risk approaches is that it relies on substantiation data as the outcome variable. As Keddell (2014) notes, "'visibility bias' affects initial notifications to child protection services … and tend to over-identify those who are poor and those overrepresented within the poor—Maori, Pasifika and women".

An example of this predictive risk approach in Aotearoa NZ was the stated intent by the Department of Corrections in 2016 to use the IDI to create actuarial risk models for the entire population which could, in the future, be used as part of decision-making at the frontline of corrections decisions (Hughes, 2016). In 2020, 52 per cent of people in prison were Māori (Department of Corrections, 2020); for women in prison, it was 63 per cent. The risks associated with the pre-emptive risk approach will clearly be disproportionately borne by Māori, particularly when criminal justice reform in Aotearoa NZ has favoured "retributive" rather than "transformative justice" (Te Uepū Hāpai i te Ora, 2019).

The era of increased data sharing and data linkage between government agencies creates new platforms for state surveillance of Indigenous peoples that extend and reify existing colonial, racialised biases (Carroll et al., 2021). In this sense, they facilitate even greater surveillance of people who are constructed as a potential future risk or liability, representing a new form of colonial angst. Increased mechanisation does not disrupt the racial logics built into the datasets or embedded within the institutions involved in surveillance. What it can do, however, is obfuscate the role of non-Indigenous decision-makers, including the state, in the lives of Indigenous peoples. The colonial white gaze is built into new modes of monitoring such that the state remains the watcher and never the watched. Simultaneously, it perpetuates conditions whereby Indigenous peoples are always known, never able to be unknown, and never the knower. Aligning with Foucault's discussion of the Panopticon and power "based on a system of permanent registration" (Foucault, 1979: 196), current modes of surveillance through the use of 'big data' reinscribe neoliberal individualism in ideas of the pre-eminence of the individualised knowable subject, who now exists as a series of linked data points, potentially indefinitely.

Māori Data Sovereignty: Resistance and Self-determination

Just as surveillance has been a constant feature of colonialism in Aotearoa NZ, so too has Māori resistance (Walker, 1990). This includes resistance to the racial logics and classificatory systems that underpinned state surveillance (Kukutai, 2012). In *Dark Matters: On the Surveillance of Blackness*, Browne (2015) proposes "dark sousveillance" as a way of conceptualising opposition to surveillance and "complicating" the notion of the Panopticon:

> As a way of knowing, dark sousveillance speaks not only to observing those in authority (the slave patroller or the plantation overseer, for instance) but also to the use of a keen and experiential insight of plantation surveillance in order to resist it. (2015: 21)

The notion of the 'watched' deploying their own forms of intelligence and agency to look and speak back (Browne, 2015) at the 'watchers' resonates. Māori have long repurposed data collected by the state as a form of counter-surveillance. For at least four decades, Māori health researchers have assembled and analysed data about Māori health inequities to monitor the impact of state actions, policies, and programmes on Māori (Robson & Harris, 2007; Pomare, 1980; Pomare & de Boer, 1988). Their efforts have served to both witness the repeated breaches of Māori rights under the Treaty of Waitangi and the UNDRIP, and provide evidence for claims against the Crown for its failures to protect Māori health (Ministry of Health, 2019).[4] In other areas, such as environmental and natural resource management, Māori researchers, organisations, and communities have developed their own Māori values-based indicators to monitor changes over time in ways that are culturally meaningful (Harmsworth & Tipa, 2006; Morgan, 2011) and to hold authorities to account[5] (Independent Māori Statutory Board, 2019).

More recently, Māori Data Sovereignty has provided a mechanism through which to articulate, and advocate for, a wider set of Māori rights and interests in Māori data—that is, any data that is about or from Māori people, Māori language, culture, resources, or environments (Te Mana Raraunga, 2018). As an approach to data, Māori Data Sovereignty requires a fundamental rethinking of how data should be collected, cared for, used, stored, shared, or restricted. Māori Data Sovereignty principles and

[4] As of June 2020, there were more than 200 claims seeking to participate in what is known as the Health Services and Outcomes Kaupapa Inquiry (Wai 2575). The claims are historical and contemporary and cover a range of issues relating to the health system, specific health services and outcomes, including health equity; primary health care; disability services; mental health; and alcohol, tobacco, and substance abuse. See: https://www.health.govt.nz/our-work/populations/maori-health/wai-2575-health-services-and-outcomes-kaupapa-inquiry

[5] In particular, see the five 'Māori Values' reports published by the Independent Māori Statutory Board at: https://www.imsb.maori.nz/value-reports/introduction/

frameworks also provide a way of thinking through surveillance, and through data practices and relations that are harmful, and to imagine alternative futures that exist beyond a colonial surveilling data system (Cormack et al., 2020; Kukutai & Cormack, 2019).

To that end Māori Data Sovereignty, and Indigenous Data Sovereignty more broadly, complicates prevailing notions of personal data and individual privacy and consent. Issues relating to personal data protection and individual privacy are well defined internationally, with some governments moving to implement stricter regulatory controls around the collection, storage, and use of personal data (e.g. EU GDPR). However, the risks of big data extend far beyond the individual, and personal data is now at "one end of a long spectrum of targets" in need of protection (Taylor et al., 2017). Many of the ways in which data surveillance occurs currently relate not only to individuals but to collectives, whether they are constructed prior to data collection or in analytical and output processes. While there is some recognition that group privacy cannot be reduced to the aggregate privacies of its members (Vis-Dunbar et al., 2011), there are few practical and operational examples of group privacy protection to counter group surveillance. One exception is in Canada, where First Nations communities that have adopted the First Nations Information Governance Centre OCAP® principles have passed their own privacy laws (First Nations Information Governance Centre, n.d.).

Māori have collective data rights—that includes a collective right to not be known by the state and to not be placed into a constructed group, particularly where those groups are manifestations of racialised colonial imaginaries. However, as 'data subjects' Māori are included in a diverse range of data aggregations, from self-defined political and social groupings (e.g. tribes) to clusters of interest defined by data analysts and controllers (e.g. children of incarcerated parents). Current regulatory approaches fail to acknowledge, let alone address, the privacy implications of these collective designations. Indigenous Data Sovereignty allows for the consideration of data rights outside of neoliberal conceptualisations of the individual, pushing us to incorporate collective rights and interests and relational understandings of data into contemporary data practices.

Since its establishment in 2015, the Māori Data Sovereignty Network, Te Mana Raraunga (TMR), has taken to task various government agencies over a range of issues including a lack of social and cultural licence to use government administrative data for census purposes, the need for Treaty-based Māori data governance over government-held Māori data, and the procurement of a facial recognition system by the Department of Internal Affairs (see: https://www.temanararaunga.maori.nz/nga-panui). A range of government initiatives has been developed to try to respond to Māori Data Sovereignty (Sporle et al., 2020), but ongoing structural inequities mean there are significant barriers to achieving the sort of transformational change needed (Kukutai & Cormack, 2020). At the international level, the Global Indigenous Data Alliance and Research Data Alliance Indigenous Data Sovereignty Interest Group have led a number of projects to try to influence government and private sector data practices. These include the development of the CARE Principles of Indigenous Data Governance (Carroll et al., 2020a) and guidelines for the use of COVID-19 data with respect to Indigenous Data Sovereignty (Carroll et al., 2021; Research Data Alliance COVID-19 Indigenous Data Working Group, 2020). The latter has been particularly important given the heightened risks that the pandemic has provided for data harms and racist surveillance (Carroll et al., 2020b, 2021).

Conclusion

In recent years, discourses of 'reconciliation' with Indigenous peoples have gained increasing prominence. However, as Pauline Wakeham (2012) discusses, at the same time as settler colonial states have been making more public calls for 'reconciliation', Indigenous peoples have been simultaneously impacted by increased state powers of surveillance. In this sense, Indigenous resistance to colonialism remains a threat to state power that needs to be managed, and surveillance continues to be a mechanism to police and manage Indigenous peoples, albeit with new and expanded technologies at play.

The expansion of state powers and reduction of civil liberties, in combination with the construction of Indigenous resistance as an ongoing threat to the safety and security of the nation-state, undermines Indigenous self-determination. In addition, surveillance reinforces colonial hierarchies of power and reinscribes dangerousness and deviance onto Indigenous peoples. While colonial logics and structures remain in place, it is not possible to have surveillance practices that operate outside of this racialised imaginary. Indigenous Data Sovereignty is enmeshed with broader anti-colonial and sovereignty movements in seeking to unsettle current harmful data practices, including surveillance, and restore social relations and practices that are relational, collective, and bounded in place.

Indigenous Data Sovereignty offers a critical approach to data and surveillance that is situated in the knowledges and lived experiences of those who are located in the "zone of non-being". In line with Simone Browne's conceptualisation of "dark sousveillance", Indigenous communities have "keen and experiential insight" (Browne, 2015: 21) into practices and relations of surveillance that illuminate potential spaces for resistance and disruption that may be unseen or unfelt by others. This (re)orientation is important to ensure that the increasing scholarly attention being paid to data colonialism does not reify the knowledge hierarchies that characterise colonial knowledge production or produce paternalistic responses misaligned with Indigenous goals of self-determination.

As Moana Jackson importantly reminds, "[N]o reality is immutable or beyond change and the centuries of indigenous resistance have always brought change in what seemed unchangeable situations. That history is part of our reality" (Jackson, 2018: 109). Indigenous Data Sovereignty is a part of this intergenerational resistance.

References

Alcoff, L. M. (1999). On judging epistemic credibility: Is social identity relevant? *Engendering Rationalities, 29*(1), 53–80.

Australian Human Rights Commission. (2008). *Social Justice Report 2007.* https://www.humanrights.gov.au/publications/social-justice-report-2007-chapter-3-northern-territory-emergency-response-intervention

Barnes, A. M., Borell, B., McCreanor, T., Nairn, R., Rankine, J., & Taiapa, K. (2012). Anti-Māori themes in New Zealand journalism—Toward alternative practice. *Pacific Journalism Review: Te Koakoa, 18*(1), 195–216.

Belich, J. (1996). *Making peoples: A history of the New Zealanders from Polynesian settlement to the end of the nineteenth century.* Penguin.

Benjamin, R. (2019a). Introduction: Discriminatory design, liberating imagination. In R. Benjamin (Ed.), *Captivating technology: Race, carceral technoscience, and liberatory imagination in everyday life* (pp. 1–22). Duke University Press.

Benjamin, R. (2019b). *Race after technology: Abolitionist tools for the New Jim Code.* Polity Press.

Berda, Y. (2013). Managing dangerous populations: Colonial legacies of security and surveillance. *Sociological Forum, 28*(3), 627–630.

Binney, J., Chaplin, G., & Wallace, C. (1979). *Miahaia. The prophet Rua Kenana and his community at Maungapohatu.* Bridget Williams Books.

Blank, A., Cram, F., Dare, T., De Haan, I., Smith, B., & Vaithinathan, R. (2015). *Ethical issues for Māori in predictive risk modelling to identify new-born children who are at high risk of future maltreatment*. https://www.msd.govt.nz/documents/about-msd-and-our-work/publications-resources/research/predictive-modelling/00-ethical-issues-for-maori-in-predictive-risk-modelling.pdf

Browne, S. (2015). *Dark Matters: On the Surveillance of Blackness*. Duke University Press.

Buchanan, R. (2018). *Ko Taranaki te Maunga*. Bridget Williams Books.

Burgess, H., & Painting, T. K. (2020). Onamata, anamata: A whakapapa perspective of Māori futurisms. In S. Walsh & A. M. Murtola (Eds.), *Whose futures?* ESRA.

Byrd, J., & Rothberg, M. (2011). Between subalternity and indigeneity. *Interventions, 13*(1), 1–12.

Capatosto, K. (2017). *Foretelling the future—A critical perspective on the use of predictive analytics in child welfare*. Kirwan Institute for the Study of Race and Ethnicity, Ohio State University.

Carroll, S. R., Rodriguez-Lonebear, D., & Martinez, A. (2019). Indigenous data governance: Strategies from United States Native Nations. *Data Science Journal, 18*(31), 1–15.

Carroll, S. R., et al. (2020a). The CARE principles for indigenous data governance. *Data Science Journal, 19*(43), 1–12. https://doi.org/10.5334/dsj-2020-043

Carroll, S. R., et al. (2020b). Indigenous data in the COVID-10 pandemic: Straddling erasure, terrorism, and sovereignty. *Social science research council, items: Insights from the Social Sciences*. https://items.ssrc.org/covid-19-and-the-social-sciences/disaster-studies/indigenous-data-in-the-covid-19-pandemic-straddling-erasure-terrorism-and-sovereignty/

Carroll, S. R., et al. (2021). Indigenous data during COVID-19: From external to internal. *Frontiers in Medical Sociology, 62*. https://doi.org/10.3389/fsoc.2021.617895

Cormack, D., Kukutai, T., & Cormack, C. (2020). Not one byte more: From data colonialism to data sovereignty. Kia kaua tētahi paita anō: mai i ngā raraunga whakatōpū ki te mana motuhake o ngā raraunga. In A. Chen (Ed.), *Shouting zeros and ones: Digital technology, ethics and policy in New Zealand* (pp. 71–83). Bridget Williams Books.

Couldry, N., & Yu, J. (2018). Deconstructing datafication's brave new world. *New Media & Society, 20*(12), 4473–4491.

de Leeuw, S., Greenwood, M., & Cameron, E. (2010). Deviant constructions: How governments preserve colonial narratives of addictions and poor mental health to intervene into the lives of Indigenous children and families in Canada. *International Journal of Mental Health and Addiction, 8*, 282–295.

Department of Corrections. (2020). Prison facts and statistics—June 2020. https://www.corrections.govt.nz/resources/statistics/quarterly_prison_statistics/prison_stats_june_2020

Dow, D. (1999). *Maori Health and Government Policy 1840–1940*. Victoria University Press.

Eubanks, V. (2018). *Automating inequality: How high-tech tools profile, police, and punish the poor*. St. Martin's Press.

Fanon, F. (1986). *Black skin, white masks*. Pluto Press.

First Nations Information Governance Centre. (n.d.). The First Nations Principles of OCAP®. https://fnigc.ca/ocap.

Fischer-Tiné, H., & Whyte, C. (2016). Introduction: Empires and emotions. In H. Fischer-Tiné (Ed.), *Anxieties, fear and panic in colonial settings: Empires on the verge of a nervous breakdown* (pp. 1–23). Palgrave Macmillan.

Foucault, M. (1979). *Discipline and punish: The birth of the prison*. Vintage Books.

Gates, K. (2011). *Our biometric future: Facial recognition technology and the culture of surveillance*. New York University Press.

Gellman, B., & Soltani, A. (2013). NSA tracking cellphone locations worldwide, Snowden documents show. *Washington Post*, 4 December. https://www.pulitzer.org/files/2014/public-service/washpost/16washpostnsa2014.pdf

Genel, K. (2006). The question of biopower: Foucault and Agamben. *Rethinking Marxism, 18*(1), 43–62.

Grosfoguel, R. (2004). Race and ethnicity or racialized ethnicities? Identities within global coloniality. *Ethnicities, 4*(3), 315–336.

Grosfoguel, R. (2016). What is racism? *Journal of World-Systems Research, 22*(1), 1–9.

Grosfoguel, R., Oso, L., & Christou, A. (2015). 'Racism', intersectionality and migration studies: Framing some theoretical reflections. *Identities, 22*(6), 635–652.

Harmsworth, G., & Tipa, G. (2006). *Maori environmental monitoring in New Zealand: Progress, concepts and future direction*. Landcare Research.

Hill, R. (2016). Māori, police and coercion in New Zealand history. In D. Keenan (Ed.), *Terror in our midst? Searching for terror in Aotearoa New Zealand* (pp. 39–61). Huia Publishers.

Hughes, T. (2016). The investment approach to justice: Taking integrated offender management to police, justice and the wider social sector. *Practice, 4*(1).

Human Rights Commission. (2012). *A fair go for all: Addressing structural discrimination in public services*. Human rights Commission.

Independent Māori Statutory Board. (2019). *Data issues of significance*. https://www.imsb.maori.nz/assets/sm/upload/pa/0z/vq/ww/IMSB%20Data%20Issues%202019_d.pdf

Ittmann, K., Cordell, D., & Maddox, G. (Eds.). (2010). *The demographics of empire: The colonial order and the creation of knowledge*. Ohio University Press.

Jackson, M. (2016). Preface: The constancy of terror. In D. Keenan (Ed.), *Terror in our midst? Searching for terror in Aotearoa New Zealand* (pp. 1–10). Huia Publishers.

Jackson, M. (2018). In the end: "The hope of decolonisation". In E. McKinley & L. Smith (Eds.), *Handbook of indigenous education* (pp. 101–111). Springer.

Jonas, M. (2018). *Ethics and the integrated data infrastructure*. Health Research Council. https://gallery.mailchimp.com/57af16fa15f95ed83e0b434a9/files/9bb5a418-ccdb-4f8b-aefe-4bc784a89b8d/Monique_Jonas_Integrated_Data_Infrastructure.02.pdf

Kak, A. (Ed.). (2020). *Regulating biometrics: Global approaches and urgent questions*. AI Now Institute. https://ainowinstitute.org/regulatingbiometrics.html

Keddell, E. (2014). The ethics of predictive risk modelling in the Aotearoa/New Zealand child welfare context: Child abuse prevention or neo-liberal tool? *Critical Social Policy, 35*(1), 69–88.

Keddell, E. (2015, June 4). Predictive risk modelling: On rights, data and politics. Re-imagining social work in Aotearoa New Zealand. *Re-Imagining Social Work in Aotearoa New Zealand* [blog]. http://www.reimaginingsocialwork.nz/2015/06/predictive-risk-modelling-on-rights-data-and-politics/

Keddell, E. (2016). Substantiation decision-making and risk prediction in child protection systems. *Policy Quarterly, 12*(2), 46–56.

Keddell, E., & Hyslop, I. (2019). Ethnic inequalities in child welfare: The role of practitioner risk perceptions. *Child & Family Social Work, 24*(4), 409–420.

Keenan, D. (2016). Introduction—Searching for terror. In D. Keenan (Ed.), *Terror in our midst? Searching for terror in Aotearoa New Zealand* (pp. 17–33). Huia Publishers.

Kukutai, T. (2012). Quantum Māori, Māori quantum: Representations of Māori identities in the census, 1857/8–2006. In R. McClean, B. Patterson, & D. Swain (Eds.), *Counting stories, moving ethnicities: Studies from Aotearoa New Zealand* (pp. 27–51). University of Waikato.

Kukutai, T., & Taylor, J. (2016). *Indigenous Data Sovereignty: Toward An Agenda*. ANU press.

Kukutai, T., & Cormack, D. (2019). Mana motuhake ā-raraunga: Datafication and social science research in Aotearoa. *Kotuitui: New Zealand Journal of Social Sciences Online, 14*(2), 201–208.

Kukutai, T., & Cormack, D. (2020). 'Pushing the space': Data sovereignty and self-determination in Aotearoa NZ. In M. Walter, T. Kukutai, S. R. Carroll, & D. Rodriguez-Lonebear (Eds.), *Indigenous data sovereignty and policy* (pp. 21–35). Routledge.

Lloyd, D., & Wolfe, P. (2016). Settler colonial logics and the neoliberal regime. *Settler Colonial Studies, 6*(2), 109–118.

Mahuika, N. (2019). A brief history of whakapapa: Māori approaches to genealogy. *Genealogy, 3*(2), 32.

Maldonado-Torres, N. (2007). On the coloniality of being. *Cultural Studies, 21*(2–3), 240–270.

Martin-Sanchez, F. J., Aguiar-Pulido, V., Lopez-Campos, G. H., Peek, N., & Sacchi, L. (2017). Secondary use and analysis of Big Data collected for patient care. *Yearbook of Medical Informatics, 26*(1), 28–37.

Meredith, P. (2006). Pakeha by law: The Europeanisation of Maori, 1912–1931. *New Zealand Universities Law Review, 22*, 103–147.

Mignolo, W. (2007). Introduction. *Cultural Studies, 21*(2–3), 155–167.

Ministry of Health. (2019). *Wai 2575 Māori Health Trends Report*. Ministry of Health.

Morgan, K. (2011). Waiora and cultural identity. Water quality assessment using the Mauri Model. *Alternative: An International Journal of Indigenous Peoples, 3*(1), 42–67.

Murphy, M. (2017). Algorithmic surveillance: The collection conundrum. *International Review of Law, Computers & Technology, 31*(2), 225–242.

New Zealand Institute of Economic Research. (2016). *Defining Social Investment, Kiwi Style* (NZIER working paper 2016/5). https://nzier.org.nz/static/media/filer_public/e8/56/e8566475-e1c7-4a2c-9f0f-bf65710b039b/wp2016-5_defining_social_investment.pdf

O'Connell, A. (2016). My entire life is online: Informed consent, big data, and decolonial knowledge. *Intersectionalities: A Global Journal of Social Work Analysis, Research, Polity, and Practice, 5*(1), 68–93.

Office of the Children's Commissioner. (2016). *State of Care 2016: What we learnt from monitoring child, youth, and family*. Office of the Children's Commissioner.

Omi, M., & Winant, H. (1986). *Racial formation in the United States*. Routledge.

Orange, C. (1987). *The Treaty of Waitangi*. Allen & Unwin.

Owen, S. (2017). Monitoring social media and protest movements: Ensuring political order through surveillance and surveillance discourse. *Social Identities, 23*(6), 688–700.

Pomare, E. (1980). *Maori standards of health: A study of the 20-year period 1955–75*. Medical Research Council of New Zealand.

Pomare, E., & de Boer, G. (1988). *Maori standards of health. A study of the years 1970–1984*. Department of Health.

Quijano, A. (2005). The challenge of the 'indigenous movement' in Latin America. *Socialism and Democracy, 19*(3), 55–78.

Quijano, A. (2007). Coloniality and modernity/rationality. *Cultural Studies, 21*(2–3), 168–178.

Research Data Alliance COVID-19 Indigenous Data Working Group. (2020). Data sharing respecting Indigenous data sovereignty. In RDA COVID-19 Working Group. *Recommendations and guidelines on data sharing*. Research Data Alliance. https://doi.org/10.15497/rda00052.

Robson, B., & Harris, R. (Eds.). (2007). *Hauora: Māori Standards of Health IV. A study of the years 2000–2005*. Te Rōpū Rangahau Hauora a Eru Pōmare.

Sa'di, A. (2012). Colonialism and surveillance. In K. Ball, K. Haggerty, & D. Lyon (Eds.), *Routledge handbook of surveillance studies* (pp. 151–158). Routledge.

SCRGSP (Steering Committee for the Review of Government Service Provision). (2018). *Report on Government Services 2018*. Productivity Commission, Canberra.

Smith, K. (2009). *Liberalism, surveillance, and resistance: Indigenous communities in Western Canada, 1877–1927*. AU Press.

Smith, L. T. (2012). *Decolonizing methodologies* (2nd ed.). Zed Books.

Sporle, A., Hudson, M., & West, K. (2020). Indigenous data and policy in Aotearoa New Zealand. In M. Walter, T. Kukutai, S. R. Carroll, & D. Rodriguez-Lonebear (Eds.), *Indigenous data sovereignty and policy* (pp. 62–80). Routledge.

Stats NZ. (2018a). *Integrated data infrastructure*. https://www.stats.govt.nz/integrated-data/integrated-data-infrastructure/

Stats NZ. (2018b). *Algorithm assessment report*. Stats NZ.

Stats NZ. (2020a). Live births by age of mother (annual June), 1981–2020. http://archive.stats.govt.nz/infoshare/

Stats NZ. (2020b). Live births by age of mother (Maori and Total population) (annual June), 1981–2020. http://archive.stats.govt.nz/infoshare/

Taylor, L., Floridi, L., & van der Sloot, B. (2017). Introduction: A new perspective on privacy. In L. Taylor, L. Floridi, & B. van der Sloot (Eds.), *Group privacy: New challenges of data technologies*. Springer.

Te Mana Raraunga. (2018). *Principles of Māori data sovereignty*. https://cdn.auckland.ac.nz/assets/psych/about/our-research/documents/TMR%2BM%C4%81ori%2BData%2BSovereignty%2BPrinciples%2BOct%2B2018.pdf

Te Uepū Hāpai i te Ora—Safe and Effective Justice Advisory Group. (2019). *Turuki! Turuki! Move together: Transforming our Criminal Justice System*. https://www.safeandeffectivejustice.govt.nz/assets/Uploads/28ce04fd87/Turuki-Turuki-Report-Interactive.pdf

Toki, V. (2017). Māori seeking self-determination or tino rangatiratanga? A note. *Journal of Māori and Indigenous Issues, 5*, 134–144.

Vis-Dunbar, M., Williams, J., & Weber Jahnke, J. H. (2011). *Indigenous and community-based notions of privacy. Informational Privacy Interdisciplinary Research Group (IPIRG) Report* No. UVic/IPIRG-2011-TR-04. : University of Victoria.

Waiton, S. (2010). The politics of surveillance: Big Brother on Prozac. *Surveillance and Society, 8*, 61–84.

Wakeham, P. (2012). Reconciling 'terror': Managing indigenous resistance in the age of apology. *American Indian Quarterly, 36*(1), 1–33.

Walker, R. (1990). *Ka whaiwhai tonu matou: Struggle without End*. Penguin.

Walter, M., Kukutai, T., Carroll, S., & Rodriguez-Lonebear, D. (Eds.). (2021). *Indigenous data sovereignty and policy*. Routledge. https://doi.org/10.4324/9780429273957

Walter, M., & Suina, M. (2019). Indigenous data, indigenous methodologies and indigenous data sovereignty. *International Journal of Social Research Methodology, 22*(3), 233–243.

Wanhalla, A. (2006). Housing un/healthy bodies: Native housing surveys and Māori health in New Zealand 1930–1945. *Health and History, 1*(1), 100–120.

Ward, A. (1974). *A show of justice: Racial 'Amalgamation' in nineteenth century New Zealand*. ANU Press.

Ware, F., Breheny, M., & Forster, M. (2017). The politics of government 'support' in Aotearoa/New Zealand: Reinforcing and reproducing the poor citizenship of young Māori parents. *Critical Social Policy, 37*(4), 499–519.

Open Access This chapter is licensed under the terms of the Creative Commons Attribution 4.0 International License (http://creativecommons.org/licenses/by/4.0/), which permits use, sharing, adaptation, distribution and reproduction in any medium or format, as long as you give appropriate credit to the original author(s) and the source, provide a link to the Creative Commons licence and indicate if changes were made.

The images or other third party material in this chapter are included in the chapter's Creative Commons licence, unless indicated otherwise in a credit line to the material. If material is not included in the chapter's Creative Commons licence and your intended use is not permitted by statutory regulation or exceeds the permitted use, you will need to obtain permission directly from the copyright holder.

PART II

State and Data Justice

The Datafied Welfare State: A Perspective from the UK

Lina Dencik

"We are witnessing the gradual disappearance of the postwar British welfare state behind a webpage and an algorithm. In its place, a digital welfare state is emerging."
—Statement on Visit to the United Kingdom by Philip Alston, United Nations Special Rapporteur on extreme poverty and human rights, 16 November 2018.

Introduction

The modern welfare state emerged out of industrialisation and the dual crises of a global recession followed by the Second World War that together created conditions for a consensus around the need to build a society better able to deal with the human costs of a largely unregulated market economy. The subsequent economic downturn of the 1970s followed by the advent of neoliberalism as a global ideology has seen the public sector shrink, labour relations shift, and financialisation take hold of the

L. Dencik (✉)
Cardiff University, Cardiff, UK
e-mail: DencikL@cardiff.ac.uk

© The Author(s) 2022

A. Hepp et al. (eds.), *New Perspectives in Critical Data Studies*,
Transforming Communications – Studies in Cross-Media Research,
https://doi.org/10.1007/978-3-030-96180-0_7

economy presenting numerous challenges for the welfare state and its continued relevance. Yet, recently the welfare state has come into renewed focus. The crisis of the COVID-19 pandemic has swiftly changed the terms of economy and state. For some, we are seeing a return of the Leviathan state, a social contract with an absolute sovereign in which the state provides the ultimate insurance against an intolerable condition (Mishra, 2020) and others see it as providing a renewed impetus for demands of universal healthcare, stable employment, and a basic income (Standing, 2020). Certainly, initial responses to the pandemic and ongoing lockdowns across the world have converged around unprecedented state interventions in the economy and a prominent rhetoric of economic planning and social security.

However, as Magalhães and Couldry (2020) note, any renewal of social welfare will be very different to how we knew it before. It will be so, in part, because the coronavirus crisis has elevated not only the role of the state but importantly that of Big Tech. They write, a renewal of social welfare "will be strongly driven by private corporations, and it will use their tools and platforms—whose ultimate goal is generating profit. Crucially, it will be based on opaque and intrusive forms of ***datafication***" (para 1, italics in original). The trend to turn more and more of social life into data points that can be collected and analysed is rapidly transforming the ways in which the provision of public services is organised with significant implications for how we might think of the welfare state. Whilst the emphasis on data infrastructures in the context of COVID-19 has made this more explicit in several different ways, the conditions for these developments were already well underway. As noted by the UN Special Rapporteur on extreme poverty and human rights, Philip Alston, the "digital welfare state" is already a reality or is emerging in many countries across the globe. In these states, "systems of social protection and assistance are increasingly driven by digital data and technologies that are used to automate, predict, identify, surveil, detect, target and punish" (Alston, 2019).

In this chapter, I elaborate on these conditions and discuss the interplay between technological infrastructures, data-driven systems, and the welfare state, focusing particularly on the UK. The welfare state in the UK follows a different trajectory than many of its European counterparts, evident also in its response to the COVID-19 pandemic, but it serves as an illuminating case for trends that are also emerging in many other contexts. The chapter draws in part on research conducted with colleagues at the

Data Justice Lab at Cardiff University that explored the uses of citizen scoring in public services as well as research carried out as part of the multi-year project DATAJUSTICE that explores the relationship between datafication and social justice. I am particularly focused on engaging with the imperatives of automation and the logics of data-driven systems in the context of the current political economy of digital technologies and how these relate to the values and visions of a society commonly associated with the welfare state. Using developments in local government and the public sector in the UK as a lens, I advance a two-part argument about the ways in which data infrastructures are transforming state-citizen relations through on the one hand advancing an actuarial logic based on personalised risk and the individualisation of social problems (what I refer to as responsibilisation) and, on the other, entrenching a dependency on an economic model that perpetuates the circulation of data accumulation (what I refer to as rentierism). These mechanisms, I argue, fundamentally shift the "matrix of social power" (Offe, 1984) that made the modern welfare state possible and position questions of data infrastructures as a core component of how we need to understand social change.

Matrix of Social Power and the Foundations of the British Welfare State

The British welfare state emerged, like elsewhere in Europe, out of the dual crises of the Great Depression and the Second World War, but it is worth noting that the foundations for a consensus around the need for the state to protect citizens from the harms of market failure, an emphasis on social solidarity, and a commitment to decommodification have earlier roots. As Thane (2013) has highlighted, demands for the state to take a permanent, as distinct from temporary and residual, responsibility for the social and economic conditions experienced by its citizens began in the 1870s in conjunction with industrial capitalism. Recognition that poverty had structural causes rather than ones that were purely moral and that responses needed to be collectivist rather than individualist grew in line with a notable increase in trade union membership and industrial conflicts in the lead up to the First World War. Yet it was only after the shocks of the Great Depression and Second World War that a government formally acknowledged that the welfare of the mass of its citizen was a major component of its activities and announced the dawning of a "welfare state"

(Thane, 2013). The arrangement saw governments, formally or informally, presiding over negotiations between capital and labour that were more or less institutionalised. Importantly, according to Judt (2007), this faith in the state—as planner, coordinator, facilitator, arbiter, provider, caretaker, and guardian—was widespread and crossed almost all political parties. It was from the outset a class compromise that was able to serve many conflicting ends and strategies simultaneously, making it attractive to a broad alliance of heterogeneous forces (Offe, 1984). "The welfare state", Judt contends, "was avowedly social, but it was far from socialist. In that sense welfare capitalism, as it unfolded in Western Europe, was truly post-ideological" (Judt, 2007: 362).

The welfare state, therefore, is more than the narrow interpretation of it as a provider of social services. Rather, as argued by Offe (1984), it can be understood as a formula that consists of the explicit obligation of the state apparatus to provide assistance and support to those citizens who suffer from specific needs and risks characteristic of the market society and is based on a recognition of the formal role of labour unions in both collective bargaining and the formation of public policy. It is, in this sense, a political solution to social contradictions that emerged out of a specific "matrix of social power": the nature of the welfare state and the agenda of any political reality is an outcome of the ways in which social classes, collective actors, and other social categories are able to shape the environment of political decision-making (Offe, 1984: 160). In Britain, whilst there was no formal 'social partnership' of the kind we see in other European countries, the labour movement was able to seek gains for the working class through social reforms to improve living conditions. Without a viable alternative solution in terms of economic policy, Hobsbawm has argued, "a reformed capitalism which recognized the importance of labour and social-democratic aspirations suited them well enough" (Hobsbawm, 1994: 272). In this sense, the British welfare state is an outcome of a widespread normative shift and a growing labour movement that was simultaneously constrained by political circumstances and an ongoing dependency on the capitalist economy.

This historical backdrop is important for any discussion of the welfare state today as it highlights the particular dynamics that informed the policy agendas being pursued. These dynamics have radically changed since the post-war period. The economic downturns of the 1970s followed by the advent of neoliberalism and globalisation as dominant ideologies across the Western world have been significant for how the welfare state has advanced. Whilst there is no consensus on how these developments intersect and responses have varied across national contexts (Genschel, 2004), the UK has been at the forefront of key transformations, rapidly transitioning to a service economy, highly dependent on global supply chains and precarious labour whilst experiencing a significant decline in trade union membership (Dencik & Wilkin, 2015). In the last decade, since the financial crisis of 2008, this has been accompanied by an austerity agenda that has weakened the public sector and overhauled welfare programmes and social care through the privatisation of services and substantial cuts (Monbiot, 2020). A recent report estimated that local authorities and councils have seen a reduction in funding of up to 60 per cent in the last ten years (Davies et al., 2019), whilst the transfer of assets from the public sector to the private sector since Thatcher in the 1980s has reduced state-owned enterprises from 10 per cent to less than 2 per cent of GDP and from 9 per cent to less than 1.5 per cent of total employment (ons.gov.uk, CPI 2016).

Technology, information and communication technologies (ICTs), in particular, have played a key role in these shifting dynamics. Instrumental in the growth of consumer capitalism, digitalisation has also been seen as a challenge to the welfare state and its ability to deliver on its promises, disrupting labour relations, undermining social security, and changing the parameters of state governance. With growing trends such as mass data collection, automation, and artificial intelligence, these tensions have only intensified, putting the welfare state into further question (Petropoulos et al., 2019). At the same time, developments in technology have also significantly shaped public administration and the way social welfare is organised through the establishment of bureaucracies and different forms of population management. The creation of databases and the monitoring of citizens were from early on key features of the welfare state and played a fundamental part in assessing population needs and determining the allocation of resources (Rule, 1973; Scott, 1994). This includes ways of advancing social engineering and discerning "deserving" and "undeserving" citizens as central features of the modern welfare state (Dencik & Kaun, 2020). In the UK, for example, the 'modernisation' of public administration in line with a growing emphasis on new public management strategies is closely linked to early forms of the digitalisation of services as a way to "rationalise" engagement with citizens (White, 2009). In addition, a perceived need to increase information gathering and sharing as a way to better manage risk has led to a growing reliance on databases that overwhelmingly pertain to vulnerable and disadvantaged groups. In what they refer to as the advent of the "database state", Anderson et al. (2009) map the myriad public sector databases that have been put in place under different government programmes in the UK, arguing that several of these do not abide by human rights and data protection laws.

These previous intersections between technology and the welfare state have paved the way to what Yeung (2018) has described as a paradigm shift in public administration from 'new public management' to 'new public analytics' organised around algorithmic regulation. In her seminal study of the welfare sector in the US, Eubanks (2018) similarly refers to a new "regime" of data analytics used to determine eligibility and assess needs across areas of housing, healthcare, and child welfare. The non-governmental organisation AlgorithmWatch (2019), meanwhile, has outlined the growing reliance on automated decision-making or decision support systems across the public sector in Europe, understood as procedures in which decisions are delegated to automatically execute decision-making models to perform an action. This might include allocating treatment for patients in the public health system in Italy, sorting the unemployed in Poland, identifying child neglect in Denmark, or detecting benefit fraud in the Netherlands. As I will go on to outline below, the UK has increasingly integrated these technologies into public services in a way that present a particular set of questions for the nature of the welfare state. These include both a concern with the epistemological and ontological premises of "dataism" (Van Dijck, 2014) and a concern with the implications of making public infrastructure subject to datafication as a "political-economic regime" (Sadowski, 2019).

The Datafication of Welfare in the UK

As part of his investigation into the UK in 2018, the UN Special Rapporteur on extreme poverty and human rights Philip Alston highlighted the important role digital technologies now play in the administration of welfare (Alston, 2018). Of particular significance is the Universal Credit system, the first 'digital-by-default' policy implemented by the UK government, designed to reform social welfare into one integrated platform for benefit claimants. A key part of this reform is the emphasis on automation as a policy goal and the processing of claims entirely through digital means. As Alston's investigation makes clear, this has contributed to entrenched inequality, exclusion, and lack of redress with significant implications for human and social rights, not least the right to social protection. Digital divides, in terms of both access and literacy, poor design, and a lack of transparency have marked a system designed to embed conditionality within the very infrastructure of welfare provision, pushing people into destitution and poverty (ibid.). This has led to calls for the Universal Credit system to be scrapped and for digital-by-default as a policy to be illegalised (see, e.g. the Labour Party manifesto of 2019).

Yet, the Universal Credit system and the turn to digital platforms as intermediaries between public administration and service users are only one part of how digital technologies are intersecting with the British welfare state. Of growing importance is the emphasis on data collection and predictive analytics as a way to inform decisions that impact people's ability to participate in society. We see this, for example, with the advent of what we describe as 'citizen scoring' in a study we carried out at the Data Justice Lab. This refers to "the use of data analytics in government *for the purposes of categorization, assessment and prediction at both individual and population level*" (Dencik et al., 2019: 3; italics in original). These practices are part of a broader trend towards organisations becoming data-driven as a way to, it is claimed, run more efficiently and, importantly, without human bias and errors. For councils and local authorities who have been facing significant cuts, the promotion of data-driven systems as a way to reduce costs and increase efficiency and effectiveness has been particularly attractive (Beer, 2019). The emphasis on the need to focus resources and advance a more strategic understanding of population needs has been a common justification for the turn to citizen scoring. In many cases this has led to the creation of what is described as 'data warehouses' or 'data lakes' in which data is collected from a range of sources and databases from across different parts of the council and are integrated as a way to get a more granular and holistic understanding of individual households and families (Dencik et al., 2019). In some instances, this has been accompanied by predictive analytics in which these data warehouses underpin further algorithmic processing designed to simulate projections of the future as a way to assess or evaluate risks and needs.

An example of this kind of practice is increasingly prevalent in policing, where a growing number of British police forces are using predictive analytics to map crime trends in neighbourhoods and to rank offenders from high to low risk of reoffending (Couchman, 2019). Such predictions draw on a range of data sources, including crime and intelligence data, missing persons data, operational data, data held by council agencies, demographic data, and even weather data (Dencik et al., 2018). At Avon and Somerset police constabulary, for example, they have contracted a software application suite from the company Qlik Sense that is used to attribute a risk profile to all existing offenders and victims of crime on record based on real-time monitoring of characteristics and behaviours. These profiles, presented as a dashboard, inform the way Avon and Somerset police organise their resources and how they decide to engage with different individuals. Similar tools are being used in child welfare where policy reforms, such as the Troubled Families programme implemented in 2012, have incentivised increased data collection and sharing on children and families. More recently, a range of tools designed to assess risk and predict potential behaviour has been implemented around the creation of these databases (Redden et al., 2020). Bristol Council, for example, has developed an in-house tool drawing on a range of social issue data-sets that are designed to attribute a risk score to all children and young people living in the city based on a prediction about the likelihood that a child falls victim to 'child exploitation'. This score is generated on the basis of the extent to which the characteristics and behaviour of a family match those of known previous victims of child exploitation. The Council of Hackney contracted a similar tool, Early Help Profiling, from the company Xantura that produces intelligence reports once a risk threshold regarding a family is passed as a way to assist decision-making by frontline staff (Dencik et al., 2018).

The uses of these kinds of technologies in the public sector are still only emerging and there is still an uneven landscape amongst local and central government in regard to how data about people is collected and used. Whilst there is a general trend towards becoming more data-driven across government, it is not obvious that there is a shared understanding of what it is appropriate to do with data. Such an interpretive vacuum is evident from the difficulty in clearly asserting where and how data-driven systems are used in government, and in the myriad tensions and negotiations that shape the implementation of such technologies within councils and local authorities (Dencik et al., 2019). However, despite the heterogeneous nature of data practices across local government and the prevalent resistance towards algorithmic decision-making from a range of stakeholders, there is a recognisable drive towards automation and predictive analytics within social welfare and the public sector in the UK at large (cf. Booth, 2019). This has only been heightened by the COVID-19 pandemic with an onus on data collection and technological solutions shaping responses to the health crisis, whether in the form of contact-tracing apps, immunity passports, or other forms of data infrastructure to track, certify, and model the coronavirus. At the same time, the transition of social and economic life to the cloud that was already well underway has been accelerated with social distancing measures (Klein, 2020; Morozov, 2020). The welfare state, therefore, in whatever form it will take following the coronavirus crisis, looks certain to be more datafied. This raises some significant questions in need of interrogation. Below, I discuss two interrelated aspects that concern, firstly, the issue of responsibilisation and, secondly, the issue of rentierism. Both of these present counter-logics to the values commonly associated with the modern welfare state.

Datafication as Responsibilisation

As noted above, the advent of 'digital by default' policy frameworks and the collection of data in welfare systems build on previous bureaucracies and emerge out of a longer history of risk management in public administration. Alston (2019) also points out that often the implementation of new technologies in public services is seen as politically neutral and void of policy implications that allows for the gradual datafication to take place without much scrutiny and public debate. Largely it is framed as a matter of efficiency and a predominantly quantitative shift: more information, processed faster. Yet the sheer scale and nature of data now collected on citizens introduce key questions about the ways in which citizens are rendered increasingly legible to the state and the use of big data to inform decisions rest on some key assumptions with significant implications for the idea of the welfare state. In this section I focus particularly on the issue of responsibilisation, understood here as associated with the neoliberal transfer of responsibilities from state to social actors. This is not to suggest that responsibilisation emerged with datafication, but rather that the advent of data-driven systems in the context of social welfare is embedded in this form of governance. The concern here is with how social problems come to be defined and, in turn, are sought to be resolved. By optimising for personalised risk, data-driven systems can construct the burden of social ills as one that belongs to individuals, addressed through behaviour and characteristics, without engaging with underlying causes and collective responses. This fundamentally challenges notions of shared social responsibility.

Data sources now stretch across a complex ecology of digital transactions that incorporates both consumer and citizen data about evermore-intimate aspects of our lives as the public sector becomes embedded within a rapidly growing data broker industry. Local authorities in the UK, for example, were found to have contracted with the credit rating agency Experian for over £2 million in 2018 (O'Brien & Williams, 2019). These developments continue a long-standing critique of the welfare state as a surveillance state that tends to target particular parts of the population. Eubanks (2018) argues, for example, that datafication is reconfiguring the traditional poorhouse in the US into the creation of "digital poorhouses" in which some parts of the population are subject to hyper-surveillance and "predatory inclusion" (Seamster & Charron-Chénier, 2017) as a condition of welfare. The issue here is not just one of privacy, but also the inherent bias of algorithmically processed data, whether because of historically skewed data-sets (e.g. arrest records), the way certain variables are weighted (e.g. the length of benefit claims), or the type of assessment that is produced (e.g. the labelling of risk) that all lead to disparate impacts of harm (Barocas & Selbst, 2016). These so-called biases have tended to align with existing social and economic inequalities often targeting and stigmatising already disadvantaged and marginalised groups (Gandy, 2010). Indeed, the very construction of a data-set emerges out of historically discriminatory practices that have implications for people's lives and can determine access to basic services and care (Ustek Spilda & Alastalo, 2020). Similarly, the ability to challenge how data about a person is collected and used is not distributed equally. In the words of Eubanks (2017), data processes "do not fall on smooth ground" and people do not share the same conditions of engagement with data-driven systems.

These concerns about surveillance, discrimination and bias, and their contingency on existing inequalities are important for discussions on the welfare state as they raise questions about how universal access and social security can be guaranteed. Of course, challenges to such values are not new. The inability of the welfare state to deliver on its promises has been a long-standing critique of it, in part due to its very reliance on a capitalist economy it is simultaneously intended to mitigate excess harm from (Offe, 1984). Often it has been precisely those at the margins bearing the brunt, whether excluded, criminalised, or neglected by the welfare systems intended to protect them. With the datafied welfare state, such critiques continue to resonate and take on further significance as these systems become embedded in "dataism", what van Dijck (2014) terms the ideological component of datafication. While the need to gather information to assess needs and risk is seen as essential in providing public services, the growing reliance on automated processing as the arbiter of social knowledge introduces some particular, and contested, epistemological and ontological assumptions for making such assessments. The "subtractive methods of understanding reality" in which information flows are reduced into numbers that can be stored and then mined produce very particular forms of informational and computational knowledge (Berry, 2011: 2). As famously noted by boyd and Crawford (2012), big data shapes the reality it measures by staking out new terrains and methods of knowing. This includes the perceived epistemic capabilities of algorithms to anticipate, conjecture, and speculate on future outcomes in a way that McQuillan (2017: 2) compares to a kind of Neo-Platonism: "a belief in a hidden mathematical order that is ontologically superior to the one available to our everyday senses". The premise is that based on enough data, correlations can predict future outcomes in such a way that facilitates pre-emption, a strategy of intervention just before an event might occur (Andrejevic et al., 2020).

With the turn to the datafied welfare state we are, therefore, confronted with some very significant assumptions about not only the neutral nature of data and technologies, but also that there is "a self-evident relationship between data and people, subsequently interpreting aggregated data to predict individual behaviour" (Van Dijck, 2014: 199). Of central importance here is the abstraction of big data in order to reduce social identities, mobilities, and practices to mere data that can be managed and sorted (Monahan, 2008). Furthermore, these "data derivatives" (Amoore, 2013) grant authority to knowledge domains based on new forms of risk calculations rooted in data science. These calculative devices, as Andrejevic (2019) argues, follow an "operative" logic in juxtaposition to one of representation. They are not concerned with why something happens, but simply that it does; it is correlations between variables that determine outcomes, not an engagement with underlying causes. In this sense, Andrejevic (2019: 108–9) contends, they not only collapse the future into the present but also threaten to lose the distinction between prediction and comprehension.

Such logics and assumptions are pertinent for understanding the nature of state-citizen relations in the datafied welfare state. They raise questions about how social ills are problematised and solved and how individuals are positioned in relation to such ills. For example, in advancing a long-standing shift towards risk management in public administration, the advent of big data expands and redefines the way we think about risks. As Poon

(2016) has highlighted, big data derives from a cultural conception of personal risk intimately connected to corporate capitalism and with roots in actuarialism. It is not technical accuracy that makes big data investment worthy or secures profits, she argues, but rather the methods for manipulating and calculating elements and definitions of risk. Importantly, these calculations derive risk from correlations between group traits in order to make predictions about individuals. We see this, for example, in data-driven systems that predict the risk of child abuse by calculating the extent to which a child matches the behaviours and characteristics of previous victims of child abuse (Dencik et al., 2019). Carrying out such risk calculations can be seen as important for targeting resources on those who might need it most. However, they also adopt a personalised understanding of risk that centres on risk factors attributed to an individual's behaviour and characteristics. This raises concerns about the ways in which responsibility for social problems might shift from the collective onto the individuals undermining values of social solidarity (Keddell, 2015; Morozov, 2015). Responses become focused on interventions targeted at individuals in a way that shift focus away from structural causes. For example, what comes to matter are measurable categories such as school attendance and number of benefit claims, rather than complex societal issues such as poverty, racism, and precarity (Dencik et al., 2019).

Furthermore, an imperative of pre-emption constructs personalised risk according to a compressed temporality. Risk is an outcome of simulated futures that draw on aggregated historical and real-time data about group traits to make predictions about an individual. In other words, it is what 'people like you' have done in the past that underpin predictions about what you might do in the future in order to inform interventions made towards you in the present. Insofar as such a temporal collapse informs decision-making, it is a form of decision-making that is intrinsically conservative (Cheney-Lippold, 2017). What is more, taken to its limit in seeking to address all possible risks and opportunities in advance, pre-emption is a-temporal, invoking a state of social stasis (Andrejevic et al., 2020). Rather than creating conditions for social mobility and human flourishing, the datafied welfare state threatens to lock individuals into their data futures and dispense with the possibility for social change (Dencik & Kaun, 2020).

In thinking about the welfare state, it, therefore, becomes imperative to consider how a growing reliance on data-driven systems constructs what counts as social knowledge and how people should be rendered legible in such a way that undermines notions of universal access, social solidarity, and human flourishing. Rather than the state being accountable to its citizens, the datafied welfare state is premised on the reverse, making citizens' lives increasingly transparent to those who are able to collect and analyse data, at the same time as knowing increasingly little about how or for what purpose that data is collected. Moreover, rather than social problems being understood as shared, the datafied welfare state advances actuarial logics that attribute risk to individuals without necessarily engaging with preventative measures for such risks. Instead, policy responses become pre-emptive, potentially shifting responsibility away from the collective whilst at the same time entrenching existing inequalities and stifling the conditions needed for social change. We therefore need to consider the turn to data infrastructures in social welfare as a form of policy intervention that is part of shaping the conditions for governance. This positions data beyond questions of bias or whether it is used for good or bad and instead requires an engagement that attends to the way problems and solutions are constructed through such infrastructures.

Datafication as Rentierism

It is important to note that the actuarial logics that are prominent in dominant processes of datafication are not an inevitable feature of digital technologies but focus our attention on the political and economic forces that shape the development of data-driven systems. As the public sector becomes increasingly intertwined with technology companies, welfare systems become embedded in global markets and infrastructures that significantly shift the terms upon which such systems can operate. In this section, I, therefore, draw attention to questions of political economy in relation to data-driven systems and consider the implications of rentierism as the operating logic of state-capital relations under datafication. Rentierism here refers to the public sector becoming dependent on a mode of capitalism in which revenue is predominantly extracted from rent (money or data) in exchange for services, with significant implications for the functioning of institutions. This relates to processes of privatisation, but the concern here is with the way the dominant business models and drivers of data-driven platforms and tools configure social practices and shape the terms upon which public institutions are able to operate. As I will go on to argue, this not only undermines a principle of decommodification by embedding public institutions in commercial operations but, furthermore, creates a relationship of dependency that threatens to displace public infrastructure with (private) computational infrastructure.

In making sense of the value of data, Zuboff's (2015, 2019) notion of surveillance capitalism has been widely used to describe the dominant business model that underpins much of today's digital technologies. This business model, she argues, relies not on a division of labour but a division of learning: between those who are able to learn and make decisions based on global data flows and those who are (often unknowingly) subject to such analyses and decisions. In this model, capital moves from a concern with incorporating labour into the market as it did under previous forms of capitalism to a concern with incorporating private experiences into the market in the form of behavioural data. This is an accumulation logic driven by data that aims to predict and modify human behaviour as a means to produce revenue and market control. Social relations under this logic are extractive rather than reciprocal and based on a formal indifference to information: it is volume rather than quality that sustains it, sourcing data from a range of infrastructures from sensors to government databases to computer-mediated economic transactions alike.

Yet in understanding the implications of this business model for the welfare state, it is worth further unpacking datafication as a "political-economic regime" (Sadowski, 2019). In doing so, Sadowski argues that we need to understand the value of data not as a commodity but as capital that propels new ways of doing business and governance. Data collection is driven by the perpetual cycle of (data) capital accumulation, which in turn drives capital to construct and rely upon a universe in which everything is made of data, including social life. The digital platform is central for this transformation in that social practices are reconfigured in such a way that enables the extraction of data (Couldry & Mejias, 2018). This matters as data in this context serves to sustain an economic process that bypasses the creation of value through production and instead relies on the capturing of value through expanding the capacity for gaining information. For Wark (2019), this presents itself as a markedly different system than how we have conventionally understood capitalism as power shifts from the owners of the means of production to the owners of the vectors along which information is gathered and used, what Wark describes as the "vectorialist class". This class controls the patents, the brands, the trademarks, the copyrights, and most importantly the logistics of the information vector. Through this, Wark argues, whilst a capitalist class owns the means of production, the means of organising labour, a vectorialist class owns the means of organising the means of production. Although Wark posits that such a shift in power relations forces us to place the vectorialist class outside a capitalist framework and as distinct from the landowning class, others have argued that understanding this organisation of power in the context of rent theory may be more fruitful (Sadowski, 2020; Srnicek, 2017).

Rent-seeking strategies are familiar in the wider shift towards financialisation that has marked advanced capitalism in Anglo-Saxon countries especially, and the drive to turn everything into a financial asset as a way to latch onto circuits of capital and consumption for the purposes of rent extraction. Whilst this logic is not new for capital, Sadowski (2020) argues what is new are the complex technologies that have been designed to extend and empower capital's abilities of assetisation, extraction, and enclosure. As Srnicek (2017) has also outlined, such expansion is driven by accumulating data as the primary revenue source for platforms that also explains the extensive acquisitions relating to big data and the significant investments in the Internet of Things (IoT) and other assets that extend data extraction. Under this analytical framework, platforms are intermediaries in the production, circulation, or consumption process and capture value from all the activities and operations that make up the platform ecosystem, extracting both monetary rent and data rent (Sadowski, 2020). That is, rentiers capture revenue from the use of digital technologies and not only rely on money as value but also treat data as a source of value. As Sadowski goes on to argue, the main strategy of these rentiers is to turn social interactions and economic transactions into 'services' that take place on their platform. This "X-as-a-service" rental model is in line with assetisation and the transformation of things and activities into resources which generate income without a sale (Birch, 2015; Sadowski, 2020).

When public sector organisations integrate tools and platforms from providers within this economy to administer the welfare state, they implement not only the systems themselves, but also a regime that propels the further datafication of social life. This matters as although rentierism can be understood as an outgrowth of capitalism, and the welfare state has always been subject to the contradictions of being dependent upon and simultaneously mitigating the harms of a capitalist economy, it configures this relationship in significant ways. With the advent of neoliberalism and globalisation, the welfare state has long been subject to forms of privatisation with a growing number of public services outsourced to private companies and large parts of the public sector commoditised and made subject to the market. The UK has been particularly prone to these trends evident in the care system, for example, where it has gone from being 95 per cent provided publicly by local authorities in 1993 to now being almost entirely provided by private companies (Monbiot, 2020), or in higher education, where commodification has grown as funding has become increasingly dependent on external and private sources (Freedman, 2011). Whilst public institutions in other advanced capitalist societies, particularly in Europe, can be said to have been more resilient to these developments, there has nevertheless been a 'convergence' in the trajectory of institutional change across national contexts that can be characterised as neoliberal (Baccaro & Howell, 2011). The turn to data-driven systems, often bound up in commercial infrastructures, across the welfare state in this sense continues the trend of privatisation and commodification. However, as I go on to argue below, under a model of rentierism, the datafied welfare state is subject to pressures that arguably move beyond binaries of de/commodification and public/private.

By plugging in to a political economy of rentier capitalism, the datafied welfare state not only advances the commodification of information about citizens and the outsourcing of service provision but also becomes locked in to a form of social ordering that restructures practices to uphold the logic of this political economy. Understanding 'welfare-as-a-service' in the context of datafication is not simply an issue of privatisation, but about establishing a set of relations that ultimately seeks to overturn public institutions as we commonly understand them. That is, by turning to data-driven systems, the welfare state reconfigures social welfare into a problem that necessarily has to be optimised computationally rather than engaged with through human experience and expertise, and embeds social welfare within an ecosystem that endlessly perpetuates this reconfiguration. Gürses et al. (2020) use the term "programmable infrastructures" to refer to this political, economic, and technological vision that advocates for the introduction of computational infrastructure onto our existing infrastructures. This vision, they argue, features the management of human behaviour, the standardisation of values, a dependency on the economic terms of technology companies, a power asymmetry of cloud providers, and an avoidance of democratic governance. As such, the datafied welfare state raises questions not just about the ways in which decisions and practices in public administration are organised, but about their contingency on a particular process that threatens to displace the very public infrastructure upon which the welfare state is built. This speaks to a particular kind of power in relation to data infrastructures that needs to be captured in our engagement with data politics.

Conclusion

At a time of global crisis, the question of how technology intersects with the welfare state has gained new significance. The COVID-19 pandemic and responses to it have shed light on not only the vulnerabilities of the welfare state but also ways in which it might be rebuilt. In many respects, it increasingly looks to do so on the pillars of Silicon Valley. The UK has been at the forefront of this trend in Europe, but the focus on contact-tracing apps, immunity passports, and location tracking has nurtured new partnerships between companies like Apple, Google, Amazon, and Palantir and governments around the world. However, the conditions for the advent of the datafied welfare state have been in the making for quite some time. Data collection and practices of citizen scoring are now prominent features of how public administration and welfare provision are organised. In the UK, austerity measures and an active shrinking of the public sector have been accompanied by a prominent shift towards the implementation of data-driven systems across key areas of the welfare state that is set to dramatically accelerate in the context of the COVID-19 crisis and its aftermath.

In order to make sense of the significance of this shift, it is important to situate the welfare state in historical and national context, understanding it as an outcome of social struggle, a political compromise, and a model of inherent contradictions. There was nothing inevitable about the emergence of the British welfare state and the values it upheld. Equally, there is nothing inevitable about the datafied welfare state we are now confronted with. Rather, it is indicative of the current matrix of social power. The ideology of dataism and the political economy of technology posit values and operational logics that are markedly different from how the welfare state has previously been understood. As I have argued here, the epistemological and ontological pillars of the datafied welfare state advance an agenda of responsibilisation that counter values of universal access, social solidarity, and human flourishing, whilst the operations of capital out of which datafication has developed position the datafied welfare state as a tenant of private cloud and service providers that threatens to undermine democratic governance and displace public infrastructure.

As the welfare state becomes further embedded in the paradigm of datafication, the question then becomes how the matrix of social power might be shifted to facilitate a different vision. This might also entail examining different models of the welfare state and the constitution of public institutions across national contexts. The COVID-19 crisis allowed for openings in demands on how society should be organised that echo those of post-war Britain at the apogee of the welfare state. This has brought hope about an opportunity to question and challenge long-standing social experiments that do not serve the majority of the population. However, in accelerating the transition to the cloud, we might find ourselves with short-term solutions that have long-term consequences for any future of the welfare state. The interrogation of power in relation to data, therefore, needs to consider not only the values and logics that are advanced through such power but, with that, the conditions of possibility for social change created by the dynamics upon which the circulation of data depends.

References

AlgorithmWatch. (2019). *Automating society: taking stock of automated decision-making in the EU.* https://algorithmwatch.org/wp-content/uploads/2019/01/Automating_Society_Report_2019.pdf

Alston, P. (2018). Statement on visit to the United Kingdom by Philip Alston, United Nations Special Rapporteur on Extreme Poverty and Human Rights, 16 November 2018.

Alston, P. (2019). Report of the Special rapporteur on extreme poverty and human rights. A/74/48037. https://www.ohchr.org/EN/NewsEvents/Pages/DisplayNews.aspx?NewsID=25

Amoore, L. (2013). *The politics of possibility: Risk and security beyond probability.* Duke University Press.

Anderson, R., et al. (2009). *The database state.* Rowntree Foundation.

Andrejevic, A., Dencik, L., & Treré, E. (2020). From pre-emption to slowness: Assessing the contrasting temporalities of data-driven predictive policing. *New Media & Society, 22*(9), 1528–1544.

Andrejevic, M. (2019). *Automated media.* Routledge.

Baccaro, L., & Howell, C. (2011). A common neoliberal trajectory: The transformation of industrial relations in advanced capitalism. *Politics and Society, 39*(4), 521–563.

Barocas, S., & Selbst, A. (2016). Big data's disparate impact. *California Law Review, 10*, 672–732.

Beer, D. (2019). *The data gaze*. Sage.

Berry, D. (2011). The computational turn: Thinking about the digital humanities.*Culture Machine, 12*, 1–22.

Birch, K. (2015). *We have never been neoliberal*. Zero Books.

Booth, R. (2019). Automating poverty. *The Guardian*. https://www.theguardian.com/technology/series/automating-poverty

boyd, d., & Crawford, K. (2012). Critical Questions for Big Data. *Information,Communication & Society, 15*(5), 662–679.

Cheney-Lippold, J. (2017). *We are data: Algorithms and the making of our digital selves*. NYU Press.

Couchman, H. (2019). *Policing by Machine: Predictive Policing and the Threat to Our Rights*. https://www.libertyhumanrights.org.uk/sites/default/files/LIB%2011%20Predictive%20Policing%20Report%20WEB.pdf

Couldry, N., & Mejias, U. (2018). Data colonialism: Rethinking big data's relation to the contemporary subject. *Television and New Media, 20*(4), 338.

Davies, G. et al. (2019). Revealed: The thousands of public spaces lost to the council funding crisis. *The Bureau of Investigative Journalism*. https://www.thebureauinvestigates.com/stories/2019-03-04/sold-from-under-you

Dencik, L., & Kaun, A. (2020). Datafication and the welfare state. *Global Perspectives*.

Dencik, L., Hintz, A., Redden, J., & Warne, H. (2018). Data Scores as Governance:investigating uses of citizen scoring in public services. Project Report, Cardiff University. https://datajustice.files.wordpress.com/2018/12/data-scores-as-governance-project-report2.pdf

Dencik, L., Redden, J., Hintz, A., & Warne, H. (2019). The 'golden view': data-driven governance in the scoring society. *Internet Policy Review, 8*(2), n.p.

Dencik, L., & Wilkin, P. (2015). *Worker resistance and media: Challenging global corporate power in the 21st century*. Peter Lang.

Eubanks, V. (2017). *Keynote at launch of Data Justice Lab*. Cardiff University, March 13 2017.

Eubanks, V. (2018). *Automating inequality: How high-tech tools profile, police, and punish the poor*. St. Martin's Press.

Gandy, O. (2010). Engaging rational discrimination: Exploring reasons for placing regulatory constraints on decision support systems. *Ethics and Information Technology, 12*, 29–42.

Genschel, P. (2004). Globalization and the welfare state: A retrospective. *Journal of European Public Policy, 11*(4), 613–636.

Gürses, S., Dobbe, R., & Poon, M. (2020). Introduction to the Programmable Infrastructures Event. TU Delft. https://www.tudelft.nl/tbm/programmable-infrastructures/

Hobsbawm, E. (1994). *The age of extremes: 1914–1991*. Abacus. https://www.bloomberg.com/opinion/articles/2020-03-17/coronavirus-will-revive-an-all-powerful-state

Judt, T. (2007). *Postwar: A history of Europe since 1945*. Pimlico.

Keddell, E. (2015). The ethics of predictive risk modelling in the Aotearoa/New Zealand child welfare context: Child abuse prevention or neo-liberal tool? *Critical Social Policy, 35*(1), 69–88.

Klein, N. (2020). How big tech plans to profit from the pandemic. *The Guardian*, 13 May 2020.

Magalhães, J. C., & Couldry, N. (2020). Tech giants are using this crisis to colonize the welfare system. *Jacobin*, 27 April 2020.

McQuillan, D. (2017). Data science as machinic neoplatonism. *Philosophy & Technology*. https://doi.org/10.1007/s13347-017-0273-3

Mishra, P. (2020). Coronavirus will revive an all-powerful state. *Bloomberg*, 17 March.

Monahan, T. (2008). Editorial: Surveillance and inequality. *Surveillance and Society, 5*(3), 217–226.

Monbiot, G. (2020). Tory privatization is at the heart of the UK's disastrous coronavirus response. *The Guardian*, 27 May.

Morozov, E. (2015). Silicon Valley: The new welfare state? Public lecture at University of Maastricht.

Morozov, E. (2020). The tech 'solutions' for coronavirus take the surveillance state to the next level. *The Guardian*, 15 April.

O'Brien, H., & Williams, O. (2019). Revealed: How citizen-scoring algorithms are being used by local government in the UK. *New Statesman*, 15 July.

Offe, C. (1984). Some contradictions of the modern welfare state. In J. Keane (Ed.), *Claus Offe: Contradictions of the welfare state*. Hutchinson.

Petropoulos, G. et al. (2019) *Digitalisation and European Welfare States*. Bruegel.

Poon, M. (2016). Corporate capitalism and the growing power of big data: Review essay. *Science, Technology & Human Values, 41*(6), 1088–1108.

Redden, J., Dencik, L., & Warne, H. (2020). Datafied child welfare services: Unpacking politics, economics and power. *Policy Studies*. https://doi.org/10.1080/01442872.2020.1724928

Rule, J. B. (1973). *Private lives and public surveillance*. Allen Lane.

Sadowski, J. (2019). When data is capital: Datafication, accumulation, and extraction. *Big Data & Society. January–June, 2019*, 1–12.

Sadowski, J. (2020). The internet of landlords: Digital platforms and new mechanisms of rentier capitalism. *Antipode, 52*(2), 562–580.

Scott, J. C. (1994). *Seeing like a state: How certain schemes to improve the human condition have failed*. Yale University Press.

Seamster, L. & Charron-Chénier, R. (2017). Predatory Inclusion and Education Debt:Rethinking the Racial Wealth Gap. *Social Currents, 4*(3), 199–207.

Srnicek, N. (2017). *Platform capitalism*. Polity Press.
Standing, G. (2020). Coronavirus has shown us why we urgently need to make a basic income a reality. *weforum.org*. Retrieved from: https://www.weforum.org/agenda/2020/04/coronavirus-made-basic-income-vital/
Thane, P. (2013). *The foundations of the welfare state* (2nd ed.). Routledge.
Ustek Spilda, F., & Alastalo, M. (2020). Software sorted exclusion of asylum seekers in Norway and Finland. *Global Perspectives*.
Van Dijck, J. (2014). Datafication, dataism and dataveillance: Big data between scientific paradigm and ideology. *Surveillance & Society, 12*(2), 197–208.
Wark, M. (2019). *Capital is dead. Is this something worse?* Verso.
Yeung, K. (2018). *Algorithmic government: Towards a New Public Analytics?* Paper presented at the Ethical and Social Challenges posed by Artificial Intelligence workshop, Cumberland Lodge, Windsor.
Zuboff, S. (2015). Big other: Surveillance capitalism and the prospect of an information civilization. *Journal of Information Technology, 30*(1), 75–89.
Zuboff, S. (2019). *The age of surveillance capitalism*. Profile Books.

Open Access This chapter is licensed under the terms of the Creative Commons Attribution 4.0 International License (http://creativecommons.org/licenses/by/4.0/), which permits use, sharing, adaptation, distribution and reproduction in any medium or format, as long as you give appropriate credit to the original author(s) and the source, provide a link to the Creative Commons licence and indicate if changes were made.

The images or other third party material in this chapter are included in the chapter's Creative Commons licence, unless indicated otherwise in a credit line to the material. If material is not included in the chapter's Creative Commons licence and your intended use is not permitted by statutory regulation or exceeds the permitted use, you will need to obtain permission directly from the copyright holder.

The Value Dynamics of Data Capitalism: Cultural Production and Consumption in a Datafied World

Göran Bolin

Introduction

The observation that 'information' and 'data' have come to the centre of capitalism has inspired a range of descriptive terms aimed at qualifying the broad concepts of "informational capitalism" (Castells, 1996; Benkler, 2006); "digital capitalism" (Schiller, 1999); "surveillance capitalism" (Zuboff, 2015); "platform capitalism" (Srnicek, 2017); or, simply, "data capitalism" (Morozov, 2015). The underlying idea of data capitalism is that data is "the new oil", a valuable asset to be extracted as a natural resource (World Economic Forum, 2011) in the process of datafication (Mayer-Schönberger & Cukier, 2013), centred on the transformation of social movement in digital space into processable, digital forms that feed on the activity of media users, consumers, and citizens. While many scholars point to the broader, general consequences of datafication for social life

G. Bolin (✉)
Södertörn University, Huddinge, Sweden
e-mail: goran.bolin@sh.se

© The Author(s) 2022

A. Hepp et al. (eds.), *New Perspectives in Critical Data Studies*, Transforming Communications – Studies in Cross-Media Research,
https://doi.org/10.1007/978-3-030-96180-0_8

(e.g. Cohen, 2012, 2018; Cheney-Lippold, 2017; Couldry & Hepp, 2016; Schäfer & van Es, 2017; Couldry & Mehijas, 2019)—transforming everything from jobs, finance, education, and power relations to intimacy and everyday sociality—we are still in need of analytical models to understand the complexity and scale of this techno-social development and the dynamics behind these transformations.

The fact that data capitalism and the principles around data capture and processing have integrated financial services, telecom providers, platform operators, media content producers, advertising and PR, retail, and consumer goods and services, as well as many public sectors that have previously operated in relative isolation from each other, brings with it the risk to mistake this increased complexity for an epochal, qualitative shift in the historical unfolding of capitalism itself. This might, of course, well be the case, but it might also be too early to empirically establish with any solidity. Several scholars have, however, pondered upon whether we are facing a fundamental shift in the workings of capitalism (Couldry, 2004; Couldry & Mehijas, 2019; Kitchin, 2014; Mosco, 2014), or if datafication is more of a myth of "big data" (boyd & Crawford, 2012, cf. Boellstorff, 2013) driven by what José van Dijck (2014) calls "dataism", or an "algorithmic ideology" (Mager, 2012). Is capitalism moving in the direction of automation, thus privileging "correlation over causation, predictability over referentiality" (Andrejevic, 2013, p. 40) where automated intelligence becomes favoured at the expense of hermeneutical interpretation, or are we witnessing an extension of old models for value generation in new clothing? And, if so, how will these value forms affect other value forms in society? Will the private value forms of the economy suppress or alter public value forms, for example, as José van Dijck et al. (2018) have recently argued? Such questions cannot easily be answered without empirical analysis, but in order to conduct such empirical studies, there is a need for robust analytical models of value—models that can handle complexity and produce indicators for the various dynamics involved.

The need for more elaborated models also stems from the fact that much research into datafication and what broadly could be considered critical data studies is dispersed over disciplines and research fields and has focused on different sectors of the data industries with a variety of methods. However, some major consequences have been established: an ongoing concentration of data capacity in the form of the centralisation of data into data centre monopolies (Rossiter, 2017; Vonderau, 2018); an increased reliance on semantics in computation to produce self-learning algorithms (Gillespie, 2014; boyd & Crawford, 2012; Kitchin, 2016); an increased metrification of digital culture through the quantifying and tracking of everyday life via apps for information, education, entertainment, health care, and work (Beer, 2016); and the rise of algorithmically based marketing practices that proliferate with increasing intensity (Turow 2011, 2017). However, even if there are new technological arrangements that seem to restructure society in its totality, there are also older power and value dynamics that shape these developments. There is thus a need to develop analytical approaches that can identify the possible rearticulations of value that result from the transformation of business models, technologies, epistemologies, and social life in the era of data capitalism. This also means taking into consideration the historical breaks in previous shifts from market, to industrial to informational capitalism, in order to judge if data capitalism represents a break from informational capitalism, or a deepening of its modes of production. A starting point for such a discussion is to focus on the value forms at the centre of data capitalism and their relation to public value forms (e.g. welfare, health, or equality), in order to contribute to the analytical toolbox of critical data studies.

The aim of this chapter is to suggest an analytical model where the composition of value can be analysed within distinct societal domains, as they are affected by datafication. In the following I will, first, define the concept of value that I will adopt for the analysis and outline its complexity and dynamics. I will then present a model of data capitalism as constituted by four different sub-dynamics—the economic, the technological, the epistemological, and the social. Following this, I will give some examples of societal domains where this model can be applied empirically before I sum up my argument in the conclusion.

Value and Values

The idea of data as the new oil is a widespread trope that gained traction towards the end of the first decade of the 2000s. It was cherished by the World Economic Forum (WEF, 2011) and became a buzz term in financial magazines and trade journals (e.g. Rotella, 2012; Toonders, 2014). A Google image search displays a multitude of slides and illustrations, and it is easy to get the perception that data is the Særimner of contemporary capitalism. In the classic Icelandic saga *The Prose Edda* by Snorri Sturluson (1916), Særimner is the mythic boar in Nordic mythology who was consumed by the Viking gods each night at Valhalla, but who each morning arose anew for perpetual consumption. It is a mythical belief of endless resources, free for all to grab and use for the benefit of eternal growth. However, the metaphor of data as oil is misleading in a number of respects, and it has been rightfully criticised (cf. Stark & Hoffmann, 2019). But, as Puschmann and Burgess (2014, p. 1699) argue, it evokes assumption that "supports the notion that data is all at once essential, valuable". And indeed, discussions on data often centre on its value—but what kind of value does data refer to? Before we take on that discussion, we need to say something about the ways in which data differs from oil as a resource.

First of all, data are not something that is discovered hidden in the ground—data, as units of information, are produced and have their root in human activity. While oil is the product of energy bound up in underground reservoirs over millions of years, data is produced in the process of datafication, that is, by the contemporary activity of human subjects. Its value is transient, and real-time processing is, therefore, necessary. Oil is a product of natural processes without the involvement of humans. Data cannot be produced without human activity—both as the generators of data through social action and as refined by statisticians and engineers in human-directed algorithmic calculation.

Second, unlike oil, data is a non-rivalrous good. The use of my data by a company (e.g. Facebook) does not infringe upon another's use of it (e.g. Google). In order for it to become valuable economically it needs to be restricted, similar to other nonmaterial and transient commodities—which basically mean all digital commodities that are spreadable via the interactive web—music, films, video clips, computer programmes, and so on. The legal regulation of digital commodities is therefore necessary—that which cannot be legally restricted, cannot be charged economic value for. So, what companies capitalise on might be the same data at their origin, but in order for it to become valuable it has to be processed in a way that makes it functional in a market.

Third, and again unlike oil, data will never be exhausted as long as there is human activity. It might in fact also survive human extinction through self-generating, autopoietic systems created through machine learning. Oil as a natural resource will eventually be exhausted. Taking these disparities together, oil is a poor metaphor for describing data as an asset within contemporary data economies, and for understanding the way in which data produces value.

The one thing that the analogy between data and oil captures is that they are both valuable assets. But how can we think about what form of value data (and oil) represent?

In his short treaty *Theory of Valuation*, John Dewey (1939) theorised value as both an essence (what it is) and a practice (how it is arrived at)—that is, both as a noun and as a verb. When we interact socially, we make value judgements on things and practices around us, and in this process of valuation we assign value to these things and practices. We could call this a practice theory of value. This means that value (noun) is the outcome of social negotiation or valuation (verb)—they are "sedimented" valuations (Sayer, 2011, p. 25) or, in analogy to Marx's labour theory of value, the reified result or product of the labour of valuation. Dewey thought of value as the outcome of all social activities that were not merely a reflex or the result of biological conditioning:

> All conduct that is not simply either blindly impulsive or mechanically routine seems to involve valuations. The problem of valuation is closely associated with the problem of the structure of the sciences of human activities and *human* relations. (Dewey, 1939, p. 3)

Underlying all social actions, then, are judgments where a social subject acts on the basis of choices rooted either in experience or in conscious evaluation of the situation. In our everyday lives we constantly valuate objects and practices in our surroundings. When we, through valuation, assign value to objects around us, we do this in the form of either nominal value (good/bad, ugly/beautiful) or ordinal value (1, 2, 3, etc.). With datafication, most values are translated into ordinal (or interval, ratio) value, because this is the way in which computers work. Datafication is thus also the process of transforming quality into quantity, or, if exemplified with the distinction between private and public value, transforming 'soft' value forms such as equality and knowledge into numerical form (e.g. equal numbers or grades).[1]

While there is a substantial amount of literature on value as a general category besides Dewey's account (e.g. Dumont 1908/2013; Graeber, 2001; Magendanz, 2003; Stark, 2009; Lamont, 2012), there is also a

[1] Datafication is thus a perfect fit for the administrative rationalities of New Public Management, as the audit culture (Strathern, 2000) of NPM also strives to quantify in order to evaluate managerial processes. The dynamics of the relations between NPM and datafication is well worth exploring further, but outside the scope of this chapter.

number of works concerning value and the digital (e.g. Skeggs, 2014; Gerlitz & Lury, 2014; Bolin, 2011; Gerlitz, 2016). Many studies are, however, empirically limited and/or theoretically restricted to the economic or commercial forms of value. In line with Boltanski and Thévenot (2006), Skeggs (2014), Stark (2009), and Heinich (2020) one can also be sceptical of the common separation between value (as an economic category) and values (as morals) and theorise the economic and the moral as integrated. A separation of value and values is especially problematic in digital economies, since the intangible character of commodities makes the valuation process foundational.

A similar distinction made in valuation studies (Helgesson & Muniesa, 2013) is between valuation (as assessment) and valorising (as the production of economic value) (Vatin, 2013). This distinction, stemming from the distinction in French between "évaluer" and "valoriser", is also problematic since, as I have shown (Bolin, 2011), the process of assessing value is also the production of value—especially in markets for non-tangible commodities that are almost entirely dependent on the *belief* in value among both producers and consumers. As John Kenneth Galbraith (1970) has argued, economics and economic reason was always founded on belief. Such arguments also underlie Bourdieu's (1993) analysis of social and cultural fields, centred on the shared conviction among relevant agents in regard to the value of a field's symbolic assets. These assets—such as money, prestige, and rank—are dependent on the shared belief among those agents competing for the field's capital. When we act in social fields, we make value judgements, and in this process of assigning value to something, we give it a meaningful and interpretable form. However, we do not assign value randomly, but within a specific *symbolic order*, set up by the field of which it is a part. Things and practices are given meaning within an ordered social context (Couldry & Hepp, 2016). Based on these meaning-making practices and the field of which we are a part, we act. This is not so much dependent on the accuracy of these beliefs, that is, whether they are 'true' or not, but more along dynamics laid out in the so-called Thomas theorem: "if men define situations as real, they are real in their consequences" (Thomas & Thomas, 1928, p. 572; cf. Merton, 1995).

By integrating the practices of valuation with its outcome, we could analytically focus on specific societal domains or fields where value is produced in processes of valuation and study the possible rearticulations of value in them. As briefly outlined above, a datafied society integrates agents that have hitherto acted in isolation from each other. As all kinds of societal production get more integrated into the process of datafication, it makes sense to situate this process historically. This is the objective of the next section.

Dynamics of Data Capitalism

In *The Rise of the Network Society*, Manuel Castells (1996) theorised the turn from industrial to informational capitalism by focusing on the shift in the modes of production. Castells relates these shifts to the technological revolutions—the steam engine, electricity, and computerisation. These were all revolutionary technologies disrupting modes of production, transforming industrial relations, and affecting all interested parties.

The historical outline and description made by Castells, however, is based in a rather narrow political economy perspective where the conception of value is first and foremost related to its market function (use value, exchange value), although he also points to the dynamic of informationalism as a specific mode of development not necessarily subsumed by economic dynamics. If we want to theorise value dynamics as based in both judgmentally based practice (value as a verb) and value as an object (value as a noun), we need to extend or complement this history by bringing in other perspectives on value formation. As already mentioned above, datafication integrates various production domains in society—financial services, telecom providers, platform operators, media content producers, advertising and PR, retail, and consumer goods and services, as well as certain public sectors. Not all of these sectors are driven by the same valuation principles, and the values at the centres of these domains are sometimes dramatically different. This becomes most obvious when it comes to the differences between industrialised commercial production and the media production generated by everyday media users when connecting on social networking media, uploading texts and images, writing blog posts, or publishing music. Such production, as I have analysed elsewhere (Bolin, 2012), very seldom has economic profit as a motive, and cherishes other value forms (e.g. social or aesthetic value).

The increased complexity of the datafied media landscape has brought with it a large number of individual and collective agents whose motives and interests produce different value forms compared to those of the commercial media industries. The complexity in terms of the number of interested parties also produces a ***complexity in relations*** and in the outcomes of

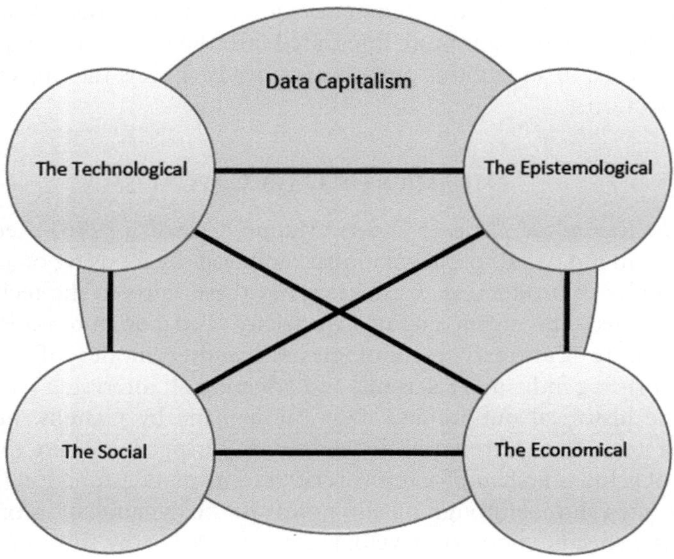

Fig. 1 Value dynamics in data capitalism

the datafication process (cf. Bolin & Hepp, 2017). This means that in this landscape there are several dynamics involved, as illustrated in Fig. 1 and further explained below.

The figure is intended to capture the relations between four different dynamics. Each dynamic is partly within but also partly outside of data capitalism as a system; not all technological dynamics are subsumed by capitalism, and not all social dynamics are drawn into profit-motivated actions (e.g. social relations within the family mainly lies outside of it). The dynamics are also related, and their relations vary if seen in relation to different empiric cases. Below they are explained in more detail, in order to then—in the next section—be discussed in relation to each other.

First and most central is, of course, the *economic dynamic* of data capitalism, the foundational strive towards return on investment, maximum profit, and so on and the constant drive towards perpetual economic growth, based on the legal regulation of private ownership (Cohen, 2019). The profit motive is of course the main motor of capitalism as such and long precedes the contemporary mode of production including its present form where data has become central. This overarching motive, of course, produces a drive to increase and speed up turnover and expand markets. Manuel Castells (1996) argues that informational capitalism has become the dominant form of capitalism since the 1980s (although the process leading up to this dominant form starts earlier). Informational capitalism is made possible through the rise of more refined communication technologies that can distribute information across society, and the question is whether what we are witnessing in data capitalism today is a deepening of informational capitalism or an abrupt break that introduces an entirely new phase. The economic dynamic can be said to be manifested in the economic dynamics of its *business models* and related financial technologies. A business model, as used here, can be described as the way in which a commercial operator organises its overall strategy for producing surplus value. For the media and communications industries, for example, we can talk of three basic business models: a text-based, an audience-based, and a service-based model.[2]

The *text-based model* has at its basis the selling of copies of media texts (written, audio-visual, auditive) in exchange for economic value (money). This model can be said to have been born with the market for books, following on from the possibilities of mass production of the written word afforded by printing technology since the mid-fifteenth century.

The *audience-based model* was then born with the advent of advertising when media producers started selling their reader's attention to advertisers (and others) who wished to reach audiences with commercial (or political) information. Many newspapers also combined these two models and based their revenues on a certain quota (say, 25% revenue from advertising and 75% from copies sold) (Gustafsson, 2009).

The *service-based model* is related not to the mass media and content-producing industries but to the telecommunications sector, where telephone companies (and, before that, postal services) traditionally offered a service for customers to communicate with distant others through their communications networks for a fee or for a subscription, or a combination of the two.

In the pre-digital world, the text-based and audience-based models, although sometimes operating in combination, were always operating independently of the service-based model. They worked on separate

[2] See Bolin (2011, chapter 3) for a more elaborate discussion of these three models and how they merged in the wake of digitisation.

markets, and there were no incentives for them to cooperate since their operations could not benefit from one another. Cinema screening, music distribution, television viewing, or print journalism could not be combined with telephone calls in any economically meaningful way. The major change that digitisation brought with it was that these three models merged. When all types of distribution of media content and all communication services such as text messaging or voice calls happen through internet networks, it becomes possible to take advantage of the user data and extract economic value, and suddenly the culture and media industries found an essential need to cooperate with the telecom business, since this is where control over the IP numbers can be found. This was necessary in order to be able to tailor advertising to consumers on an individual level. This development was what was needed for developing new business models based on consumer data.

Second, data capitalism and the datafication of society presume a technologically driven process centred on the "quantification and potential tracking of all kinds of human behaviour and sociality through online media technologies" (van Dijck, 2014, p. 198). Technological advancement has, naturally, always played a major role in the shifts in capitalism over the years: market capitalism was intimately tied to new means of transportation and navigation, industrial capitalism was centred around the invention of the steam engine, and informational capitalism built on the ability to process and disseminate information. We can of course debate how technologies are born and developed, to paraphrase Brian Winston (1995), and whether invention or societal need comes first, but there is no denying the centrality of technology for capitalism's developments as a system. Technology is, however, not only subsumed capitalism, but has its own *technological dynamic*, driven by functionality and efficiency. Technological dynamics have been central to human development since the invention of the wheel, which means that technology precedes capitalism. In the words of Heidegger, technology is also intimately related to knowledge and truth through the Greek notion of technē (τέχνη), that is, art or craft, and -logía (-λογία), that is, the study of something: "Technology is a mode of revealing. Technology comes to presence [West] in the realm where revealing and unconcealment take place, where alētheia, truth, happens" (Heidegger, 1954/1977, p. 13; square parenthesis in original). Technology and knowledge are thus intimately related (cf. Braman, 2012), and knowledge has always been a central feature in the advancement of capitalism across its many modes.

Third, and related to the technological dynamics of data capitalism, knowledge follows an *epistemological dynamic* which is centred on increasingly sophisticated means of gaining intelligence about social subjects as consumers of products and services, including media and cultural products. Epistemology, and ways of gaining knowledge, has always been a central component in the development of new techniques for controlling the environment and enters the scene with the turn of the Second Industrial Revolution in the mid-1800s (Castells, 1996, p. 34), a point after which scientific knowledge becomes one of the main drivers of technological and commercial development. One of the most important developments in this area is the advancement of statistical technologies (Hacking, 1975/2006; Porter, 1986) which facilitated more refined ways of controlling the environment, including mapping social behaviour and attitudes to anticipate future events. The increased interest in gaining knowledge about consumers and media audiences intensifies with the rise of consumer society in the mid-twentieth century when economists such as John Kenneth Galbraith (1958/1964) observed that advertising interferes with traditional economic explanation. These ideas are picked up and further developed by, among others, Jean Baudrillard, who, in a series of books in the late 1960s and 1970s, discussed the rearticulated value forms that come into view alongside consumer society, when design components and symbolic features add to the exchange value of commercial commodities (Baudrillard, 1968, 1970/1998, 1972/1981, 1973/1975). Baudrillard proposed the concept of sign value to add to the already existing use and exchange value in political-economic analysis. When consumption, rather than production, becomes the motor of capitalism, as Baudrillard argues, knowledge about consumer behaviour, attitudes, tastes, and lifestyles becomes increasingly important.

Fourth, and with consumption and the drive for intelligence about consumers, the epistemological and the technological are directed towards social life, which means that they become confronted with a *social dynamic* emanating from those who generate data at various instances in practices of production and consumption. While digital media since the rise of the interactive web has made everyone a potential producer, although on a different level and on a different scale compared to the global media and communications giants—traditional mass media corporations as well as online platform companies—the social becomes tightly integrated into those practices of production and consumption, becoming a fundamental feature of data capitalism. The domain of the social, however, is not primarily driven by the endeavour to maximise profit or economic value, but is centred on social values such as belonging, recognition, identity, and sociality. One can regard these as two types of productive activity, where the traditional professional media are producing content and services within a market economy based on a profit motivation, producing surplus value, whereas the production made by everyday media users is non-profit motivated, and within a social and cultural economy, resulting in social difference and identities (Bolin, 2012).

The four dynamics described above are each based on their specific *value regimes*, centred on a specific value form (e.g. economic, social, aesthetic, or technological). The value regimes, the type of order that is produced by the common social belief in the same value forms, are what constitute these areas as societal domains, with their own internal dynamics. A societal domain with its own value regime is, in a way, similar to social fields as theorised by Bourdieu (1993).

Fields are also centred on a common value but, in contrast to societal domains, a field is a space for competition over the central value, or capital, as Bourdieu defines it. A societal domain is more than its competition, although competition also exists in the domains. The domains discussed above, such as technology, cannot be considered a field, since its main value form—functionality—cannot be competed over. The same goes for the domain of the social. Rather, domains are areas of social life that are assembled through a common evaluative regime. Sometimes these domains work harmoniously in relation to one another. But sometimes there arise tensions between them when the specific value regimes are conflicting or incompatible. The following section will set out to describe these tensions.

Relations Between Value Forms

In order to empirically study the specific relations between value regimes, one has to operationalise them into specific societal domains, for example the domain of cultural production and consumption, or the domain of education, both of which involve relations between epistemological, technological, social, and economic value forms.

Firstly, within the domain of cultural production and consumption, knowledge about the social behaviour of customers and media users is central and results in lifestyle segmentations and audience profiles and the development of personalised recommender systems that serve individually tailored recommendations on films, books, news items, and other cultural commodities based on previous use, geographical location or access through connections on social media, and so on. In a datafied society, the combinations of such technological advancements with new forms of business models have significantly expanded on the ways in which the systematic tracking and mapping of audience behaviour occur for the purpose of automated personalised targeting to the benefit of capital accumulation and corporate profits. Take, for example, the personalised recommender systems that many platforms for the distribution of media content build on. Recommender systems are at the heart of platform services such as Spotify, YouTube, Netflix, or Amazon and direct the user towards specific types of content depending on previous choices. The most obvious are in the form of overt recommendations—"Customers who viewed this item also viewed"—while other recommendation types work in subtler ways. Recommender systems build on filtering techniques of different types, where the two main techniques are "content-based filtering" (recommendations based on the similarity of content) and "collaborative filtering" (recommendations based on similarities in user profiles) (Burke et al., 2011; Hildén, 2021).

Recommender systems are at the heart of the datafied economy's business models and they build on the ability of large-scale data processing where different types of proximity between content and behaviour are analysed. Its present uses are invariably profit-motivated. However, if we trace the invention of recommendation systems back to their original formulations, we see that the main driver of this invention is not capital accumulation, but the technological ability to measure document proximity in large databases in order for people to be able to make "serendipitous discoveries" and "to give users a greater chance of finding documents they did not know to look for" (Karlgren, 1990, p. 1).

Recommender systems were thus developed in the early 1990s in order to make it easier to discover texts in large textual archives. At that time, the technology could not be used for commercial purposes. Today, however, the interactive web allows recommender systems to be combined with business models that can generate personalised ad targeting. This is how value regimes sometimes can work alongside each other while also producing conflicting tensions.

A second example concerns the domain of education where epistemological dynamics are confronted with technological (and other) dynamics. The COVID-19 situation during 2020 has, for example, significantly sped up developments of distance education, with a range of *EdTech* solutions being developed for teachers and school managers working at all levels of the field. As has been observed by many within the field of critical studies of educational technologies (e.g. Breiter, 2014; Selwyn, 2016; Williamson, 2017; Jarke & Breiter, 2019), the digitisation of education brings with it some value registers that have not previously been found within educational systems and that stem from digital management systems as well as from digital learning material. José van Dijck et al. (2018) make important notes on how 'platformisation' (including datafication, personalisation, and commodification), in the long term, could transform traditional educational values such as *Bildung* and education into instrumental skills and what Biesta (2010) calls "learnification" and the move from "teacher's autonomy" to "automated data analytics" (van Dijck et al., 2018, p. 118f). Much of the critical discussion around the datafication of education is policy-oriented (e.g. Williamson, 2017), on privacy issues (Lindh & Nolin, 2016), or build on corporate discourse on EdTech (Yu & Couldry, 2020), but there is yet very little research into the valuation processes related to these technologies or on the actual implementation of educational technologies in schools (but see e.g. Sjödén, 2015). It should, therefore, be of importance for future research to more carefully study the tension between values as they appear in educational settings in order to judge the possible consequences for traditional educational value forms. Such long-term consequences require longer-term studies in order to understand possible rearticulations of value registers, and one can hope that in the near future knowledge about these processes will be at hand.

This is perhaps most important for understanding the tension between public and private value forms. Epistemological dynamics can be beneficial for fostering public values such as *Bildung* and understanding, but can also be used instrumentally for economic ends (i.e. subsumed by the economic dynamics of capitalism), boosting private value such as economic profits. The subsumption of epistemological value need not necessarily foster private value—it can also be subsumed by social dynamics and used as ends for boosting social value forms, such as welfare.

Conclusions

Critical data studies is a new and rapidly growing field of interdisciplinary research. As all such fields, it seeks its analytical forms and models that can aid in the understanding of the technological, social, economic, and epistemic complexities of its research object. As researchers approach the phenomenon from their respective vantage points, this means that some will focus on technological affordances, others will focus on business models, while others will look to the social apprehensions of technology and everyday use, just to single out some possible approaches. In trying to gain a holistic understanding of the phenomenon this complexity is a challenge in itself, but it also opens up a space for valuable cross-disciplinary theoretical and analytic syntheses of previous research.

Above, I have suggested such a synthesis in the form of a model for analysing the implications of datafication within the framework of data capitalism, I have specifically argued for the benefits of using the concept of value and valuation practices as a prismatic focus for studying the dynamics of this specific form of capitalism. I have tried to nuance previous research that has, quite naturally, pointed to the implications that the commercial nature of the datafication process has on the social by pointing to the additional dynamics involved for the benefit of future empiric studies. I have argued that one such approach for conducting empirical studies is to regard the various domains involved in the datafication process from the perspective of the tensions that exist between different forms of value that are drawn into the datafication process. To use value as a prismatic focus can lay bare the social dynamics of the various domains involved, as value is both produced and perceived socially in practices of evaluation.

I have identified four types of dynamics, centred on their own value regimes, that underlie data capitalism: the economical, technological, epistemological, and social dynamics. No doubt other domains can be added to this list, and future research will most likely identify these. I then provided some examples of how tensions within specific domains can play out in relation to the interface between different value forms. My examples concerned the meetings between technological dynamics and new business models, and the tensions between epistemological, technological, social, and economic dynamics. Exploring the negotiations of value within various social domains, and empirically studying and analytically theorising value dynamics can also help determine whether datafication is a true game-changer or whether it is merely a slight adjustment to informational capitalism, as theorised by Manuel Castells (1996) and others.

Acknowledgements This chapter started as a presentation during my stay as a research fellow under the ZeMKI fellowship programme in November 2018. I am grateful for all the comments provided by the ZeMKI researchers back then and especially to Andreas Hepp for hosting me. My thanks also extend to Jockum Hildén for discussions on the typologies of recommender systems.

References

Andrejevic, M. (2013). *Infoglut. How too much information is changing the way we think and know*. Routledge.

Baudrillard, J. (1968). *Le système des objets*. Gallimard.

Baudrillard, J. (1970/1998). *The consumer society: Myths and structures*. Sage.

Baudrillard, J. (1972/1981). *For a critique of the political economy of the sign*. Telos.

Baudrillard, J. (1973/1975). *The mirror of production*. Telos.

Beer, D. (2016). *Metric power*. Palgrave Macmillan.

Benkler, Y. (2006). *The wealth of networks. How social production transforms markets and freedom*. Yale University Press.

Biesta, G. (2010). *Good education in an age of measurement: Ethics, politics, democracy*. Routledge.

Boellstorff, T. (2013). Making big data, in theory. *First Monday, 18*(10). https://firstmonday.org/ojs/index.php/fm/article/view/4869

Bolin, G. (2011). *Value and the media: Cultural production and consumption in digital markets*. Ashgate.

Bolin, G. (2012). The labour of media use: The two active audiences. *Information, Communication and Society, 15*(6), 796–814.

Bolin, G., & Hepp, A. (2017). The complexities of mediatization: Charting the road ahead. In O. Driessens, G. Bolin, A. Hepp, & S. Hjarvard (Eds.), *Dynamics of mediatization* (pp. 315–331). Palgrave.

Boltanski, L., & Thévenot, L. (1991/2006). *On justification. Economies of worth*. Princeton & Oxford: Princeton University Press.

Bourdieu, P. (1993). *The field of cultural production. Essays on art and literature*. Polity.

boyd, d., & Crawford, K. (2012). Critical questions for big data. *Information, Communication & Society, 15*(5), 662–679.

Braman, S. (2012). Technology and epistemology: Information policy and desire. In G. Bolin (Ed.), *Cultural technologies. The shaping of culture in media and society* (pp. 133–150). Routledge.

Breiter, A. (2014). Schools as mediatized worlds from a cross-cultural perspective. In A. Hepp & F. Krotz (Eds.), *Mediatized worlds: Culture and society in a media age*. Palgrave Macmillan.

Burke, R., Felfernig, A., & Göker, M. H. (2011). Recommender systems: An overview. *Ai Magazine, 32*(3), 13–18.

Castells, M. (1996). *The rise of the network society: The information age: Economy, society and culture vol. I*. Black well.

Cheney-Lippold, J. (2017). *We are data: Algorithms and the making of our digital selves*. NYU Press.
Cohen, J. E. (2012). *Configuring the networked self: Law, code, and the play of everyday life*. Yale University Press.
Cohen, J. E. (2018). The biopolitical public domain. The legal construction of the surveillance economy. *Philosophy and Technology, 31*, 213–233. https://doi.org/10.1007/s13347-017-0258-2
Cohen, J. E. (2019). *Between truth and power: The legal constructions of informational capitalism*. Oxford University Press.
Couldry, N. (2004). Theorising media as practice. *Social Semiotics, 14*(2), 115–132.
Couldry, N., & Hepp, A. (2016). *The mediated construction of society*. Polity Press.
Couldry, N., & Mehijas, N. (2019). *The costs of connection: How data is colonizing human life and appropriating it for capitalism*. Stanford University Press.
Dewey, J. (1939). *Theory of valuation*. Chicago University Press.
Dumont, L. (1980/2013). On value: The Radcliffe-Brown lecture in social anthropology, 1980. *HUA: Journal of Ethnographic Theory, 3*(1): 287–315.
Galbraith, J. K. (1958/1964). *The affluent society*. Hamish Hamilton.
Galbraith, J. K. (1970). Economics as a system of belief. *American Economic Review, 60*(2), 469–478.
Gerlitz, C. (2016). What counts? Reflections on the multivalence of social media data. *Digital Culture & Society, 2*(2), 19–38.
Gerlitz, C., & Lury, C. (2014). Social media and self-evaluating assemblages: On numbers, orderings and values. *Distinktion: Scandinavian Journal of Social Theory, 15*(2), 174–188.
Gillespie, T. (2014). The relevance of algorithms. In T. Gillespie, P. Boczkowski, & K. Foot (Eds.), *Media technologies: Essays on communication, materiality, and society* (pp. 167–194). MIT Press.
Graeber, D. (2001). *Toward an anthropological theory of value: The false coin of our dreams*. Palgrave.
Gustafsson, K. E. (2009). Dagspressens affärsmodeller. *Ekonomiska samfundets tidskrift, 1*, 23–28.
Hacking, I. (1975/2006). *The emergence of probability: A philosophical study of early ideas about probability, induction and statistical inference* (2nd ed.). Cambridge University Press.
Heidegger, M. (1954/1977). *The question concerning technology, and other essays*. Garland Publishers.
Heinich, N. (2020). a pragmatic redefinition of value(s): Toward a general model of valuation. *Theory, Culture & Society*. https://doi.org/10.1177/0263276420915993
Helgesson, C. F., & Muniesa, F. (2013). For what it's worth: An introduction to valuation studies. *Valuation Studies, 1*(1), 1–10.

Hildén, J. (2021). The public service approach to recommender systems: Cultivating diversity. *Television & New Media*, Ahead of print. https://doi.org/10.1177/15274764211020106

Jarke, J., & Breiter, A. (2019). Editorial: The datafication of education. *Learning, Media and Technology, 44*(1), 1–6.

Karlgren, J. (1990). *An algebra for recommendations: Using reader data as a basis for measuring document proximity.* Working Paper No 179, The Systems Development and Artificial Intelligence Laboratory, Department of Computer and Systems Sciences, Royal Institute of Technology https://jussikarlgren.files.wordpress.com/1990/09/algebrawp.pdf

Kitchin, R. (2014). *The data revolution: Big data, open data, data infrastructures and their consequences.* Sage.

Kitchin, R. (2016). Thinking critically about and researching algorithms. *Information, Communication & Society, 20*(1), 14–29.

Lamont, M. (2012). Toward a comparative sociology of valuation and evaluation. *Annual Review of Sociology, 38*, 201–221.

Lindh, M., & Nolin, J. (2016). Information we collect: Surveillance and privacy in the implementation of google apps for education. *European Educational Research Journal, 15*(6), 644–663.

Magendanz, D. (2003). Conflict and complexity in value theory. *The Journal of Value Inquiry, 37*, 443–453.

Mager, A. (2012). Algorithmic ideology. *Information, Communication & Society, 15*(5), 769–787.

Mayer-Schönberger, V., & Cukier, K. (2013). *Big data: A revolution that will transform how we live, work, and think.* Eamon Dolan/Houghton Mifflin Harcourt.

Merton, R. K. (1995). The Thomas theorem and the Matthew effect. *Social Forces, 74*(2), 379–424.

Morozov, E. (2015). Digital technologies and the future of data capitalism. *Social Europe.* https://www.socialeurope.eu/digital-technologies-and-the-future-of-data-capitalism.

Mosco, V. (2014). *To the cloud: Big data in a turbulent world.* Paradigm Publishers.

Porter, T. M. (1986). *The rise of statistical thinking 1820–1900.* Princeton University Press.

Puschmann, C., & Burgess, J. (2014). Metaphors of big data. *International Journal of Communication, 8*. www.ijoc.org

Rossiter, N. (2017). Imperial infrastructures and Asia beyond Asia: Data centres, state formation and the territoriality of logistical media. *The FibreCulture Journal* (FCJ-220).

Rotella, P. (2012, April 2). Is data the new oil? *Forbes.* http://www.forbes.com/sites/perryrotella/2012/04/02/is-data-the-new-oil

Sayer, A. (2011). *Why things matter to people: Social science, values and ethical life*. Cambridge University Press.

Schäfer, M. T., & van Es, K. (2017). *The datafied society: Studying culture through data*. Amsterdam University Press.

Schiller, D. (1999). *Digital capitalism: Networking the global market system*. MIT Press.

Selwyn, N. (2016). *Is technology good for education?* Polity.

Sjödén, B. (2015). *What makes good educational software?* Lund University.

Skeggs, B. (2014). values beyond value? Is anything beyond the logic of capital? *British Journal of Sociology, 65*(1), 1–20.

Srnicek, N. (2017). *Platform capitalism*. Polity.

Stark, D. (2009). *The sense of dissonance: Accounts of worth in economic life*. Princeton University Press.

Stark, L., & Hoffmann, A. L. (2019, May 2). Data is the new what? popular metaphors & professional ethics in emerging data culture. *Journal of Cultural Analytics*. https://doi.org/10.31235/osf.io/2xguw

Strathern, M. (2000). *Audit cultures: Anthropological studies in accountability, ethics and the academy*. Routledge.

Sturluson, S. (ca. 1200/1916). *The prose Edda*. Translated by Arthur Gilchrist Brodeur. Internet Sacred Text Archive. https://www.sacred-texts.com/neu/pre/index.htm

Thomas, W. I., & Thomas, D. S. (1928). *The child in America: Behavior problems and programs*. A.A. Knopf.

Toonders, J. (2014). Data is the new oil of the digital economy. *Wired* no. 7/2014. https://www.wired.com/insights/2014/07/data-new-oil-digital-economy/

Turow, J. (2011). *The daily you. How the new advertising industry is defining your identity and your worth*. Yale University Press.

Turow, J. (2017). *The aisles have eyes: How retailers track your shopping, strip your privacy, and define your power*. Yale University Press.

van Dijck, J. (2014). Datafication, dataism and dataveillance: Big data between scientific paradigm and ideology. *Surveillance & Society, 12*(2), 197–208.

van Dijck, J., de Poell, T., & Waal, M. (2018). *The platform society*. Oxford University Press.

Vatin, F. (2013). Valuation as evaluating and valorizing. *Valuation Studies, 1*(1), 31–50.

Vonderau, A. (2018). Scaling the cloud: Making state and infrastructure in Sweden. *Ethnos*, 1–21. https://doi.org/10.1080/00141844.2018.1471513

Williamson, B. (2017). *Big data in education: The digital future of learning, policy and practice*. Sage.

Winston, B. (1995). How are media born and developed? In J. Downing, A. Mohammadi, & A. Sreberni-Mohammadi (Eds.), *Questioning the media: A critical introduction*. Sage.

World Economic Forum (WEF). (2011, January). *Personal data: The emergence of a new asset class*. Report. http://www3.weforum.org/docs/WEF_ITTC_PersonalDataNewAsset_Report_2011.pdf

Yu, J., & Couldry, N. (2020). Education as a domain of natural data extraction: Analysing corporate discourse about educational tracking. *Information, Communication & Society*. https://doi.org/10.1080/1369118X.2020.1764604

Zuboff, S. (2015). Big other: Surveillance capitalism and the prospects of an information civilization. *Journal of Information Technology, 30*(1), 75–89.

Open Access This chapter is licensed under the terms of the Creative Commons Attribution 4.0 International License (http://creativecommons.org/licenses/by/4.0/), which permits use, sharing, adaptation, distribution and reproduction in any medium or format, as long as you give appropriate credit to the original author(s) and the source, provide a link to the Creative Commons licence and indicate if changes were made.

The images or other third party material in this chapter are included in the chapter's Creative Commons licence, unless indicated otherwise in a credit line to the material. If material is not included in the chapter's Creative Commons licence and your intended use is not permitted by statutory regulation or exceeds the permitted use, you will need to obtain permission directly from the copyright holder.

Mapping Data Justice as a Multidimensional Concept Through Feminist and Legal Perspectives

Claude Draude, Gerrit Hornung, and Goda Klumbytė

Introduction

"Data justice" is a broad paradigm that encompasses and supersedes legal issues of data ownership and privacy regulations to account for complex power imbalances and injustices that are brought about by big data collection and use (Taylor, 2017). The concept is multifaceted in its applications. Data justice can aim towards "just decisions" and "just procedures" in administrative, legal, contractual, and other situations, where these decisions and procedures are made based on increasingly larger amounts of data. It can also mean aiming towards just decisions in sociotechnical systems' design when it relies on and pertains to data. Broader understandings of data justice can refer to justice on the level of policies, institutions, and societal structures pertaining to (big) data collection and use.

C. Draude (✉) • G. Hornung • G. Klumbytė
University of Kassel, Kassel, Germany
e-mail: claude.draude@uni-kassel.de; gerrit.hornung@uni-kassel.de; goda.klumbyte@uni-kassel.de

© The Author(s) 2022

A. Hepp et al. (eds.), *New Perspectives in Critical Data Studies*, Transforming Communications – Studies in Cross-Media Research,
https://doi.org/10.1007/978-3-030-96180-0_9

The relationship between data and justice as separate components is marked by complex considerations. For one, drawing on the metaphor of the classical image of *Justitia* as blindfolded, justice is envisioned through *not having* or *not using* certain types of data, such as those related to categories of race, class, gender, and so on. Simultaneously, just decisions require other types of data related to a person's societal attributes and actions. *Justitia's* blindfold is not meant to prevent her from seeing the relevant facts of a specific case. On the contrary, covering only one decisive fact may lead to arbitrary and unjust results. Thus, *having* and *using* certain types of data is seen as a prerequisite of justice, invoking a need to distinguish between relevant and irrelevant data to make just decisions.

Navigating the tension between *having/using* and *not having/using* certain types of data is crucial for more equitable, just, non-discriminatory futures. The underrepresentation of certain groups, people, contexts in datasets has problematic effects in proving injustice because inequalities need to be made visible. Not having data about diverse populations in engineering, computing, medicine, design, and architecture also leads to services and products that are unusable or inaccessible for some (Criado-Perez, 2019). Data collection, however, can be problematic, especially for vulnerable and marginalised groups. Concerns of privacy as well as underrepresentation and over-surveillance have been raised by people of colour, feminists, and LGBTQI communities (Browne, 2015; Shephard, 2016, 2018; Weinberg, 2017). The question of decision power over what counts as relevant and irrelevant data for making just decisions, whether answered "by design" or left open, is nonetheless particularly significant.

This points to aspects of data justice that concern well-being, participation, and broader contextual and societal aspects—what is commonly known as social justice. Reconfiguring data justice to include social justice means drawing attention to deeper structural imbalances in a given society and using data justice as a guide for legal and policy frameworks that actively work towards eliminating injustices. For algorithmic decision-making systems, this would entail paying attention to who is unduly over-served and underserved in regard to the use of these systems (Benjamin, 2019). This ties data justice to equity, equality, and fairness, as well as to "design justice", a concept promoting participation, power in design processes, and justice built into scenarios, systems, and infrastructures (Costanza-Chock, 2020). Such an understanding of justice has been prominent in intersectional feminist thinking that points out interlacing effects of structural inequalities.

Digitalisation has an enormous impact on all of the above-mentioned aspects of data justice. Algorithms and data processing have always been at the core of computing but today's pervasiveness of information technology and, especially the rise of machine learning and algorithmic decision-making (ADM) technologies with their reliance on large data sets, pose new threats and provide new opportunities. Support in decision-making is one of the most prominent and controversially discussed applications of machine learning. While this chapter pertains to information systems in general, we specifically developed the arguments with ADM technologies in mind.

A key threat posed by ADM technologies is that they stimulate further adverse effects on vulnerable and marginalised populations, increasing opacity and challenges in controlling the use of data in decision-making, and increasing difficulty in locating and ascribing accountability and responsibility. Critical research in systems design has shown that the pervasive universalisation of technological, legal, and policy solutions might also hinder context-specific and community-attuned approaches to data justice (Couldry & Mejias, 2019; Dourish & Mainwaring, 2012; Say Chan, 2018; Thatcher et al., 2016). These risks call for a concept of data justice that is both normatively robust to provide benchmarks for legal decision-making, practically applicable to address design justice concerns as well as conceptually flexible enough to accommodate or at least serve as common ground to bring in broader questions of social justice that go beyond regulation and appeal to historically formed sociopolitical structures and contexts. This chapter proposes just such an account of data justice as an interdisciplinary, multidimensional concept. First, we interrogate data justice through the lenses of feminist and legal studies and second, we investigate pathways towards realising data justice as a multidimensional practice in IT-design aided by these perspectives. We start by looking at how data justice is framed in feminist research and feminist-informed critical data and design perspectives. We then describe how it is conceptualised in law, particularly in the context of the GDPR and legal debates around privacy in Europe. On this basis we provide suggestions for what kind of concepts and tools are required to implement design justice in systems design and legal frameworks. We show that legal understandings of data justice can help generate data justice tools through regulatory frameworks, while feminist critique and design approaches can provide fruitful interventions towards more just information systems that take structural inequalities and context into account. Lastly, we discuss the extent to which data justice as a normative concept can be operationalised and highlight the potential challenges an interdisciplinary approach poses to the implementation of data justice. This includes questioning the limits of such implementations and tracing the convergences and divergences between feminist and legal approaches in regard to possible interventions towards data justice.

Ultimately, this chapter sets out to advance the design of data-driven computational systems that are most just for all. Our approach means rethinking basic assumptions of data justice from feminist and legal perspectives and suggests how this could impact IT-design and open up pathways towards more interdisciplinary approaches in critical data studies. Understanding data justice as multidimensional and interdisciplinary provides a conceptual ground that serves both the needs of legal formalisation and the feminist imperatives of contextualisation and specificity. While this chapter concludes with specific recommendations for design, its main aim is to rework the theoretical basis of where those recommendations should stem from.

We acknowledge that our perspective on data justice is unavoidably affected by our situatedness in a Western European educational institution and our observations pertain mostly to a European legal framework and justice-related issues that emerge in technology design and use.

"What's in a Name?"—Data Justice as a Concept in Feminist and Legal Scholarship

Debates around digitalisation, its effects, and the process of digital technologies becoming essential infrastructure, form a backdrop to understanding and implementing data justice as a normative concept and as a practical concern. As a broader normative concept, data justice points to understanding data as necessarily situated in sociopolitical context and practices (Dencik et al., 2019). This expands concerns regarding data beyond the domains of security, privacy, and data protection to incorporate imbalances of power. This also accounts for effects of digitalisation and datafication and relating those to social justice, citizenship, and political participation (ibid.).

Different definitions of data justice point to different contexts within which data justice can be discussed and applied. Dencik et al. (2019) distinguish between academic/conceptual definition (data justice as analysis of data that pays specific attention to structural inequalities and the different implications of datafication for different parts of society), the design dimension (data justice as pertaining to design conditions and processes through which data infrastructures emerge), and activist definition (whereby data justice is employed at the intersection of political and social activism and technology that challenges the status quo).

As a normative concept, we propose an understanding of data justice that zooms in on the imbalances of power within IT-design and use and opens a space for deliberations on how such injustices could be ameliorated. Such a concept would incorporate both legal and design perspectives as well as provide space to include concerns around data collection and use as well as the design and use of data-intensive technological systems, such as algorithmic decision-making systems.

Feminist Accounts of Data Justice

Data justice points to power discrepancies in data-related systems. Feminist approaches to data generally argue that data is entangled and laden with politics, which feminist analyses and tools can unveil and intervene in (D'Ignazio & Klein, 2019). Here, both data and justice, as well as data justice, connect several important aspects. In "Data Feminism", D'Ignazio and Klein argue that both the embodiment of data subjects and the effects of data on differently embodied people are important; that data are historical, contextual, and political as well as always processed and interpreted and that their presumed neutrality and objectivity should be interrogated; that what kind of data is collected, for which purposes, and how data is used is a political matter which should be critically discussed.

Such concerns regarding data orient justice towards embodied and contextualised understandings of it. This is particularly important not only for feminist but also anti-racist and social-justice oriented concepts. Anti-racist feminist legal scholar Kimberlé Crenshaw—following many black scholars and activists outside of a legal discipline such as Audre Lorde, bell hooks, Combahee River Collective, and others—using the metaphor of traffic intersections, coined the term "intersectionality" and pointed out that specific positions that index access to rights, resources as well as the possibilities of claiming justice are affected not by one but often by several social categories at the same time (1989). Intersectionality allows the rethinking of discrimination through multiple layers of meaning.

Following anti-racism activism and thought is particularly important for understanding social justice as well as data-related initiatives such as Data For Black Lives.[1] Among others, Eubanks (2018), Benjamin (2019), and Noble (2018) have pointed out that data inequalities and the lack of data justice have detrimental effects on communities of colour and other marginalised populations. People of colour are over-surveilled and generally over-served in terms of data accumulation for the purposes of evaluation and control, including predictive policing and other algorithmic prediction systems (Singelnstein, 2018; Thurn & Egbert, 2019; Hofmann, 2020), while they are simultaneously underserved when it comes to benefiting from algorithmic and data-driven systems (West et al., 2019; Hart, 2017; Yarger et al., 2019). Calls for justice from collectives such as Data For Black Lives or authors of Feminist Data Manifest-No[2] centre contextualised justice that is attentive to differential conditions, impacts, and effects experienced by different populations, and that requires a broad scope of socio-culturally embedded, nuanced, and specific solutions that pertain to legal, policy, and design measures.

Legal Framework: Justice, Data, and the Challenges of Digitalisation

While data justice is debated within a feminist framework in relation to contextualisation and differential conditions, in law, as it pertains to digitalisation, the expression of justice is centred around two essential elements: the principle of equality and procedural justice. The principle of equality requires that the state treat things equally if they are the same or at least substantially the same. Despite many problems around the question of what is "substantially the same", this notion of equality in Western legal tradition is probably the oldest and most agreed part of justice (Willoweit, 2012), dating back to ancient Greece; expressed in Ulpian's Digest I. 1.10: "*honeste vivere, neminem laedere, suum cuique tribuere*". Procedural justice assumes that if a transparent, inclusive procedure is used to make a decision, then that procedure ensures that a sufficiently fair and legitimate decision will be made (Luhmann, 2001).

Digitalisation has ambiguous effects on justice (Schliesky, 2019; Härtel, 2019). It could improve decision-making, leading to better and fairer

[1] See http://d4bl.org/

[2] See https://www.manifestno.com/

decisions, if the decision-maker is provided with more data which is relevant for a decision, if there are new algorithms which are able to better deal with these relevant data (e.g. accelerate processing, visualise results or individualise data use through personalisation), or if there are new algorithms which help to distinguish between relevant and irrelevant data (e.g. addressing the problem of human biases).

But digitalisation poses new risks for legal concepts of justice and legal proceedings aiming at just decisions. This may be due to (1) enormous amounts of personal data about data subjects, specific group characteristics such as race, class, and gender, or data on individual behaviour (Hoffmann-Riem, 2018). This poses the risk of enabling decision-makers to discriminate (more efficiently) against specific persons, specific groups, or specific types of behaviour if they wish to do so or are unconsciously driven by prejudice. Big data analysis will also not be a tool for everybody, as those equipped with the necessary hardware and software will be able to derive additional knowledge (raising e.g. questions of antitrust law—Körber, 2016; Louven, 2018), leading to additional power imbalances (Taeger, 2019).

Digitalisation may (2) also create conditions for data-powered findings, which lead to the conclusion that certain characteristics are (statistically) connected with certain groups or behaviours. Depending on the level of statistical significance, this may also influence decisions, without encouraging investigation into possible deeper structural factors that led to such statistical connection and thus divesting attention away from structural solutions. These types of empirical findings as well as new algorithms used in current and future legal proceedings also carry (3) inherent risks of non-transparency. The difficulties with the interpretation, explainability, and transparency of algorithmic systems make it very difficult or at times impossible, at least within current legal procedures, to monitor the results of such systems.

Data Protection Law

Basic constitutional law principles such as democracy and the rule of law as well as fundamental rights need to be freshly concretised to better address the aforementioned risks (Unger & von Ungern-Sternberg, 2019). In the German legal tradition, concepts of democracy and justice have been connected to the fundamental right to self-determination since the population census decision of the German Federal Constitutional

Court of 1983 (Bundesverfassungsgericht, 1983), whereas other legal systems have a concept of privacy that focuses solely on the individual and her "right to be let alone", which was first described by Warren and Brandeis in 1890 and lacks the society-oriented part of its European counterparts (Klar & Kühling, 2016; Gusy, 2018).

Being enshrined in fundamental rights (e.g. Art. 20–26 of the Charter of Fundamental Rights of the EU, CFR), equality and fairness are also essential parts of the regulatory framework on personal data. Concepts of data justice may, then, build on existing regulations, namely the GDPR, the EU e-Privacy Directive, and applicable national laws. In particular, the GDPR has taken up the idea that data protection does not only protect personal privacy, but also autonomy and self-determination. At the same time, we need to address the loopholes which the non-regulation of non-personal data creates.

According to Art. 4 (1) GDPR, "personal data" means any information relating to an identified or identifiable natural person (the "data subject"); an identifiable natural person is one who can be identified, directly or indirectly, in particular by reference to an identifier such as a name, an identification number, location data, an online identifier, or to one or more factors specific to the physical, physiological, genetic, mental, economic, cultural, or social identity of that natural person. This is a very broad concept (confirmed by the European Court of Justice, 2016) and the main reason for the ever-growing applicability of data protection law.

According to Art. 1 (2) GDPR, the regulation protects fundamental rights and freedoms of natural persons and "in particular their right to the protection of personal data". While the latter refers to Art. 8 CFR, it is important to note that data protection law also protects other fundamental rights (Hornung & Spiecker gen. Döhmann, 2019), particularly equality before the law (Art. 20 CFR), non-discrimination (Art. 21 CFR), cultural, religious, and linguistic diversity (Art. 22 CFR), and equality between women and men (Art. 23 CFR), as well as the rights of the child and the elderly (Art. 24 and Art. 25 CFR, respectively). These rights must be considered when applying the GDPR, which is particularly relevant when there is a need for balancing conflicting interests.

Issues of non-discrimination also play a role in the protection of "special categories of personal data" (Art. 9 GDPR, including personal data revealing racial or ethnic origin, political opinions, religious or philosophical beliefs, and data concerning health or a natural person's sex life or sexual orientation). Although experience shows that the sensitivity of personal data depends, to a large extent, on the purpose and circumstances of data processing (Simitis, 1990) and there is thus "no insignificant data anymore" (Bundesverfassungsgericht, 1983), the data categories in Art. 9 GDPR are specially protected as experience shows that there is a high risk of discriminating against people on this basis.

The notion of "procedural justice" is expressed in the GDPR by extensive procedural requirements for the processor. Following the principle of accountability (Art. 5 (2) GDPR), the processor is obliged to comprehensively inform the data subject, to implement appropriate technical and organisational measures to demonstrate compliance with the GDPR, to maintain records of processing activities, and to carry out a data protection impact assessment. The GDPR has considerably increased procedural requirements, calling for accountable and just data processing procedures as well as data protection by design and by default (Art. 25 GDPR).

The Justice Aspects of Non-personal Data

While non-personal data may be regulated by other fields of law, current data protection law has nothing to say on this type of data. At first glance this appears unproblematic as anonymity works as a shield against discrimination. This shield is however becoming weaker as big data analysis allows for (sudden or subtle) re-identification (Hornung & Wagner, 2019). Looking closer, identifiability is in no way a requirement for unjust decisions. It will often suffice to discriminate if the decision-maker or the ADM processes note the fact that people belong to certain groups or show certain attributes. While belonging to bigger anonymous groups may work as an effective shield against the risks of personalised data processing, the discrimination of these groups also poses risks for judicial remedy. At least in some cases it will be hard to prove that specific individuals are adversely affected. If equality laws do not include mechanisms of collective redress, these cases of "victimless discrimination" remain invisible to the legal system or at least not addressed properly (Lahuerta, 2018).

Furthermore, examples such as racial profiling (Angwin et al., 2016) and other developments within digitalisation, big data, and AI (Boehme-Neßler, 2008; Unger & von Ungern-Sternberg, 2019) have brought about new risks. Algorithmic personal pricing may be based on information about upscale IT equipment or behavioural analysis, teasing out the last bit of individual willingness to pay (Paal, 2019; Zuiderveen Borgesius & Poort, 2017). Using specific search terms (such as "maternity leave") in internet search engines may add to personal profiles on which discriminatory decisions may be based without knowing the specific person that is being discriminated. Safeguards such as anonymous job applications may not protect against these mechanisms, as big data and AI algorithms could still produce power imbalances to the disadvantage of non-identifiable persons. In addition, many protected categories of personal data can be coded by proxy (Dwork et al., 2012; Barocas & Selbst, 2016), as is the case, for example, with ethnicity and postal codes or specific search terms and maternity status.

These cases show that belonging (or even only being assigned) to a certain group as well as showing a certain feature or behaviour suffices to enable unjust decisions. Current data protection law, therefore, may be an important building block in a regulatory data justice framework but there is a need to address these further issues as well.

Expanding Data Justice Through Feminist and Legal Perspectives

We have identified how (data) justice has been approached conceptually and normatively from feminist and social justice perspectives that highlight context, structural inequalities, and the differential distribution of the negative and positive effects of digitalisation. We showed that justice is put into practice in law through substantive justice (the notion of equality) and procedural justice, and how these concepts are put into place in the specific European context of data protection law. Furthermore, we identified that non-personal data can be a basis for causing injustice and discrimination. In this section, we delve deeper into possibilities regarding data justice from feminist and legal perspectives, namely what kinds of alterations to the understanding and implementation of data justice are offered by both.

Feminist Avenues Towards Rethinking Data Justice

Feminist scholarship has generated important critiques of systems of oppression and inequality. One major take is that the interlinking of social categories and power relations expands through societies concerning individual, structural, and symbolic dimensions (Harding, 1986). Feminist research problematises knowledge structures such as categorisation and classification (Bowker & Star, 1999), which relates to the above-mentioned ambiguous character of *having* and *not having* data. In this way, feminist critique is at least twofold: On an empirical level, it provides a lens that shows how social inequalities are interwoven in the society's socio-economic-political fabric, making threads of injustice visible. This visibility is crucial for informing IT-design for social change (see section "An Overview of Technical Education, Higher Education, and Unemployability in India"). On a conceptual level, feminist thought highlights the productive and performative power of social categorisation itself. Cutting through these levels is a tension between relying on categories in order to point out injustices, criticising existing categories, expanding categorisations, and also challenging the very notion of stable categorisations (Benhabib et al., 1995).

This tension has been explored in feminist and queer legal thinking as it relates to perspectives on difference, equality, and justice. Feminist legal scholars have expressed varying positions on what constitutes gender equality in law, which is defined through gender neutrality or through gender specificity (Baer, 2011; Fineman, 2009). Both have faced critique: the former for the lack of acknowledgement of the different social, material, and cultural positions of women, while the latter is criticised for presenting universalist, essentialising, white- and Euro-centric views on women (ibid.). Queer scholars have also pointed out that any sort of categorisation fails those that fall outside of the normative definitions of legal subjects, as is often the case with transgender, intersex, and non-binary people (Spade, 2015). Relatedly, an ongoing debate in feminist and queer legal scholarship entails questioning the possibilities and the limits of identity-based discourses that focus on individual rights and binary categories of gender (Spade, 2015; Fineman, 2009).

Acknowledging that feminist thinking and feminist and queer legal theorising is diverse, we nonetheless highlight key avenues offered by research in these fields towards rethinking data justice:

Intersectionality: Intersectional analysis understands the position of a person to be the node of different socio-cultural categories (gender, race, class) that index their access to resources as well as their experiences of disenfranchisement (Crenshaw, 2019). Intersectionality links structural oppression, individual experience, and the symbolic order. Building upon black feminist activism, writing, and scholarship (see hooks, 1981, 1990), intersectionality as a concept informed (and informs) international policy making and discussions on human rights at the UN and for NGOs;

historically, in countering violence against women of colour (Yuval-Davis, 2006). Intersectionality shows why making injustice visible while simultaneously interrogating social categories with which injustice is commonly addressed is so important. Intersectionality remains a crucial and still largely absent perspective that could contribute to understanding and designing more just information systems through the interrogation of intersecting structural causes, effects, and the processes of discrimination and oppression.

Contextualisation: Contextualisation refers to the understanding of social categories as interdependent with the context they are produced in or in which they have meaning (Lykke, 2010). Different contexts, situations, and locations hold different consequences depending on how a person is categorised (Davis, 1983; Evans & Lépinard, 2019). Justice is taken to acquire meaning in the broader context of social relations, histories of violence/oppression, and relations of power. This means that the definition of justice, or what might constitute just action, is context-dependent, re-orienting the definition of data justice towards social justice that is concerned with overall just and fair relations in society in regard to the distribution of wealth, power, and other resources (Gangadharan, 2020).

Relationality: Interconnectedness and relationality are inherent in understanding justice as social justice. Critical scholarship and activism (including feminist thinking—Sander-Staudt, 2016, de la Bellacasa, 2017, and disability justice activism and theory—e.g. Mingus, 2010), as well as feminist legal scholarship (see Hunter, 2018) have highlighted that viewing the subject as abstract, individual, independent, autonomous, and rational is tied to the history of Enlightenment and its formation of the modern state. This legacy has until relatively recently denied subjectivity to women and children, racialised, and naturalised subjects as well as disabled people (Braidotti, 2007), based on the idea that they lack the rational, individualist, and self-defining attributes of subjecthood. Feminist thought understands individuals as shaped by relations. This is particularly relevant when thinking about discrimination that arises not out of specific personal conditions but because of being ascribed to a certain community. Feminist legal scholars have explored how feminist epistemology can introduce transformative shifts in legal reasoning or methodology in general through stressing contextualisation and relationality (Baer, 1999; Szablewska & Bachmann, 2015).

Distributed relational responsibility and accountability: Contextuality and relationality already imply a reconfigured understanding of responsibility and accountability. Feminist philosophy, particularly posthumanist new materialism, conceptualises agency as relational and emerging from within specific situational contexts (Braidotti, 2013; Barad, 2007). For data justice this would mean regarding responsibility and accountability as embedded within IT-design and sociotechnical infrastructures (Draude, 2020; Busch, 2018). From this follows a relational understanding of accountability that—given the distributed character of data collection, processing, use—could help implement a more realistic and just take on accountability in information systems that acknowledges the role of design, ownership, and use, *as well as* the unequal power that different actors possess. This is important because distributed responsibility should not lead to an avoidance of accountability or the offloading of responsibility from corporations to users, for example. Data protection law has recently made steps towards incorporating this idea, stipulating in Art. 5 (2) GDPR that data controllers shall be responsible for, and be able to demonstrate compliance with the data protection principles in Art. 5 (1) GDPR (principle of accountability). Data protection by design is evenly tied to these principles in Art. 25 (1) GDPR, building a strong nexus between accountability and system design.

Deconstructing the private/public distinction: From a Euro-Western perspective the struggle to open up the private as a realm for political struggle has been at the core of feminism (e.g., Elshtain, 1993), for example, in cases of sexual and domestic violence—a field which for a long time was not addressed by the legal system because it was seen as a "private" domain, and which, to this day, often pressures women who raise claims of sexual harassment to present their personal sexual history as proof of character. Similarly, cases of online sexual harassment are often distributed along the lines of sexual and racial marginalisation (Shephard, 2016). However, the deconstruction of the public/private dichotomy when it also wants to do justice to people of colour and other marginalised communities must interrogate its context and history and relate to intersectional perspectives (Collins, 1991).

Feminist Suggestions for IT-Design Towards Data Justice

The concepts outlined above invite the re-positioning of justice towards a more structural, expanded, and social understanding of justice, contextualising it and tying it to questions of equality and fairness, beyond individual rights and towards the concerns of communities. This resonates with questions of algorithmic and data justice such as: How is access to just outcomes and just procedure different for communities? How does big data-based decision-making disproportionately affect already marginalised communities? In what ways does the understanding of privacy as an individual domain impede the understanding of discrimination and disenfranchisement through categorisation and inference? The feminist considerations outlined above can provide a conceptual reorientation of data justice. Here, we provide some recommendations on how these conceptual reorientations can be applied towards the implementation of data justice as a pragmatic approach within IT-design. The idea is to not create foolproof design approaches but, rather, to treat data justice as a normative direction, along which processes of technology design could be refocused in order to design more accountable and contextualised systems.

First, the contextualisation of justice towards intersectional and social justice would allow those involved to pay attention to both issues around non-personal data and related power imbalances, and to ways in which collective decision-making and the power of self-determination could be enhanced. The diversification of data sets is often recommended to counter social bias in ADM and machine learning systems (Zou & Schiebinger, 2018). Intersectionality brings to the fore the challenges of such diversification. One problem of dealing with data and intersectionality is data disaggregation. According to Crenshaw (1989), the segmentation of discrimination has a counter-beneficial effect for women of colour. Since a lot of data is gathered or broken down into separate categories, intersectional positioning in the world is not represented. Here, the explicit sampling of intersections could be recommended. If diversification of data is pursued it is not enough to bring in diverse groups or features. Instead, the active oversampling of marginalised groups is recommended. Also, while the diversification of data might enhance the performance of the machine learning system, how people are affected by the system outcomes or what it means to become visible in a data set must also be considered.

This leads to the second point: To design more accountable sociotechnical systems it is important to *situate* the information systems and their design processes (Draude et al., 2020). This means that it is important to understand such systems as embedded in political, social, and other relevant contexts as well as in structural power relations. Draude et al. propose several guidelines for how such situatedness could be achieved. One of them is the systematic and iterative attention to a "4P set of questions", which stands for *people* affected and involved; *place*, for example, of data collection, of use, of affected areas; *power* relations; and *participation* of human/non-human actors (ibid., pp. 336–337).

Third, for the implementation of feminist values into data systems, approaches that question power relations and foster collaboration are promising. Participatory Design (PD) has a long tradition in information systems and human-computer interaction (Simonsen & Robertson, 2013; Sundblad, 2011). In contrast to its historical origin focusing on industry workers, today's PD approaches are mostly used in local, often smaller-scale projects addressing community building, neighbourhood projects with local youth or the elderly as user groups. Inherent to PD is interrogating power setting in the development process, such as between developer and user, democratic and emancipatory values, and participation throughout all stages of development, ideally by all people affected. To meet challenges brought upon by AI and big data, existing PD research and methodology must be updated. This is increasingly recognised in computing (Bannon et al., 2018). For data justice, PD offers the potential to transfer the claim of not just having a voice but, furthermore, of "having a say" (Kensing & Greenbaum, 2013) in overall system design to the new challenges brought about through (big) data collection, gathering, labelling, use, and storage. Juliane Jarke (2019a, 2019b), in her work on co-creating digital public services with older adults, has shown how participatory practices can be employed for data curation, self-determination in data ownership, and the shifting of power relations through collaborative methods such as "data walkshops".

In computing, norms, standardisations, and algorithms have so far acted as processes of exclusion when it comes to implementing social diversity. But they also provide the gateway for bringing in normative claims (Friedman & Kahn Jr, 2003; Roßnagel et al., 2018), such as non-discrimination or gender equality. To be implementable in information system development, critical knowledge has to be made operationalisable. Some approaches to design or to software engineering have taken this up. To name just two: Anti-Oppressive Design translates Patricia Hill Collins' racial justice work into a framework for HCI (Smyth & Dimond, 2014); the Gender-Extended Research and Development Model (GERD) takes up intersectional gender studies and reconstructs software engineering cycles enriched with feminist reflections and examples (Draude & Maaß, 2018).

To sum up, feminist perspectives offer some promising avenues for implementing data justice through conceptual and design means. Data justice here gets reconfigured as a more contextual, situated, and systemic approach to justice. These conceptual considerations can then be translated into systems design approaches that prioritise contextualisation through situating, participatory design methods and methods that are sensitive to power relations. These approaches are both resonating as well as contrasting with specifically legal approaches. In the next section, we will explore how data justice can be intervened into and implemented from a legal perspective in the current legal framework, particularly within the GDPR.

Legal Interventions for Data Justice in the Current Legal Framework

While feminist scholarship allows for structural design interventions, applying existing laws first of all requires the utilisation of data justice as a pragmatic and normative tool. In the GDPR, data justice questions play out regarding having/using and not having/using specific personal data. The GDPR provides space for interpretation when it comes to data justice questions, and it is, therefore, important to discuss how the GDPR and the tools that it offers *could* be used.

First, it is possible to understand several provisions of the GDPR as being duties to respect personal autonomy.[3] Art. 5 (1) (a) GDPR emphasises that personal data shall be processed "fairly". Little use has been made of this requirement so far (Roßnagel, 2019), but it could work as a gateway to incorporate notions of justice, equity, and equality—in particular, factoring in vulnerable groups and social inequalities. As there is so far no legal precedent at all, the term "fairly" offers space for interventions, including perceptions from feminist research on IT-design (cf. above). To this end, there is the need to translate into legal practice an understanding of intersectionality (and intersectional discrimination) as well as the relationality and situatedness this brings. This could mean giving up an absolute, fixed identity-based concept of discrimination in favour of

[3] The following ideas relate to data protection and privacy to personal autonomy, arguing that autonomy is one of their very foundations. Some scholars (e.g. Mokrosinska, 2018) argue that building privacy on the grounds of autonomy is not only insufficient, but even risky, because it connects data protection "only" to the person and is, therefore, blind to the political/democratic value of privacy. While we acknowledge these concerns, we believe that this depends on the context and legal tradition, as, for example, the German legal tradition connects privacy to both autonomy and democracy.

understanding discrimination as related to location, context, time, and changing with societal circumstances.

Autonomy is also emphasised in the definition of the data subject's consent in Art. 4 (11) and Art. 7 GDPR. Such consent is only valid if it is a freely given, specific, informed, and unambiguous indication of the data subject's wishes, prohibiting pre-checked checkboxes which the user must deselect to refuse her consent (European Court of Justice, 2019). Feminist scholarship has established the concept of "relational autonomy" (Mackenzie & Stoljar, 2000). As elaborated upon in section "Introduction", feminist thought emphasises contextualisation and relationality. The data subject's consent, the capability for self-determination and assessing the impact of one's consent become highly challenging in times of complex, networked, distributed information systems.

Data justice also relates to and is operationalised through several rights that are detailed in the GDPR, namely in the rights of the data subject to access (Art. 15), rectification (Art. 16 GDPR), erasure (Art. 17 GDPR), restriction of processing (Art. 18 GDPR), data portability (Art. 20 GDPR), and object (Art. 21 GDPR). The right to data portability, an important innovation of the GDPR (de Hert et al., 2018), provides for the right of the data subjects to receive the personal data concerning them which they have provided to a controller, in a structured, commonly used, and machine-readable format and the right to transmit those data to another controller without hindrance by the controller to which the personal data have been provided. Aiming at reducing lock-in effects, particularly in applications with high network effects, could in the future be a powerful tool to decrease the data power of these providers. As already mentioned, the GDPR considers the sensitivity of the data to be processed (Art. 9 GDPR). The Regulation also addresses the risks of automated individual decision-making, including profiling, in Art. 22 GDPR (Malgieri, 2019). Such decision-making shall only take place upon explicit consent of the data subject, following a contract with her or if authorised by Union or Member State law, if this law lays down suitable measures to safeguard the data subject's rights and freedoms, and legitimate interests.

Several procedural requirements should also be noted. The rights of the data subject are tools for procedural interventions for the data subject. The general principle of transparency (Art. 5 (1) (a) GDPR) is specified in many new and extended provisions. This is not restricted to reactive duties, such as granting access to data upon the request of the data subject. There are also obligations to proactively inform the data subject (in general, before processing is initiated) in Art. 12 ff. GDPR, and to notify personal data breaches to both the supervisory authority and the data subject (Art. 33, 34 GDPR). These duties are of utmost importance because transparency is a *conditio* sine qua non for almost every other right: Without knowing the controller and the details of the data processing, data subjects will not feel the need to control the actions of powerful controllers. As the German Federal Constitutional Court put it in its famous population census decision of 1983 (cf. Hornung & Schnabel, 2009), if individuals cannot oversee and control which or even what kind of information about them is openly accessible in their social environment and if they cannot even appraise the knowledge of possible communication partners, they may be inhibited in making use of their freedom.

Other procedural requirements form direct interventions within organisations processing personal data. Each controller must maintain a record of processing activities (Art. 30 GDPR), including the purposes of the processing, the categories of personal data and recipients, and even, where possible, the envisaged time limits for erasure. Art. 32 GDPR contains the duty to implement appropriate technical and organisational measures to ensure a level of security appropriate to the risk evocated by data processing. If that risk is likely to be high, Art. 35 GDPR imposes the duty to carry out a data protection impact assessment, including measures envisaged to address the risks (Bieker et al., 2016; Raab, 2020). Data protection officers (Art. 37, 38 GDPR) are important tools of mandatory internal self-control. The GDPR has also strengthened self-regulation in data protection law by voluntary codes of conduct (Art. 40, 41 GDPR), and certification (Art. 42, 43 GDPR). All in all, these requirements enforce, and the voluntary instruments offer important self-learning mechanisms for controllers and processors.

Towards Operationalising Data Justice

Following the aim of protecting fundamental rights of equality (Art. 1 (2) GDPR, cf. above), the idea of data justice—including the knowledge social sciences and humanities can add to this concept—should be kept in mind when interpreting the GDPR, particularly its sweeping clauses. Examples include the principle that personal data shall be processed "fairly" (Art 5 (1) (a) GDPR), the question of if there are "interests or fundamental rights and freedoms of the data subject" which override legitimate interests pursued by the controller (Art. 6 (1) (f) GDPR), the "appropriate measures to provide information" (Art. 12 (1) GDPR), the implementation of "suitable measures to safeguard the data subject's rights and freedoms and legitimate interests" when automated decision-making takes place (Art. 22 (3) GDPR), and every provision which forces controllers or processors to assess the specific risk of the data processing (e.g. appropriate technical and organisational measures, Art. 24 (1) GDPR; data protection by design, Art. 25 (1) GDPR; security of processing, Art. 32 (1) GDPR; data breach notification, Art. 33, 34 GDPR; data protection impact assessment, Art. 35 GDPR).

Our general assumption is that there are sufficient possibilities to address issues of data justice within the current legal framework. The main task will be to make decision-makers familiar with this idea, including the knowledge provided by feminist research, such as the specific discriminatory risks that intersectionality brings forth (cf. section "Introduction") as well as the broader understanding of data justice including structural policy measures and best practices for technology design. Data protection practitioners could profit from the feminist perceptions that any realistic concept of discrimination must include not only individual but also structural understandings of injustice—and that, following the idea of intersectionality and situatedness, we need to shift the understanding of a person's identity as a fixed position in society towards a more interactive, dynamic understanding.

Particular attention should be drawn to issues of profiling, which have been identified as being particularly risky not only for the individual, but also from the perspectives of social inequalities and social justice (Büchi et al., 2020). Furthermore, industry ethical standards as well as possible policy-oriented regulations could take the aforementioned legal as well as more conceptual feminist re-framings of data justice into account.

Some of these considerations are already implemented through ideas such as addressing quality and diversity of data as well as employing various sociotechnical design approaches. Nonetheless, a more explicit challenge to stark structural inequalities related to digitalisation could be considered as normative elements of data justice that could play a role in industry standards. Future policy making should also include issues of intersectionality, perhaps. Through procedural requirements for the integrated work of the respective bodies supervising issues of sectional equality (such as gender equality officers and disabled-employee officers).

Considering data justice and the rights of the data subject, it seems unclear how effective they are in practice. This is related to the general problems of the respective rights (such as procedural obstacles, unwillingness of controllers and processors to answer and to react to complaints), but also to issues of group-related discrimination, which "law in the books"-thinking is not able to address. Data protection rights form a comprehensive set of claims that will work as valuable tools for those who are fit and willing to fight against unjust processing and discriminatory effects. Those who lack these capabilities or who are (or feel) deterred because of their situation may face considerable difficulties in doing so.

While supervisory authorities usually report to the public the number of complaints they receive in their activity reports (Art. 59 GDPR), there is no information about the sociodemographic characteristics of those making use of this right. Regarding the rights against controllers and processors (access, rectification, erasure, restriction of processing, data portability, and object), there is not even any general statistics. There also appears to be no research at all on the question of which groups use their data protection rights in practice and whether the current procedural rules and requirements disfavour vulnerable groups.

The idea of making a data subject's rights workable for everyone also calls for digital literacy and education, a task that yet again poses problems of equally addressing different social groups. Data controllers should be urged to commit themselves to implementing measures to make rights workable particularly for vulnerable groups. These measures should then be subject to review by the supervisory authorities, as they have, inter alia, the tasks of promoting public awareness of data processing risks (Art. 57 (1) (b) GDPR) and providing information to data subjects concerning the exercise of their rights (e). In fulfilling these tasks, it would be possible to connect to representatives from vulnerable groups and enable forms of citizen participation.

Ideally, these activities should—together with research on the implementation of data subject's rights—lead to specific best practices and technological tools that are usable and affordable (for all) and enable to effectively exercise the respective rights. The need for such tools will become even more urgent, as ADM poses serious additional risks for the practical use of rights aiming at transparency. Best practices and tools could later form part of codes of conduct for specific processing sectors (Art. 40 GDPR).

The GDPR has seriously increased the duties of controllers to assist data subjects: Art. 12 (1) GDPR stipulates that the controller shall take appropriate measures to provide any information and communication in a concise, transparent, intelligible, and easily accessible form, using clear and plain language. According to Art. 12 (2) GDPR, the controller shall also "facilitate the exercise of data subject rights under Articles 15 to 22". This obligation could be interpreted to include a duty of controllers who implement ADM systems to also ensure that they implement effective technical tools for data subjects who want to make use of their rights.

Regarding applicable law, data protection law is restricted to personal data. Even considering the wide definition given in Art. 4 (1) GDPR, many cases of using data and the power of algorithmic decision-making are not covered by the GDPR. When considering appropriate possibilities for addressing this issue, regulators may be able to learn from the instruments data protection law provides—an approach which the German data ethics commission (Datenethikkommission, 2019) appears to follow in its expert opinion of 2019. Using justice and solidarity as ethical and legal principles (pp. 46 f.), it attempts to formulate data rights and corresponding data responsibilities (*Datenrechte und korrespondierende Datenpflichten*), arguing that a right to digital self-determination (*Recht auf digitale Selbstbestimmung*) should apply to legal persons as well as collective groups (85 ff.).

Many rules in the GDPR relate to specific data subjects and may not be used where those subjects do not exist. The general notion of a data subject's rights could however be transferred, for example, by granting persons affected by automated anonymous decisions a right not to access their personal data, but to access the algorithms and training rationales of AI decision-making. Following the idea that clandestine data power is particularly dangerous for data justice, new laws for mandatory pro-active transparency could be introduced as well. A recent example is the duty to inform consumers before distance and off-premises contracts about the fact that the price was personalised based on automated decision-making.[4] Such rules surrounding transparency could also include feminist research's critique that social inequalities and injustice are often not visible enough (cf. section "Introduction").

Many procedural instruments are also transferable to non-personal data and its algorithmic use. Records of processing activities could enhance transparency and allow for external administrative and judicial control. "Data justice impact assessments" or "AI impact assessments" could initiate the awareness within organisations, particularly in regard to the impact

[4] Art. 6 (1) (ea) of Directive 2011/83/EU, as amended by Directive (EU) 2019/2161.

on vulnerable groups. There could even be an obligation to involve and consult representatives of these groups. Such participatory approaches would not only be a way of taking into account suggestions for inclusive participatory design and co-creation (cf. section "State of Research: The QS and Maker Movements' Organisational Elites"), but also strengthen the legitimacy of design decisions. In bigger organisations, an ombudsman ("data justice protection officer") could oversee risky data processing and serve as a contact person for those affected by automated decision-making.

Certification and audits could be used to demonstrate compliance of existing products and procedures—either voluntary or in the case of risky algorithms and use cases, compulsory. They may disburden individuals from having to understand IT systems and data processing structures in detail in order to understand risks and personal implications (Hornung & Hartl, 2014; Rodrigues & Papakonstantinou, 2018; Hornung & Bauer, 2019). Given the ever-increasing complexity of ADM, this preventative control by experts will become more important and could again strengthen accountability. It could also become part of a wider system of external control, including elements like an obligation to disclose training data.

Potentials and Limitations of Integrating Feminist and Legal Perspectives for Data Justice in IT-Design

In lieu of a conclusion, we follow the potentials and limits of integrating and operationalising feminist and legal perspectives towards data justice in IT-design and what this means for critical data studies. Instead of combining the multiple perspectives provided into one unified framework, we map data justice as a multidimensional concept that has its normative-legal and pragmatic-design aspects, both of which can be intervened in from the perspectives of feminist and legal scholarship. This we see as both an advantage and a limitation. It is a potential drawback because a unified normative framework would be useful for IT-system design. On the other hand, the lack of such a unification allows for both interpretative flexibility regarding the concept of data justice as well as space for specific, situated, and contextualised translations of what data justice might mean in concrete cases, communities, and situations.

Feminist perspectives push the concept of (data) justice away from universalism and towards relational contextualised justice. This repositions data justice as a more systemic understanding of justice where the tools needed to ensure data justice are also systemic and broad, encompassing pragmatic legal interventions but also measures for policy and design requirements. The legal perspective meanwhile operates within the idea of a universal, formal definition of justice that ensures the broadest common denominator of justice for all, resonating with similar ideas of universality and broad applicability in IT-design.

This tension can be productive and does not necessarily need to be immediately resolved. Both perspectives are needed, particularly because they can be operationalised differently. The former, more conceptually inclined, feminist perspective can be particularly well suited for instigating broader conceptual and structural change when it comes to understanding what data justice might mean and also what should be included within normative and value-based considerations in technology design. More pragmatic legal approaches instead point to possible changes in the existing legal framework and can perhaps be integrated more easily with formal and model-based approaches prevalent in design.

In this chapter, we have outlined the broad conceptual changes needed with a feminist perspective, more concrete interventions from a legal perspective, and what kind of implications they both might bear on IT-design. Nonetheless, some questions remain that are significant both for IT-design and critical data studies as an interdisciplinary research field that purports to not only research but also to intervene in data-based systems.

First, how can we envision what "data protection" and "data justice" in IT-design entails? For one, this requires the more precise investigation into how the translations can be made between a more conceptual normative level and the more pragmatic levels of design—or put differently, what norms actually entail and how normative aspects are realised in IT-design. There is so far no common European understanding in regard to the exact content of the fundamental right to data protection enshrined in Art. 8 CFR (Marsch, 2018). This means that besides the new methods of information system design, fundamental rights innovations (Hornung, 2015) could also play a role in the shaping of data justice. Here feminist legal scholarship is still important, as it continues to investigate what "gender equality", "equity", and "privacy" might mean for different legal subjects.

Second, and relatedly, the question remains: At which points might we need to put effort into translating feminist critique (Simitis 1990) into formalisable and generalisable legal regulations? This is not a new question but has been explored by feminist computer scientists as well as feminist and queer legal scholars as pointed out earlier. Nonetheless, it highlights the need to investigate the conceptual possibilities of "translation" as well as existing best practices and to provide the space for activist voices and the voices of those constituencies that are directly suffering from "data injustice".

Last but not least, it is important to note that there already exists a plethora of sociotechnical approaches that can be used in designing more just data systems (such as participatory design). However, they are employed relatively rarely. This means that the question of how to make sure that feminist and legal justice-oriented design recommendations (particularly if they are not formalised in law) are taken up more extensively remains open as well.

To conclude, new perspectives in critical data studies require closer interdisciplinary collaborations. It is not only the analysis but pragmatic interventions that can originate in critical data studies that require understanding and work into the tensions that interdisciplinary perspectives bring. Our invitation is to embrace those perspectives and continue to expand their reach through possible structural regulation, policy regulations, voluntary industry standards, and other measures, with the hope that these different approaches can be brought into "strategic resonance" with each other: a kind of resonance that leaves space for tensions as sources of conceptual possibility. We hope that this chapter contributes to the creation of such a methodological and conceptual open space for considerations of data justice and interdisciplinary approaches in critical data studies.

References

Angwin, J., Larson, J., Mattu, S., & Kirchner, L. (2016, May 23). Machine bias. *ProPublica*. https://www.propublica.org/article/machine-bias-risk-assessments-in-criminal-sentencing?token=lZ_nPrh6oVJEnMzc-TH1Jr59Ibe3K8XZC

Baer, J. A. (1999). *Our lives before the law: Constructing a feminist jurisprudence*. Princeton University Press.

Baer, J. A. (2011). Feminist theory and the law. In R. E. Goodin (Ed.), *The Oxford handbook of political science*. Oxford University Press.

Bannon, L., Bardzell, J., & Bødker, S. (2018). Reimagining participatory design. *Interactions, 26*(1), 26–32.

Barad, K. (2007). *Meeting the universe halfway: Quantum physics and the entanglement of matter and meaning*. Duke University Press.

Barocas, S., & Selbst, A. D. (2016). Big data's disparate impact essay. *California Law Review, 104*(3), 671–732.

Benhabib, S., Butler, J., Cornell, D., & Fraser, N. (1995). *Feminist contentions: A philosophical exchange*. Routledge.

Benjamin, R. (2019). *Race after technology: Abolitionist tools for the new Jim code*. Polity Press.

Bieker, F., Friedewald, M., Hansen, M., Obersteller, H., & Rost, M. (2016). A process for data protection impact assessment under the European general data protection regulation. In S. Schiffner, J. Serna, D. Ikonomou, & K. Rannenberg (Eds.), *Privacy Technologies and Policy. APF 2016. Lecture Notes in Computer Science* (Vol. 9857). Springer.

Boehme-Neßler, V. (2008). *Unscharfes Recht. Überlegungen zur Relativierung des Rechts in der digitalisierten Welt*. Duncker & Humblot.

Bowker, G. C., & Star, S. L. (1999). *sorting things out: Classification and its consequences*. MIT Press.

Braidotti, R. (2007). Feminist epistemology after postmodernism: Critiquing science, technology and globalisation. *Interdisciplinary Science Reviews, 32*(1), 65–74.

Braidotti, R. (2013). *The posthuman*. Polity Press.

Browne, S. (2015). *Dark matters: On the surveillance of blackness*. Duke University Press.

Büchi, M., Fosch-Villaronga, E., Lutz, C., Tamò-Larrieux, A., Velidi, S., & Viljoen, S. (2020). The chilling effects of algorithmic profiling: Mapping the issues. *Computer Law & Security Review, 36*, 105367.

Bundesverfassungsgericht. (1983). Decision of 15. December 1983 (1 BvR 209, 269, 362, 420, 440, 484/83—"Volkszählung"), *65*, 1–71.

Busch, C. (2018). *Algorithmic accountability* [Gutachten]. ABIDA-Projekt.

Collins, P. H. (1991a). *Black feminist thought: Knowledge, consciousness, and the politics of empowerment*. Routledge.

Costanza-Chock, S. (2020). *Design justice: Community-led practices to build the worlds we need*. MIT Press.

Couldry, N., & Mejias, U. A. (2019). Data colonialism: Rethinking big data's relation to the contemporary subject. *Television & New Media, 20*(4), 336–349.

Crenshaw, K. (1989). Demarginalizing the intersection of race and sex: A black feminist critique of antidiscrimination doctrine, feminist theory and antiracist politics. *University of Chicago Legal Forum, 1989*, 139–168.

Crenshaw, K. W. (2019). *On intersectionality: Essential writings*. New Press.

Criado-Perez, C. (2019). Invisible women. Exposing data bias in a world designed for men. Chatto & Windus.

D'Ignazio, C., & Klein, L. F. (2019). *Data feminism*. MIT Press.

Datenethikkommission. (2019). *Gutachten der Datenethikkommission*. BMI & BMJV. https://www.bmi.bund.de/SharedDocs/downloads/DE/publikationen/themen/it-digitalpolitik/gutachten-datenethikkommission.pdf;jsessionid=2FBF166716BEFD690C724CCD8039936A.2_cid287?__blob=publicationFile&v=6

Davis, A. Y. (1983). *Women, race, & class*. Vintage Books.

de Hert, P., Papakonstantinou, V., Malgieri, G., Beslay, L., & Sanchez, I. (2018). The right to data portability in the GDPR: Towards user-centric interoperability of digital services. *Computer Law & Security Review, 34*, 193–203.

de la Bellacasa, M. P. (2017). *Matters of care: Speculative ethics in more than human worlds*. University of Minnesota Press.

Dencik, L., Hintz, A., Redden, J., & Treré, E. (2019). Exploring data justice: Conceptions, applications and directions. *Information, Communication & Society, 22*(7), 873–881.

Dourish, P., & Mainwaring, S. D. (2012). Ubicomp's colonial impulse. *Proceedings of the 2012 ACM Conference on Ubiquitous Computing*, 133–142.

Draude, C. (2020). "Boundaries Do Not Sit Still" from Interaction to Agential Intra-action in HCI. In M. Kurosu (Ed.), *Human-computer interaction. design and user experience* (pp. 20–32). Springer International Publishing.

Draude, C., Klumbyte, G., Lücking, P., & Treusch, P. (2020). Situated algorithms: A sociotechnical systemic approach to bias. *Online Information Review, 44*(2), 325–342.

Draude, C., & Maaß, S. (2018). Making IT work: Integrating gender research in computing through a process model. *Proceedings of the 4th Conference on Gender & IT*, 43–50.

Dwork, C., Hardt, M., Pitassi, T., Reingold, O., & Zemel, R. (2012). Fairness through awareness. *Proceedings of the 3rd Innovations in Theoretical Computer Science Conference*, 214–226.

Elshtain, J. B. (1993). *Public man, private woman: Women in social and political thought* (2nd ed.). Princeton University Press.

Eubanks, V. (2018). *Automating inequality: How high-tech tools profile, police, and punish the poor*. St. Martin's Publishing Group. 7.

European Court of Justice. (2016). Decision of 19 October 2016 (C-582/14—Breyer), ECLI:EU:C:2016:779.

European Court of Justice. (2019). Decision of 11 October 2019 (C-673/17—Planet 49), *ECLI*:EU:C:2019:801.

Evans, E., & Lépinard, E. (2019). *Intersectionality in feminist and queer movements: Confronting privileges*. Routledge.

Fineman, M. A. (2009). Introduction: Feminist and queer legal theory. In M. A. Fineman, J. E. Jackson, & A. P. Romero (Eds.), *Feminist and queer legal theory: Intimate encounters, uncomfortable conversations*. Ashgate.

Friedman, B., & Kahn, P. H., Jr. (2003). Human values, ethics, and design. In A. Sears & J. A. Jacko (Eds.), *The human-computer interaction handbook: Fundamentals, evolving technologies, and emerging applications* (pp. 1177–1201). Lawrence Erlbaum Assoc Inc.

Gangadharan, S. P. (2020). Context, research, refusal: Perspectives on abstract problem-solving. *Our Data Bodies: Human Rights and Data Justice*. Retrieved April 30, 2020, from https://www.odbproject.org/2020/04/30/context-research-refusal-perspectives-on-abstract-problem-solving/

Gusy, C. (2018). Datenschutz als Privatheitsschutz oder Datenschutz statt Privatheitsschutz? *Europäische Grundrechte-Zeitschrift, 45*(9), 244–255.

Harding, S. G. (1986). *The science question in feminism*. Cornell University Press.

Hart, R. D. (2017, July 10). If you're not a white male, artificial intelligence's use in healthcare could be dangerous. *Quartz*. https://qz.com/1023448/if-youre-not-a-white-male-artificial-intelligences-use-in-healthcare-could-be-dangerous/

Härtel, I. (2019). Digitalisierung im Lichte des Verfassungsrechts–Algorithmen, Predictive Policing, autonomes Fahren. *Landes- Und Kommunalverwaltung*, 49–60.

Hoffmann-Riem, W. (Ed.). (2018). *Big data—Regulative Herausforderungen*. Nomos Verlag.

Hofmann, H. (2020). *Predictive policing: Methodologie, Systematisierung und rechtliche Würdigung der algorithmusbasierten Kriminalitätsprognose durch die Polizeibehörden*. Duncker & Humblot.

hooks, b. (1981). *Ain't I a woman. Black women and feminism*. South End Press.

hooks, b. (1990). Homeplace: A site of resistance. In *Yearning: Race, gender and cultural politics*. South End Press.

Hornung, G. (2015). *Grundrechtsinnovationen*. Mohr Siebeck.

Hornung, G., & Bauer, S. (2019). Privacy through certification? The new Certification Scheme of the General Data Protection Regulation. In P. Rott (Ed.), *Certification—Trust, accountability, liability*. Springer.

Hornung, G., & Hartl, K. (2014). Datenschutz durch Marktanreize—auch in Europa? Stand der Diskussion zu Datenschutzzertifizierung und -audit. *Zeitschrift für Datenschutz*, 219–225.

Hornung, G., & Schnabel, C. (2009). Data protection in Germany I: The population census decision and the right to informational self-determination. *Computer Law and Security Review, 25*, 84–88.

Hornung, G., & Spiecker gen. Döhmann, I. (2019). Commentary on Art. 1 GDPR. In S. Simitis, G. Hornung, I. Spiecker Döhmann (Eds.), *Datenschutzrecht. DSGVO mit BDSG*. Nomos Verlag.

Hornung, G., & Wagner, B. (2019). Der schleichende Personenbezug. Die Zwickmühle der Re-Identifizierbarkeit in Zeiten von Big Data und Ubiquitous Computing. *Computer und Recht, 9*(2019), 565–574.

Hunter, R. (2018). Contesting the dominant paradigm: Feminist critique of liberal legalism. In M. Davies & V. E. Murno (Eds.), *The Ashgate research companion to feminist legal theory*. Routledge.

Jarke, J. (2019a). Co-creating digital citizenship: Considering the reconfiguration of participation in digital public service design. *Mensch Und Computer 2019: Workshopband*. Bonn. Gesellschaft für Informatik e.V.

Jarke, J. (2019b). Open government for all? Co-creating digital public services for older adults through data walks. *Online Information Review, 43*(6), 1003–1020.

Kensing, F., & Greenbaum, J. (2013). Heritage: Having a say. In J. Simonsen & T. Robertson (Eds.), *Routledge international handbook of participatory design* (pp. 21–36). Routledge.

Klar, M., & Kühling, J. (2016). Privatheit und Datenschutz in der EU und den USA–Kollision zweier Welten? *Archiv des öffentlichen Rechts, 141*(2), 165–224.

Körber, T. (2016). Ist Wissen Marktmacht? Überlegungen zum Verhältnis von Datenschutz, „Datenmacht“ und Kartellrecht. *Neue Zeitschrift für Kartellrecht*, 303–309; 348–356.

Lahuerta, S. B. (2018). Enforcing EU equality law through collective redress: Lagging behind? *Common Market Law Review, 55*, 783–818.

Louven, S. (2018). Datenmacht und Zugang zu Daten. *Neue Zeitschrift für Kartellrecht*, 217–222.

Luhmann, N. (2001). *Legitimation durch Verfahren* (6th ed.). Suhrkamp.

Lykke, N. (2010). *Feminist studies: A guide to intersectional theory, methodology and writing*. Routledge.

Mackenzie, C., & Stoljar, N. (Eds.). (2000). *Relational autonomy. Feminist perspectives on automony, agency, and the social self.* Oxford University Press.

Malgieri, G. (2019). Automated decision-making in the EU Member States: The right to explanation and other "suitable safeguards" in the national legislations. *Computer Law & Security Review, 35*, 105327.

Marsch, N. (2018). *Das europäische Datenschutzgrundrecht*. Mohr Siebeck.

Mingus, M. (2010, January 22). Interdependency (excerpts from several talks). *Leaving Evidence*. https://leavingevidence.wordpress.com/2010/01/22/interdependency-exerpts-from-several-talks/

Mokrosinska, D. (2018). Privacy and autonomy: On some misconceptions concerning the political dimensions of privacy. *Law and Philosophy, 37*(2), 117–143.

Noble, S. U. (2018). *Algorithms of oppression: How search engines reinforce racism*. NYU Press.

Paal, B. (2019). Missbrauchstatbestand und Algorithmic Pricing. Dynamische und individualisierte Preise im virtuellen Wettbewerb. *Gewerblicher Rechtsschutz und Urheberrecht*, 43–53.

Raab, C. D. (2020). Information privacy, impact assessment, and the place of ethics. *Computer Law & Security Review, 37*, 105404.

Rodrigues, R., & Papakonstantinou, V. (Eds.). (2018). *Privacy and data protection seals*. Springer.

Roßnagel, A. (2019). Commentary on Art. 5 GDPR. In S. Simitis, G. Hornung, & I. Spiecker gen. Döhmann (Eds.), *Datenschutzrecht. DSGVO mit BDSG*. Nomos.

Roßnagel, A., Hornung, G., & Geminn, C. L. (Eds.). (2018). *Rechtsverträgliche Technikgestaltung und technikadäquate Rechtsentwicklung* (Vol. 7). Kassel University Press.

Sander-Staudt, M. (2016). Care ethics. *Internet encyclopedia of philosophy*. https://iep.utm.edu/care-eth/

Say Chan, A. (2018). Decolonial computing and networking beyond digital universalism. *Catalyst: Feminism, Theory, Technoscience, 4*(2), 1–5.

Schliesky, U. (2019). Digitale Ethik und Recht. *Neue Juristische Wochenschrift*, 3692–3697.

Shephard, N. (2016). *Big data and sexual surveillance*. Association for Progressive Communications. https://www.apc.org/sites/default/files/BigDataSexualSurveillance_0_0.pdf

Shephard, N. (2018). *Queer enough? Categories and norms at the border*. XYZ. https://xyz.informationactivism.org/en/queer-enough/

Simitis, S. (1990). "Sensitive Daten" – Zur Geschichte und Wirkung einer Fiktion. In E. Brem, J. N. Druey, E. A. Kramer & I. Schwander (Eds.), Festschrift zum 65. Geburtstag von Mario M. Pedrazzini (pp. 469–493). Stämpfli.

Simonsen, J., & Robertson, T. (Eds.). (2013). *Routledge international handbook of participatory design*. Routledge.

Singelnstein, T. (2018). Predictive policing: Algorithmenbasierte Straftatprognosen zur vorausschauenden Kriminalintervention. *Neue Zeitschrift für Strafrecht*, 1–9.

Smyth, T., & Dimond, J. (2014). Anti-oppressive design. *Interactions, 21*, 68–71. https://doi.org/10.1145/2668969

Spade, D. (2015). *Normal life: Administrative violence, critical trans politics, and the limits of law*. Duke University Press.

Sundblad, Y. (2011). UTOPIA: participatory design from Scandinavia to the world. In J. Impagliazzo, P. Lundin, & B. Wangler (Eds.), *History of Nordic computing 3. HiNC 2010. IFIP advances in information and communication technology* (Vol. 350). Springer.

Szablewska, N., & Bachmann, S.-D. (Eds.). (2015). *Current issues in transitional justice: Towards a more holistic approach*. Springer International Publishing.

Taeger, J. (Ed.). (2019). *Die Macht der Daten und der Algorithmen: Regulierung von IT, IoT und KI*. OlWIR Verlag für Wirtschaft, Informatik und Recht.

Taylor, L. (2017). What is data justice? The case for connecting digital rights and freedoms globally. *Big Data & Society, 4*(2).

Thatcher, J., O'Sullivan, D., & Mahmoudi, D. (2016). Data colonialism through accumulation by dispossession: New metaphors for daily data. *Environment and Planning D: Society and Space, 34*(6), 990–1006.

Thurn, R., & Egbert, S. (2019). Predictive Policing: Die Algorithmisierung der Polizei als Risiko für die Bürgerrechte? *Vorgänge, 227*, 71–84.

Unger, S., & von Ungern-Sternberg, A. (Eds.). (2019). *Demokratie und künstliche Intelligenz*. Mohr Siebeck.

Warren, S., & Brandeis, L. (1890). The right to privacy. *Harvard Law Review, 4*(5), 193–220.

Weinberg, L. (2017). Rethinking privacy: A feminist approach to privacy rights after snowden. *Westminster Papers in Communication and Culture, 12*(3), 5–20.

West, S. M., Whittaker, M., & Crawford, K. (2019). Discriminating systems: Gender, race and power in AI. *AI Now Institute*, 1–33.

Willoweit, D. (2012). Recht und Willkür. Rechtsgeschichtliche Annäherungen an den Begriff des Rechts. *Rechtstheorie, 43*(2), 143–158.

Yarger, L., Fay, C. P., & Neupane, B. (2019). Algorithmic equity in the hiring of underrepresented IT job candidates. *Online Information Review, 44*(2), 383–395.

Yuval-Davis, N. (2006). Intersectionality and feminist politics. *European Journal of Women's Studies, 13*(3), 193–209.

Zou, J., & Schiebinger, L. (2018). AI can be sexist and racist—It's time to make it fair. *Nature, 559*(7714), 324–326. https://doi.org/10.1038/d41586-018-05707-8

Zuiderveen Borgesius, F., & Poort, J. (2017). Online price discrimination and EU data privacy law. *Journal of Consumer Policy, 40*, 347–366.

Open Access This chapter is licensed under the terms of the Creative Commons Attribution 4.0 International License (http://creativecommons.org/licenses/by/4.0/), which permits use, sharing, adaptation, distribution and reproduction in any medium or format, as long as you give appropriate credit to the original author(s) and the source, provide a link to the Creative Commons licence and indicate if changes were made.

The images or other third party material in this chapter are included in the chapter's Creative Commons licence, unless indicated otherwise in a credit line to the material. If material is not included in the chapter's Creative Commons licence and your intended use is not permitted by statutory regulation or exceeds the permitted use, you will need to obtain permission directly from the copyright holder.

Reconfiguring Education Through Data: How Data Practices Reconfigure Teacher Professionalism and Curriculum

Lyndsay Grant

Introduction

'Data power' permeates nearly every aspect of educational policy and practice, governing not only how educational institutions and individuals are made accountable, but also how we come to think about what 'education' is, and what counts as 'good' educational practice (Grant, 2017). Education is becoming increasingly 'datafied and digitised' (Jarke & Breiter, 2019; Williamson, 2017) and data has become a primary mode of governing education (Fenwick et al., 2014; Ozga, 2016). Educational performance is measured, analysed, visualised and applied at every scale—from international benchmarking to individual student assessments—and used to create comparisons, evaluations and interventions across the educational landscape (Gorur, 2015; Grek & Ozga, 2010; Hamilton, 2017;

L. Grant (✉)
School of Education, University of Bristol, Bristol, UK
e-mail: Lyndsay.Grant@bristol.ac.uk

© The Author(s) 2022

A. Hepp et al. (eds.), *New Perspectives in Critical Data Studies*, Transforming Communications – Studies in Cross-Media Research,
https://doi.org/10.1007/978-3-030-96180-0_10

Hamilton et al., 2015). This turn to data has permeated deeply into schools' everyday practices, with England's education system characterised as particularly 'advanced' in terms of its extensive production and use of data (Ball, 2015; Bradbury & Roberts-Holmes, 2017; Ozga, 2009).[1] While schools in many education systems have long been required to produce and report some form of quantitative data about their operations (Lawn, 2013), the late twentieth and early twenty-first centuries have seen a sharp increase in the use of pupil performance data as an accountability measure for school and teacher performance (Ozga, 2009). Concurrently, in the last decade, the growth of digital data infrastructures has created new, networked forms of governing education (Williamson, 2016) with an accompanying intensification in practices of generating, analysing, visualising and intervening in education with digital data (Sellar, 2015; Selwyn, 2016).

The datafication of educational practice and policy has far-reaching political implications for how education is governed at every level and scale. It also raises significant implications for how we—as a society—think about what education is, and what it is for. These are political questions that go beyond questions of educational effectiveness to questions of the social purpose of education. Increase in educational datafication, focused around efforts to more precisely monitor and predict pupil performance, risks creating an education that functions as a machine for reproducing existing knowledge and social orders rather than a more *educational* process of creating possibilities for the development of new knowledge and the formation of new subjectivities (Biesta, 2010, 2013).

It can seem that numbers have become an all-powerful and encompassing force governing every aspect of school life. Data policies, discourses and technologies do not, however, have straightforwardly predictable 'effects' on educational practice but are themselves the product of fragile assemblages, performed through political work, and are potentially incoherent and inconsistent (Piattoeva & Boden, 2020). In-depth explorations of how educational data practices work 'on the ground' are therefore needed in order to understand the complexities of how data power works in and through specific people, practices, policies, discourses, and digital

[1] I refer to 'England' throughout to indicate the focus of this chapter on the national education system of England, which, while sharing some similarities, is distinct from the education systems in the other, devolved nations of the United Kingdom (Scotland, Wales and Northern Ireland).

and material resources that come together. This chapter takes an in-depth approach to the study of educational data practices in an educational setting of a secondary school in England in order to understand the specificities, complexities and ambivalences of the workings of data power.

The study reported in this chapter was a critical exploration of how data was made and what it 'did' in one school in England, tracing the 'social life of data' (Beer & Burrows, 2013) and how it worked to (re)configure educational practices. This 'up close' approach to tracing data showed the constraining effects as well as the complexity, contestations and ambivalences of data power as it played out in practice. Datafication was evident in many different aspects of school life, including pupil—and teacher—attendance monitoring, educational and financial accountability processes, as well as pupil performance. While there are many ways in which datafication acts to reconfigure education, in this chapter I focus primarily on how data reconfigured two key aspects of the field: the English curriculum and teachers' professional judgements.

Since the 1990s, tighter specification of curriculum content and regular testing and reporting of results have become key policy technologies in the political control of classroom teaching and pedagogy in England (Moss, 2017; Ozga, 2009). This high-stakes testing regime has led to a situation in which assessment requirements drive both the content and the pace of delivery of the curriculum, rather than identifying the best ways in which to assess a curriculum built with wider aims in mind (Moss, 2014, 2017). Frequent testing to monitor children's 'expected progress' through a tightly defined curriculum reflects a limited view of how children learn, in which children are seen as "functional machines" who should all automatically progress at the same rate (Llewellyn, 2016). Such measures have significant educational consequences. The process of standardising curricula and monitoring progress can obscure teachers' ability to develop a more situated understanding of their pupils' learning and to adapt content, pace and the approach to teaching in relation to the specific learning needs of the pupils in their class (Llewellyn, 2016; Moss, 2017). In this chapter, I explore how new processes of datafication associated with frequent and high-stakes testing of progress, worked to reconfigure the English curriculum around the demands to evidence particular kinds of learning data, and the consequences of this for pupils' access to a broad curriculum.

A large proportion of teachers' work now includes facilitating the production and capture of pupil performance data, and incorporating it into their daily practice (Selwyn et al., 2016). Demand to produce an increasing volume of data in order to monitor, anticipate and intervene in pupil performance has led to a significant reshaping of teachers' subjectivities and professional practice. Data has become an important part of teachers' sense of self-worth and understanding of their own effectiveness (Bradbury, 2019; Lewis & Holloway, 2019), leading to cynical compliance with performative processes of datafication at the expense of building relationships or interrogating data for educational value (Hardy & Lewis, 2016). Within regimes of performative accountability, teachers' work and effectiveness are made visible, subject to comparison and evaluation, and they may internalise data logics as a new sense of professional purpose (Ball, 2003; Bradbury, 2019; Hardy, 2015; Lewis & Holloway, 2019). Contributing further to these accounts of the governing of teachers' work and subjectivities through data, I explore how moves towards 'objective' data as the basis for decision-making orientated teachers' judgements towards data in ways that worked to standardise judgement and exclude more multifaceted, situated and values-driven modes of professional knowledge that were characterised as 'human' and therefore inevitably biased.

Data Practices Reconfiguring Education

Data technologies and practices do not simply measure and represent aspects of social life but rather need to be understood as actively participating in producing new social practices (Barad, 2007; Savage, 2013). They can be usefully understood as part of a world-making practice, performing and reconfiguring the ongoing emergence of the world (Barad, 2007). The question of how data comes to reconfigure education is therefore not primarily about whether data accurately measures educational performance or whether that data is correctly interpreted or applied—points which are usefully addressed by statistical critiques (e.g. Leckie & Goldstein, 2016). The more pertinent question is how educational data practices configure and perform what 'education' is. For example, as Gorur (2015) shows, the data practices within the OECD's Programme for International Student Assessment (PISA) comprise a complex chain of translations and negotiations that condense a selected set of knowledge and skills with the varied experience of millions of school students over several school years to produce a seemingly coherent ranking of national school systems, which then drives further national policy reforms in attempts to move up the rankings. Educational data then, must be understood as both material and discursive, part of the ongoing sociomaterial reconfiguration of education. The question this then raises is what kinds of education are being configured through data, and how compatible these are with our social values of what constitutes a 'good' education.

One of the common concerns about increasing datafication is around its potentially reductive effects, in which complex social and human relations are reduced to only those areas that can be easily measured and opportunities for the individual are restricted. This has been a longstanding critique of high-stakes educational assessment, addressing the way that education systems can render invisible important aspects of education that are harder to measure at scale (Hardy, 2015; Thedvall, 2015). The turn to data in education has also been used to make complex, contestable and ultimately political decisions appear easily resolvable and outside the scope of more democratic and deliberative processes (Amoore, 2019). Such data-driven approaches can also be seen as reducing the scope for valuing un-quantified and unquantifiable social and human factors in education, including pupils' and teachers' voices and experiences.

For the educational philosopher Gert Biesta (2010, 2013), determining educational decisions through data is not only potentially reductive of social complexity in the ways discussed above, but also profoundly uneducational. He notes that we must start from an understanding of the purposes and values of education before we can decide what should be measured or how to measure it, and that establishing these purposes must be open to democratic debate. While government and school policies that mobilise pupil performance data might aim to 'raise standards', it is not at all clear how these measurements reflect wider purposes or values about the purpose of these standards, or education more broadly.

While discussions of educational purposes are often focused on the domains of qualification (knowledge and skills) and socialisation (induction into social orders), for Biesta (2010), the domain of subjectification must also be considered as an essential domain in any truly 'educational' project. Subjectification here, drawing on Arendt, refers to the development of human freedom in relation to others, and in education allows for the possibility of learners developing their own ideas, subjectivities and agency. Subjectification—which emphasises the emergence of *new* subjectivities and *new* knowledge—requires the possibility for pupils to enter into new relations with others' knowledge and subjectivities with necessarily unpredictable results. The drive towards predicting and determining educational outcomes through data, potentially threatens these kinds of encounters, and without which, our educational systems will continue to only reproduce the knowledge and subjectivities of the past.

While data practices are likely to shape opportunities for subjectification in education in many ways, the English curriculum and professional judgement provide useful examples of how this is working in two key domains of educational practice. These elements directly touch on the importance of creating possibilities for the formation of new knowledge and new subjectivities.

Data practices are impacting on how pupils encounter, experience and are assessed on new knowledge as they attempt to precisely monitor pupils' progress through curricular content. Data practices come to shape teachers' understanding of pupils' learning in more standardised ways that may offer fewer opportunities for the unique strengths and contributions of pupils to be recognised and responded to.

Following the Social Life of Data in an English Secondary School

To understand these questions of how data practices come to reconfigure the ways that 'education' is understood, thought about, and enacted, it is important to explore how they are performed and experienced within specific educational settings, as well as understanding how these settings themselves are shaped by their participation in networks of discourses, policies and technologies. By following the social life of data practices within an English secondary school over the course of a school year, I was able to track the specificity of how data practices worked to produce particular reconfigurations of education. Rather than positioning the school as a 'case study' that might be representative of similar cases, or as an illustration of the 'effects' wrought by global and national policies, discourses and technologies, the school site is conceptualised as a point of articulation within multiple intersecting networks and flows, as an entry point into "an assemblage of material, semiotic and social flows and practices" (Sellar, 2015).

Ridgewood School,[2] in which the fieldwork took place was a large, comprehensive, suburban, secondary (age 11–16) academy school in England. While I am not aiming to show that the findings from this school

[2] Names of institutions, individuals, titles and locations have been given pseudonyms to preserve anonymity.

are representative of all state secondary schools in England, Ridgewood was not unusual in its overall constitution, or the ways in which it responded to the demands of data. To trace data practices within this school I took an ethnographic approach in order to follow how data was created, circulated, processed, visualised, and represented and brought into relation with other people, discourses and objects. Over the course of December 2014–May 2015, I spent three periods of around one week each within the school, observing data practices, interviewing key members of staff and collecting key documents, displays, and technologies, allowing me to get a sense of the overall yearly cycle of data production. My entry point for fieldwork within the school was the 'data office', in which three members of staff worked: Sarah, the Head of Improving Achievement (HIA) and Chris, the Data Manager (who were also teaching staff) and Jenny, a full-time Data Administrator. I also followed the flows of data—digital and printed documents, conversations, school staff—into and out of this office, tracing the connections back to classrooms (including resources and teachers) and forwards to school-level decision-making processes about targeting and resource allocation. Following the data back to classrooms, I observed and interviewed two English teachers, Joe and Sophie, as they engaged with the digital and physical sociomaterial resource in their classrooms to generate, input and interpret pupil performance data.

In this chapter, I focus primarily on interview data from two key participants, Sarah and Joe, alongside fieldnotes and collected documents from the data office and an English classroom. I focus on Sarah as the senior architect of the school's data systems, as she was in a position to articulate the rationale behind their development and use and how they were intended to integrate into school-wide approaches and strategies. She was a maths teacher and the head of the data office and was engaged in bringing many disparate elements of the school's work into data-driven control, thereby increasing the scope of the school's activities that fell within the power of the data office. I focus on Joe as he was an English teacher closely involved in translating school-wide data strategies into classroom practices within the English department. His interview generated a particularly compelling account of the ambivalence of data in teaching as he was both reflectively questioning of the effect of data on pupils' learning while also being an enthusiastic proponent of the power of data to improve standards at the school. Both teachers allied themselves with the development of new data systems in the school, in the process claiming privileged positions for themselves as legitimate arbiters of data driven knowledge, interpretation and application. They are chosen here, therefore, not as 'representative' of staff within the school, but because their position enabled them to give deep insights into the logics and power of the school's data practices.

Prioritising Pupils: The Data Drop Machinery

Data practices can be understood as part of a material-discursive apparatus that works to not just measure but perform the very thing that it measures (Barad, 2007). In this case, data practices including disparate elements of assessment regimes, curriculum policies, children taking tests, software platforms, reports in which data is presented and communicated, league tables published in local newspapers, and so on, work to perform particular ideas, practices and materialisation of what education is and what it should be. These data practices work, together, to make education and schooling both known, and knowable, through data. The data apparatus can thus be understood as a sprawling, extensive arrangement that extended well beyond the school walls.

Staff in the data office alongside teachers in different subject areas had worked to develop a school-wide system for data collection, analysis and decision-making. An important part of this system was the regular generation of pupil progress and attainment data, known informally by staff as "data drops". To create these data drops, every pupil was assessed to measure their attainment and progress, six times per year—about every six weeks. This data was sent to the data office, where it was collated, processed and displayed as part of a 'data wall' in the data office. This data wall displayed postcard sized print-outs of pupils' photographs and data, arranged in a series of rows. Sarah explained to me that the arrangement of postcards on the data wall represented which pupils were targeted as priorities to receive 'interventions' (booster classes) to improve their performance.

Pupils' priority was determined using a bespoke algorithm that the school had devised themselves and calculated using simple coding in an Excel spreadsheet. This '"priority coefficient"' calculation, as Sarah termed it, assigned scores to pupils based on their performance data in English and mathematics, teachers' forecasts for their future performance, and their socioeconomic status, compared against national targets for attainment and progress. Scores were then ranked in order to indicate the level of priority for pupils to be assigned to intervention classes.

While the formula devised by the school may be bespoke to this school, the use of data-driven algorithms to take a diverse set of attributes and data and derive a single, actionable output, is not. As Amoore (2020, p. 4) notes, "what matters to the algorithm, and what the algorithm makes matter, is the capacity to generate an actionable output from a set of attributes". Whereas with more complex, proprietary and machine-learning algorithms it is difficult to 'open the black box' to see exactly how they work, this school's relatively simple algorithmic calculations provided an opportunity to explore in more depth the assumptions, decisions and values that went into the making up of this calculation.

Reconfiguring Access to and Delivery of the Curriculum

Starting with the data drop system and the data wall, I traced the data practices at work in Ridgewood School back to where pupil performance data was produced in classrooms, and forwards to how it was used to make decisions about pupils, staff and the allocation of resources, in the process reshaping the ways that staff and pupils thought about and practised 'education'. One of the notable reconfigurations was how the school's data practices performed differential access to the curriculum for different pupils and restructured how teachers organised the pace and delivery of curricular material.

An Algorithmic Triage Device Determining Curricular Access

The data practices instantiated in the pupil priority coefficient and materialised in the data wall functioned as an algorithmic triage device that produced differential access to the full curriculum for different kinds of pupils. Pupils who were identified through this calculation as high priority were assigned to attend additional English or mathematics booster classes, or both, aimed at improving their performance data to meet school and national attainment and progress targets. This was a calculated trade-off on the part of the school: some pupils would study a narrower curriculum in return for more pupils achieving higher grades in English and mathematics exams, which counted more highly towards the school's accountability targets. In this way, the pupil priority coefficient worked as an algorithmic triage device that determined pupils' access to the full curriculum, as well as eligibility for interventions.

An important element of this algorithm was that pupils who were closest to meeting targets, that is, those pupils who were forecast a 'near miss', were assigned a higher priority than pupils who were a long way from reaching their targets. Sarah described this to me as, "it's all about intervening with the right children", that is, identifying those children whose performance would be more likely to meet school targets with the aid of interventions, rather than those whose performance was so far from targets that even with additional support they may not improve enough. This can be seen as a continuation of the long-familiar process of institutions triaging access to limited resources, automated for an algorithmic age. As Gillborn and Youdell (2000) showed twenty years ago, triage processes in schools targeted resources to pupils seen as 'treatable', or borderline cases, where pupils were just a short distance from meeting attainment targets while ignoring pupils seen as either 'safe' or 'hopeless'. More significantly, in their study, triage processes discriminated against Black and minority ethnic pupils, as teachers' perceptions of pupil's potential were shaped by racism.

In Ridgewood School, data office staff saw the use of pupil performance data and algorithmic calculations as a way of avoiding such teacher bias and ensuring that triage decisions were based solely on objective assessments. This framing of data-driven decisions as objectively fair, however, conceals the ways that inequality and discrimination are already present within the data. While the priority coefficient algorithm did not include data on pupil ethnicity, gender or social class, pupil performance data already reflects unequal educational outcomes between these different groups of pupils (Department for Education, 2018; UNICEF Office of Research, 2018). The algorithm also included an additional weighting for economically disadvantaged pupils in receipt of welfare benefits, meaning that the likelihood of these pupils being assigned to intervention classes was higher than for their more advantaged classmates. The claims of objectivity and fairness were made simply on the basis of an algorithmic data-driven decisions removing the possibility of human bias, thereby overlooking the extent to which bias is already present in data sets.

While the priority coefficient algorithm determined which pupils should be assigned to English or maths intervention groups, it was Sarah who decided which subjects students would be withdrawn from to free up time for these additional classes. She usually withdrew pupils from arts and sports subjects such as photography, drama or physical education that were not part of accountability metrics. Pupils and parents were not routinely involved in these decisions or even made aware that there was a decision to be made: Sarah explained that she took these decisions on the basis of pupils' best interests. Yet, as pupils' interests were defined by the same accountability metrics as evaluated by the schools' performance, any differences between pupils' and the school's interests were elided. It is of course possible that dropping these subjects in order to gain higher grades in English and maths *was* in the best interests of some pupils, but importantly, this data-driven approach did not allow for pupils', subject teachers' and parents' voices, aims and ambitions to be included. These exclusions limited the scope for more democratic debates about when, why and for whom trade-offs between wider curricular access and individual or school performance might be an ethical—and educational—choice, including those pupils ineligible for interventions because their performance levels were deemed to be too low or high for these efforts to be worthwhile to the school. Thus, data power operated through this triage device by subsuming open questions of different and potentially competing interests with closed answers determined through data. In these ways, a 'good education' for any individual pupil was simply that which produced 'good data' for the school.

Disaggregating the Curriculum to Calculate 'Progress' Data

As well as differentially determining pupils' access to the curriculum, the demands of data practices to show frequent and fine-grained pupil progress data reconfigured how curricular content was organised and assessed. Producing data to measure and predict pupils' progress required teachers to assess pupils six times per year. In English, teachers adopted a pre-and-post-test model in which pupils were tested against four objectives in reading and writing, at the beginning and end of each of six terms—an increase from six to twenty-four tests over the year. As well as occupying a significant amount of the available class time, this regime resulted in far-reaching changes to how curricular content was taught. The objectives precisely specified the skills that pupils were required to demonstrate, yet were generic in terms of the knowledge or content. For example Joe, an English teacher, referred to his marking process against an objective in reference to the assessment criteria as "[i]f there's short quotations then that's Level

Five as well, if they're paraphrasing then that's a Level Six skill and so on and so forth". To coach students to perform well in these tests, each lesson was organised around the explicit teaching and practice of these generic objectives, with pupils required to attach a sticker to their workbooks summarising the assessment criteria for their target level objective, and to assess themselves against these criteria at the end of each lesson. In these ways, the requirements to produce detailed pupil progress data had become materialised in daily classroom practice, shaping each lesson around the practice and performance of skills against their assessment criteria.

Some teachers had questioned the requirements to explicitly focus every lesson on specified assessment criteria, wanting to go beyond the objectives to engage with English literature more widely. Joe, who had devised the use of target stickers, responded that,

> if you're not doing it, you're not doing English. I think what they were trying to do is something other than the skills-based practice that I see English as being [...] if you're trying to measure a child against something other than what's on this tracker then that something that you're measuring them against isn't English, as the government wants it to be taught.

While the national curriculum, specified centrally by government, does specify objectives to be taught and the criteria against which they were to be assessed, it does not stipulate that these are the entirety of what should be taught. In the telling quote above, Joe shows how the demands for detailed and frequent data updates had produced a curriculum that was entirely determined by assessment, instead of one embodying wider educational aims and responding to pupils' specific learning needs (Moss, 2017). Educational tests are more usually understood as a proxy for a pupils' learning; even the best tests can only assess a limited selection of the total knowledge, skills and understanding that have been learned. By requiring all teachers to use target stickers in every lesson in order to feed into the demands of the data drop process, however, the entirety of the English literature curriculum had effectively become reduced to its assessment criteria. A focus on pupils' current and target levels had become the only possible way of thinking, doing and talking about pupils' educational journeys (Livingstone & Sefton-Green, 2016).

Averaging 'Progress' Scores

While data drop requirements had given rise to the generation of large volumes of data per pupil, converting these into a single score for each pupil was a complex process in which the result was a compromise between representing coverage of curricular content with pupils' improvement over time. To create a single score for each pupil for the data drop spreadsheet, the English department calculated the mean average of all the 'post' test scores that a pupil had completed to date. This meant combining pupils' scores from tests of entirely separate objectives, effectively amalgamating snapshots of performance across different content areas of the curriculum. By including all scores across the year, the average score also suppressed out any improvement pupils had made over time. In order to feed the demands of the data drop system for simple progress data that could be plotted as a linear and predictable path (Llewellyn, 2016), these measures conflated coverage of the curriculum with improvement in performance. In the English department itself, these compromises were well understood, with Joe commenting, "I know that that's not their attainment, it's an average", and indeed, it remained a live issue as the department actively considered alternative possible compromises. Yet, once the single progress measure had been entered into the data drop spreadsheet, it was treated in subsequent calculations as an objective measure of attainment and used as the basis for the consequential decisions about curricular access and interventions discussed above.

Reconfiguring the Possibilities of Qualification, Socialisation and Subjectification

In a high-stakes accountability system such as in England, it is not surprising that schools' response to policy levers is to focus incessantly on improving measures that will improve their ranking. League tables of pupil performance are published in local and national newspapers, driving 'consumer choice' in the form of parents choosing schools for their child, with funding cuts following if lower numbers of pupils attend. Schools that are deemed as inadequate or requiring improvement in data-driven inspections may have their leadership staff replaced, be taken over by an (often private sector) academy sponsor or re-brokered to a new academy sponsor, and be subjected to more frequent monitoring and inspection until they meet the required standards (Ofsted, 2017).

The school's data practices worked to reorganise how pupils' progress through the curriculum was measured, organised and delivered. The data drop process, designed to monitor and intervene in pupils' performance and progress in order to produce the right kind of data for the school, also reconfigured access to and organisation of the curriculum, with consequences for the kinds of knowledge and learning that pupils were able to encounter.

Returning to Biesta's (2013) domains of educational purposes—qualification, socialisation and subjectification—can help to explore just what is at stake and what is made to matter through these data practices. The domain of qualification—that is, the knowledge and skills in which students must show themselves to be competent—is most clearly related to this reconfiguration of the curriculum. The algorithmic triage processes that restricted some students' access to a wider curriculum in return for higher scores in English and mathematics clearly prioritised qualification in high-stakes subjects that counted more in accountability targets, and for some pupils over others. Importantly, pupils who were identified as disadvantaged were more likely to have to make this trade-off. While there is a worthwhile educational question about whether achieving higher qualifications in English and mathematics is more important than engaging in a wider curriculum, the data practices at work presented these outcomes as the only option rather than open debates, eliding the interests of the school with those of pupils and excluding the voices of pupils, parents and teachers. Yet the data practices did not stop at narrowing the range of subjects in which pupils were able to become qualified. The domain of qualification had been stripped back to a form of 'credentialism', in which the knowledge, skills, understandings, dispositions and judgements that allow someone to be truly qualified to achieve something were limited to performance of skills that could be measured against assessment criteria.

The reconfiguration of the curriculum through data practices also had implications for the educational domain of subjectification. Subjectification—the unpredictable process of creating independent and novel ways of being through encounters with diversity and plurality—was limited by a curriculum focused on performing disaggregated, generic skills, leaving less scope for pupils and teachers to deeply engage with the content and meanings of the literature they read or considering the diverse ways that pupils related to it. Opportunities for more open-ended and expansive educational encounters were replaced with a curriculum organised around reproducing precisely predictable outcomes instead of more risky, open-ended educational encounters that might lead to learners developing their own, unique ideas and subjectivities.

Reconfiguring Teachers' Educational Knowledge and Judgements

In elevating data as the primary mode of knowing about education, data practices in the school also worked to reconfigure teachers' epistemological orientations and their judgements about pupils, learning and education. These reconfigurations, however, had not become accepted consensus, but were a matter of some debate and tension within the school, indicating how data epistemologies also play a role in reconfiguring power relationships in institutions like schools.

The privileging of data as the primary way of understanding education, reflects an 'evidence-based' approach to teaching which seeks to identify straightforward causal connections between teaching 'input' and educational 'outcomes' in the form of pupil data (Biesta, 2013). Such an approach aims to identify 'what works' in order to replicate its subsequent situations, which fails to consider how education is dynamic, subject to recursive effects and social interpretations, meaning that repeated inputs do not necessarily lead to predictable outcomes. Teachers are never faced with exactly the same situation twice, and must use their wider professional experience and values as well as any evidence to make situated, informed, and normative judgements that consider the desirability as well as the effectiveness of their actions and decisions at any one time (ibid.). Professional judgement, therefore, is necessary to remain open towards the possibility of emerging new knowledge and subjectivities, as Biesta writes, "we need judgment rather than recipes in order to be able to engage with this openness and do so in an educational way" (ibid. 2013, p. 137).

Excluding Professional Judgement

Frequent, formal tests at Ridgewood School to generate pupil data were part of the data office's attempts to create a more objective and accurate model for monitoring, intervening in and predicting pupil and school performance. In the process, other forms of understanding pupils' achievement through teachers' professional judgements were excluded as inherently 'biased'. Sarah described her frustration with some teachers who did not always enter the direct results of pupil tests into the spreadsheet but instead entered a score that reflected their professional view of a pupils' overall capabilities. In my fieldnotes, I recorded a discussion in which Chris, a maths teacher and the data manager, and Sarah discussed the problem of an English teacher who did not use the spreadsheet to calculate an average level for the pupil from assessed results, but simply input her overall judgement of the pupils' level. A second teacher was described as max valuing", that is, entering the pupil's highest level rather than their averaged score to date. Chris objected that such "holistic" approaches made it impossible for others to know on what "evidence" the score was based or to compare it to other teachers' grades, indicating the importance of a data trail to creating accountability for teachers' judgements of pupil performance. Sarah expressed frustration that these teachers did not understand that the average level was calculated from their own assessment data, and that they seemed to think the averaged test scores were the result of Chris claiming to "know the child better" or the grade just "magically appearing". In so doing, she framed their approaches as innumerate and illegitimate.

As a result of Sarah's and Chris's concerns, teachers were issued with an explicit instruction by Sarah to strictly limit themselves to assessed test results in reporting pupils' levels: "don't use professional judgement, use the actual number". For Joe, who had worked with Sarah and Chris to develop a system of assessment proformas for the English department to standardise marking approaches, this was necessary to "help [...] the teacher to gauge what that child is actually achieving". Joe drew an explicit contrast between untrustworthy human judgements and more objective data-driven assessments: "people are untrustworthy, just by being human, we make errors and so we test them [pupils]". Joe's phrase "actual achievement" equates to Sarah's "actual number", in which judgement of a pupil's level is legitimately defined and determined only through very specific test events, in contrast to illegitimate "holistic" approaches that took account of teacher's interpersonal and more informal knowledge of pupils learning.

This raises significant questions about the forms of professional knowledge that were made legitimate and illegitimate through the data practices at work. I asked Sarah whether there might be a legitimate reason for a teacher to give a different level to that calculated by averaged assessments; her argument was that unless higher levels could be evidenced through written work within two days of the initial assessment, then there could be no legitimate reason for teachers to have a different view of a pupil's level. In other words, a pupil's achievement could only be legitimately known through standardised assessments, and any professional knowledge derived from other sources, such as class discussions or work that was not formally assessed, was rendered illegitimate.

While the other teachers I spoke to did not openly discuss concerns about this process with me (in part, perhaps, due to the politics of voicing disagreement with school policy), Sarah's expressions of frustration and issuing of directives to use the "actual number" in response to teachers' alternative methods, indicated that there was far from a complete consensus within the school about how pupil performance data should be generated. This debate perhaps reflects questions about what, exactly, the data was thought to represent and how it fitted into the overall purpose of the data drop system. This tension between approaches to educational data—"professional judgement" or "the actual number"—highlights two points of tension in educational data epistemologies, with implications for what it means to be a professional teacher.

The first tension concerns whether data is being considered at the level of the individual pupil or in aggregate. Teachers who were using their professional judgement as a source of knowledge alongside test results were working with a more personal approach to data that reflected their understanding of a pupil as an individual learner, taking account of their wider strengths, weaknesses and capabilities. Sarah and Chris in the data office had a more statistically driven approach, in which it was more important to standardise data, to be able to compare pupils and compile aggregate data sets from standardised tests. Aggregating data into large sets is also, of course, a key statistical technique in which small, random margins of error in individual data points are cancelled out, allowing an overall pattern to be more clearly seen. The picture of "holistic" teachers focusing on individual pupils and the data office using test data to drive aggregate statistical analyses, was not the whole story, however, as individual pupils' data points were still the basis on which decisions were made about priority for interventions.

The implications for teacher professionalism can be better understood by considering a second tension: whether the data analysis was primarily concerned with generating insights about pupils' learning, or with anticipating and intervening in future pupil and school performance. The concern to reflect a pupil's capabilities more broadly than test results amongst the "holistic" approach suggests that data was seen primarily as a way of understanding pupils' learning strengths and weaknesses; an understanding that could potentially inform judgements around approaches to teaching and learning. Analysis of pupil performance data certainly has the potential to yield insights about differential performance, such as between different pupils with different characteristics, which could potentially be used to inform cohort or collective-level responses by the school. For example, data analysis might indicate that pupils from some ethnic groups were outperformed by others, which could lead to an investigation of how and why this might be within this school, and the development of new approaches that better met the needs of all pupils. Following Biesta's (2013) insistence on the importance of open-ended approaches to education, this would be an *educational* approach, in which teachers were also learners, exploring new questions about their pupils' learning, and creating new knowledge to inform professional judgements and decisions.

The purpose of analysing pupil performance data was, however, not primarily focused on opening up new questions and possible responses, but on anticipating and intervening in the production of future high-stakes data—pupils' final exams. Joe was explicit about how tests of pupil progress were designed primarily to replicate final exams rather than give a broader picture of pupils' learning, "if they're being tested at the end of their five years by doing a GCSE [end of school exams] we're just getting them ready for that". The primary emphasis on using data to anticipate and intervene in future performance can be seen in how the school applied the results of their data analysis. Rather than using insights generated from data analysis to explore, understand, and respond by adapting different educational approaches, responses focused on a more limited approach of simply increasing the quantity of instruction that pupils would receive in English or mathematics. This can be seen as a mode of anticipatory governance, in which data is analysed to identify and quantify future risks in order to determine actions in the present to respond as if those risks were already here (Amoore, 2011); in this case as risks to the school's performance targets. Pupils were produced through their data as a "risk subject" (Adams et al., 2009), and thereby made subject to further intervention in order to mitigate the impact of those risks. From an anticipatory perspective, data based on test scores alone would be a better indicator of future performance in similar conditions than "holistic" data that tried to include pupils' wider capabilities and achievements. As an anticipatory regime, the data practices in the school worked not to open up new questions and develop new responses, but to predict and then incorporate them in the ever-present possible future outcomes.

Becoming a Data-orientated Teacher and School

The data practices at work in the school demanded a re-orientation of teachers' professionalism towards the production and management of pupil data as a core professional practice. For Joe, an orientation towards data seemed to give him a sense of greater control over his own work and pupils' progress, as well as a sense of confidence that he was teaching the subject "as the government intended". Yet, although he was able to show me which pupils were colour coded red on his spreadsheet to indicate insufficient progress, he had not used the data he described as "ridiculously powerful" to investigate *why* some pupils might be struggling and others succeeding, or to inform different approaches to his teaching. Rather, this orientation towards data appeared to be primarily a way of performing his professionalism as a teacher whose sense of self was informed through data and whose professional knowledge was constituted through accountability measures (Hardy & Lewis, 2016; Lewis & Holloway, 2019). The use of mandated resources and procedures such as target stickers, assessment proformas and frequent tests also worked to define and standardise teachers' classroom practice towards producing the data demanded by the school's data drop systems. This worked in the favour of teachers such as Joe, whose alignment with a data-driven approach meant that he was given the additional responsibility of Deputy Head of Department in order to develop and implement new data systems which shaped the practice of his colleagues.

This re-orientation of teacher professionalism towards working with data, served as a form of exercising power by determining the scope of teachers' professional roles, as they became primarily accountable for producing pupil data. Those with the legitimacy to make data-driven claims, such as the data office teachers, held considerable authority within the school. For example, Sarah told me how she "hauled teachers in" to look at the data displays in the data office to show them "which pupils they should be working on", evaluated teachers' performance, and used school data analysis to drive policy-making. Sarah and Chris's approach in the data office was about driving system change throughout the school, as she described to me: "It isn't just about 'we'll crunch some numbers and give you some answers' [...] how I see the work of this office is, actually, we find a problem, something that's not being done very well, we find a process and a system to make it be done better, we give that system back to that person and that person then does that better". Diverse aspects of the management of school life, from teacher attendance to school performance, were increasingly becoming drawn into a data-driven systems approach in which processes and evidence for decision-making were driven by the data office.

Standardising Judgement and Practice

This elevation of data-derived knowledge and decision-making above teachers' holistic or situated professional judgements can be seen as a part of an overall process of standardisation. As with national and international benchmarking and policy-making, standardised performance metrics are a governing technology that allows for the comparative evaluation of different teachers and pupils performance (Fenwick et al., 2014). Such standardisation is a key part of an 'evidence-based' approach to teaching in which causal, rather than interpretative, relationships are assumed between teaching 'inputs' and pupil performance 'outcomes' (Biesta, 2013). Standardised metrics are employed in order to quantify and codify this relationship and to make data directly actionable.

The data practices in Ridgewood School made education knowable, accountable, and actionable through a data apparatus that also included international comparisons, national accountability frameworks and school data practices. The logics of this data apparatus reconfigured teachers' practice, performing data-driven, standardised, codified and quantified forms of teacher knowledge and decision-making as objective and legitimate, while more holistic professional knowledge and judgements were framed as inherently subjective and therefore illegitimate. The data practices within Ridgewood School were part of an ongoing re-orientation of teachers' professionalism towards producing, managing and responding to pupil data. Teachers such as Joe, Sarah and the data office team who aligned themselves with these practices were able to exercise considerable influence and power within the school, performing themselves as 'good' teachers, able to make legitimate claims about pupils' learning and using their knowledge to shape the work of other teachers within the school.

Conclusion: What Is Made to Matter and What Is Excluded from Mattering?

Educational data practices are shaping education in many different ways, from international policy-making through benchmarking to the subject positions of pupils in and out of the classroom. The organisation of the curriculum and the role of teachers' professional knowledge and judgements are two aspects of this process in which it is possible to closely follow in some depth how data works to make a difference to education. The analysis of these two aspects in this chapter helps to shed some light on the ways that data practices are not only making some elements of education more visible, but how they reconfigure the ways in which it is possible to think about and practice education, and the social purposes enacted through our education systems. These domains help to show how data practices of monitoring, standardising and predicting educational practice and outcomes can undermine the emergence of new subjectivities and knowledge, potentially limiting educational possibilities to the reproduction of existing knowledge and social orders.

Educational data practices are reconfiguring how pupils are assessed on, access, and experience new knowledge as they attempt to precisely monitor pupils' progress through curricular content. Through an algorithmic triage device, decisions were taken about which pupils were deemed priorities to receive interventions, in the process excluding them from participating in the wider curriculum. Importantly, these decisions were designed to prioritise pupils who made the biggest impact on the school's accountability metrics. The use of a data-driven algorithm framed these decisions as objective outcomes in the best interests of both pupils and the school, in the process eliding these potentially different interests and excluding pupils', parents' and teachers' voices from mattering in this debate. In pursuit of more predictable performance data the English literature curriculum had become focused around practising generic skills that could be measured against assessment criteria, resulting in a curriculum that failed to engage meaningfully with the new knowledge pupils may be creating through their engagement with the literature they studied.

As these data practices reached out beyond the system of data collection itself, they also worked to reconfigure teachers' professionalism, away from contextualised and multifaceted ways of knowing and making judgements, towards a more standardised and data-orientated form of professionalism. As data practices shape teachers' understanding of pupils' learning in more standardised ways, they may offer fewer opportunities for teachers to recognise and respond to the unique strengths and contributions of pupils. These new modes of teacher professionalism were matters of some controversy, but those teachers with the legitimacy to make data-driven claims were able to exercise considerable influence throughout the school.

This exploration of how data practices reconfigured the curriculum and teacher judgement in an English secondary school also serves to open up questions about the implications of data power in its attempts to minimise risk in other areas of social life beyond education. As data power attempts to more precisely predict social outcomes and standardise modes of judgement, it has consequences for how far we are able to engage with the openness, risk, and unpredictability that are necessary to create new knowledge and subjectivities to deal with new challenges and build new worlds.

References

Adams, V., Murphy, M., & Clarke, A. (2009). Anticipation: Technoscience, life, affect, temporality. *Subjectivity, 28*, 246–265. https://doi.org/10.1057/sub.2009.18

Amoore, L. (2011). Data derivatives: On the emergence of a security risk calculus for our times. *Theory, Culture & Society, 28*(6), 24–43. https://doi.org/10.1177/0263276411417430

Amoore, L. (2019). Doubt and the algorithm: On the partial accounts of machine learning. *Theory, Culture & Society*. https://doi.org/10.1177/0263276419851846

Amoore, L. (2020). *Cloud ethics: Algorithms and the attributes of ourselves and others*. Duke University Press.

Ball, S. J. (2003). The teacher's soul and the terrors of performativity. *Journal of Education Policy, 18*(2), 215–228. https://doi.org/10.1080/0268093022000043065

Ball, S. J. (2015). Education, governance and the tyranny of numbers. *Journal of Education Policy, 30*(3), 299–301. https://doi.org/10.1080/02680939.2015.1013271

Barad, K. (2007). *Meeting the universe halfway: Quantum physics and the entanglement of matter and meaning*. Duke University Press.

Beer, D., & Burrows, R. (2013). Popular culture, digital archives and the new social life of data. *Theory, Culture & Society, 30*(4), 47–71. https://doi.org/10.1177/0263276413476542

Biesta, G. (2010). *Good education in an age of measurement*. Routledge.

Biesta, G. (2013). *The beautiful risk of education.* Routledge.

Bradbury, A. (2019). Datafied at four: The role of data in the 'schoolification' of early childhood education in England. *Learning, Media and Technology, 44*(1), 7–21. https://doi.org/10.1080/17439884.2018.1511577

Bradbury, A., & Roberts-Holmes, G. (2017). Creating an Ofsted story: The role of early years assessment data in schools' narratives of progress. *British Journal of Sociology of Education, 38*(7), 943–955. https://doi.org/10.1080/01425692.2016.1202748

Department for Education. (2018). Revised GCSE and equivalent results in England: 2016 to 2017 [Statistical first release]. London. https://assets.publishing.service.gov.uk/government/uploads/system/uploads/attachment_data/file/676596/SFR01_2018.pdf

Fenwick, T. J., Mangez, E., & Ozga, J. (2014). Governing knowledge: Comparison, knowledge-based technologies and expertise in the regulation of education. In T. J. Fenwick, E. Mangez, & J. Ozga (Eds.), *World Yearbook of Education 2014.* Routledge.

Gillborn, D., & Youdell, D. (2000). *Rationing education: Policy, practice, reform, and equity.* Open University Press.

Gorur, R. (2015). Producing calculable worlds: Education at a glance. *Discourse: Studies in the Cultural Politics of Education, 36*(4), 578–595. https://doi.org/10.1080/01596306.2015.974942

Grant, L. (2017). *"Don't use professional judgement, use the actual number": The production and performance of educational data practice in an English secondary school* (Doctoral thesis). University of Bristol.

Grek, S., & Ozga, J. (2010). Re-inventing public education: The new role of knowledge in education policy making. *Public Policy and Administration, 25*(3), 271–288. https://doi.org/10.1177/0952076709356870

Hamilton, M. (2017). How international large-scale skills assessments engage with national actors: Mobilising networks through policy, media and public knowledge. *Critical Studies in Education, 58*(3), 280–294. https://doi.org/10.1080/17508487.2017.1330761

Hamilton, M., Maddox, B., & Addey, C. (2015). *Literacy as numbers.* Cambridge University Press.

Hardy, I. (2015). Data, numbers and accountability: The complexity, nature and effects of data use in schools. *British Journal of Educational Studies, 63*(4), 467–486. https://doi.org/10.1080/00071005.2015.1066489

Hardy, I., & Lewis, S. (2016). The 'doublethink' of data: Educational performativity and the field of schooling practices. *British Journal of Sociology of Education, 38*(5), 671–685. https://doi.org/10.1080/01425692.2016.1150155

Jarke, J., & Breiter, A. (2016). Datafying education: How digital assessment practices reconfigure the organisation of learning (Research Network "Communicative Figurations" No. 11). https://doi.org/10.13140/RG.2.1.2866.5686

Jarke, J., & Breiter, A. (2019). Editorial: The datafication of education. *Learning, Media and Technology, 44*(1), 1–6. https://doi.org/10.1080/17439884.2019.1573833

Lawn, M. (2013). *The rise of data in education systems: Collection, visualisation and use* (M. Lawn, Ed.). Symposium Books.

Leckie, G., & Goldstein, H. (2016). *The evolution of school league tables in England 1992–2016: 'contextual value-added', 'expected progress' and 'progress 8'* (Bristol Working Papers in Education No. #02/2016). Bristol. bristol.ac.uk/education/research/publications

Lewis, S., & Holloway, J. (2019). Datafying the teaching 'profession': Remaking the professional teacher in the image of data. *Cambridge Journal of Education, 49*(1), 35–51. https://doi.org/10.1080/0305764X.2018.1441373

Livingstone, S., & Sefton-Green, J. (2016). *The class: Living and learning in the digital age [E-Book]*. New York University Press. http://opensquare.nyupress.org/books/9781479824243/

Llewellyn, A. (2016). Problematising the pursuit of progress in mathematics education. *Educational Studies in Mathematics, 92*(3), 299–314. https://doi.org/10.1007/s10649-015-9645-8

Moss, G. (2014). Putting literacy attainment data in context: Examining the past in search of the present. *Comparative Education, 50*(3), 357–373. https://doi.org/10.1080/03050068.2014.921369

Moss, G. (2017). Assessment, accountability and the literacy curriculum: Reimagining the future in the light of the past. *Literacy, 51*(2), 56–64. https://doi.org/10.1111/lit.12104

Ofsted. (2017). *School inspection handbook.* Ofsted School Inspection Handbook. https://doi.org/10.4324/9780203416242_chapter_2

Ozga, J. (2009). Governing education through data in England: From regulation to self-evaluation. *Journal of Education Policy, 24*(2), 149–162. https://doi.org/10.1080/02680930902733121

Ozga, J. (2016). Trust in numbers? Digital Education Governance and the inspection process. *European Educational Research Journal, 15*(1), 69–81. https://doi.org/10.1177/1474904115616629

Piattoeva, N., & Boden, R. (2020). Escaping numbers? The ambiguities of the governance of education through data. *International Studies in Sociology of Education, 00*(00), 1–18. https://doi.org/10.1080/09620214.2020.1725590

Savage, M. (2013). The "Social Life of Methods": A Critical Introduction. *Theory, Culture & Society, 30*(4), 3–21. https://doi.org/10.1177/0263276413486160

Sellar, S. (2015). Data infrastructure: A review of expanding accountability systems and large-scale assessments in education. *Discourse: Studies in the Cultural Politics of Education, 36*(5), 765–777. https://doi.org/10.1080/01596306.2014.931117

Selwyn, N. (2016). "There's so much data": Exploring the realities of data-based school governance. *European Educational Research Journal, 15*(1), 54–68. https://doi.org/10.1177/1474904115602909

Selwyn, N., Nemorin, S., & Johnson, N. (2016). High-tech, hard work: An investigation of teachers' work in the digital age. *Learning, Media and Technology* (Online First). https://doi.org/10.1080/17439884.2016.1252770

Thedvall, R. (2015). Managing preschool the Lean way. Evaluating work processes by numbers and colours. *Social Anthropology, 23*(1), 42–52. https://doi.org/10.1111/1469-8676.12105

UNICEF Office of Research. (2018). An unfair start: Inequality in children's education in rich countries: Innocenti report card 15. Florence.

Williamson, B. (2016). Digital education governance: An introduction. *European Educational Research Journal, 15*(1), 3–13. https://doi.org/10.1177/1474904115616630

Williamson, B. (2017). *Big data in education: The digital future of learning, policy and practice.* Sage.

Open Access This chapter is licensed under the terms of the Creative Commons Attribution 4.0 International License (http://creativecommons.org/licenses/by/4.0/), which permits use, sharing, adaptation, distribution and reproduction in any medium or format, as long as you give appropriate credit to the original author(s) and the source, provide a link to the Creative Commons licence and indicate if changes were made.

The images or other third party material in this chapter are included in the chapter's Creative Commons licence, unless indicated otherwise in a credit line to the material. If material is not included in the chapter's Creative Commons licence and your intended use is not permitted by statutory regulation or exceeds the permitted use, you will need to obtain permission directly from the copyright holder.

Public Values and Technological Change: Mapping how Municipalities Grapple with Data Ethics

Lotje Siffels, David van den Berg, Mirko Tobias Schäfer, and Iris Muis

Introduction

Increasingly, Dutch municipalities use novel data practices for public management. These range from data analysis for more efficient waste management, to generating novel data sources for analysing criminal activities, to combining various data sources for predicting social welfare fraud (Redactie Gemeente.nu, 2016; van Ark, 2018). A process of decentralisation has delegated many tasks from central government to municipalities without giving them more resources and capacities. Local governments often see data practices as the most efficient way to deal with additional tasks and to

L. Siffels (✉)
Radboud University, Nijmegen, The Netherlands
e-mail: l.siffels@ftr.ru.nl

D. van den Berg • M. T. Schäfer • I. Muis
Utrecht Data School, Utrecht University, Utrecht, The Netherlands
e-mail: m.t.schaefer@uu.nl; i.m.muis@uu.nl

© The Author(s) 2022

A. Hepp et al. (eds.), *New Perspectives in Critical Data Studies*, Transforming Communications – Studies in Cross-Media Research, https://doi.org/10.1007/978-3-030-96180-0_11

distribute limited resources in a just and effective way. These data practices (such as predictive analysis, the automatic collection of records, the use of dashboards, combining various datasets and the capturing or digitisation of previously inaccessible or unavailable records) are not just replacing or innovating older practices, but are seen as a welcome solution to a shortage of resources and capabilities (Vermeulen, 2015: 139; Maarse & Jeurissen, 2016: 224).[1] However, data projects are not free from ethical issues and there is a real possibility that such a project affects public values. A recent example of this played out in the city of Rotterdam when discriminatorily processed data from residents of two entire neighbourhoods were used to detect a risk population of citizens that *might* commit social welfare fraud. The backlash to this activity has placed the ethical issues surrounding data practices within the purview of political debate and journalistic scrutiny (Redactie Nieuws Digitaleoverheid.nl, 2020).

There is an emerging debate in critical data studies on the use of algorithms in public management (e.g. Eubanks, 2018; O'Neil, 2016), which indicates that data practices are changing citizenship and democracy (e.g. Hintz et al., 2019). The argument in this debate emphasises that technology, in this case data, models and their automated analysis through algorithms, carry and transform values. Many scholars have focused on this relationship between (public) values and (emerging) technologies (Bannister & Connolly, 2014; Bertot et al., 2010). However, in this debate and in the broader public debate on the data practices of government organisations, there is little to no attention paid to how these practices are influenced by their resources, their experience and knowledge about data practices and their thinking about public values. Very little empirical research has been carried out on the subject. Evaluating data projects in public administration calls for a method to structurally connect these values to the data projects, the municipalities' operational capacities and how they legitimate their actions.

Over the past few years, the authors of this chapter have immersed themselves within this field to help municipalities detect possible ethical issues in their data projects and to gather research insights. We used an

[1] Examples for these data projects include a monitor for predicting foundation rot (City of Zaanstad), a model for predicting early school leavers (City of Dordrecht), automatic number plate scanning for parking space management (The Hague, Leiden, Utrecht and others), predictive analysis for waste management, social benefit fraud, and housing violations, the use of software for simulating traffic flows, construction, water management and policy effects.

ethical deliberation tool called the Data Ethics Decision Aid (DEDA) which helps participants working on a data project to become aware of and discuss the ethical aspects that are relevant to the project.[2] The Utrecht Data school provides this impact assessment as a service to companies, government organisations and NGOs. This chapter describes how the assessment of data projects with the DEDA enables participatory observation, granting insight into an organisation's data practices, their awareness of data ethics and how policy objectives are translated into data projects, which then carry or transform public values. Simultaneously, the Data Ethics Decision Aid enables municipalities to review their data projects while considering the ethical issues at stake. Our assessments with DEDA therefore have a dual-use function: they serve as a process through which municipalities can establish possible ethical problems within data projects and adapt the design accordingly, but the process is also used by the authors as a participatory observation method.

We will analyse our findings through the framework provided by Mark H. Moore's 'strategic triangle'. Mark Moore introduced the strategic triangle in his book *Creating Public Value* (Moore, 1995); it is a framework for understanding governmental public value creation. Moore argues that in order to create public value an iterative process is needed, where public management needs to move back and forth between their operational capacities, the authorising environment (which includes the political sphere as well as civil society) and the public value they aim to create. Moore's triangle provides a way of broadening the debate about public data practices to discuss how the decisions that are taken within local government on a managerial level are embedded within their operational capacities and their practices of legitimation.

In the first section, we describe how the DEDA works and how it gives us insight into the current data practices of Dutch local governments. The second section introduces Mark H. Moore's strategic triangle as a lens through which we are able to map the relations between legitimation, operational capacities and public values as they appear in the data project assessments. We use data obtained through our DEDA workshops to show how public value creation, operational capacities and the authorising environment interrelate when data projects are set up in local government. The aim is to understand how data practices affect our understanding of

[2] For more information about DEDA, See Utrecht Data School, DEDA: <https://dataschool.nl/deda/?lang=en>

citizenship and democracy, and how they transform government organisations and their practices. Our chapter shows that the Data Ethics Decision Aid is an effective way of immersing deeply into the local government sector and collecting rich data on the organisations' data projects, their operational capacities, and how they address ethical issues and value questions. By introducing our approach, we hope to provide a new perspective for critical data studies, which focuses on the practice rather than the theory of doing data ethics. This provides empirical richness to the data justice debate. It also provides a more nuanced perspective on the widely heterogeneous data practices and the responses to the challenges raised by them. It also allows for identifying possibilities for intervention that has the potential of lasting social impact, rather than maintaining an analytical distance and merely commenting on technological and social transformation. From our analysis we draw conclusions on the ways in which ethical deliberation sometimes fails, and how the political sphere and civil society can sometimes be excluded from decision making surrounding data projects.

Method: Participatory Observation with DEDA

The Data Ethics Decision Aid (DEDA) was developed by the Utrecht Data School (UDS) at Utrecht University.[3] In 2016, Aline Franzke, Mirko Tobias Schäfer, and a group of Applied Ethics students collaborated with data analysts, project managers, the data protection officer and policy advisors from the City of Utrecht to develop a process for reviewing data projects in view of ethical issues. This resulted in the first version of DEDA, which since then has undergone several revisions and updates.[4] At its foundation lies a broad understanding of data ethics, as phrased by Luciano Floridi: "the branch of ethics that studies and evaluates moral problems related to data [..], algorithms [...] and corresponding practices" (Floridi & Taddeo, 2016). DEDA actively contributes to increasing awareness of

[3] Utrecht Data School is a research and teaching platform investigating the impact of datafication on citizenship and democracy. The researchers look specifically at how datafication affects public management, transforms the public sphere and manifests in public space. Insights are gathered through research projects with external partners being active in either one or more of these areas.

[4] See Utrecht Data School, DEDA: <https://dataschool.nl/deda/?lang=en>

data ethics and how data practices can carry and transform values.[5] These types of values[6] can be organisational, individual, public and anything in between. It helps participants to recognise values embedded in the design of the project or values that could be affected by the project. They reflect how their own actions, the policies of their organisation, and regulation affect both their project design and its impact on various stakeholders. The dialogical process reveals a great deal of unexpected and easily overlooked issues that could not have been tackled when checking the boxes of a guideline or the many AI and data ethics manifestos listing broad ethical principles.[7] Through this process, our research yields a direct impact even before we have analysed the data from our observation.[8] (Image 1)

The purpose of the workshop is, therefore, twofold. First, the workshops function as a way of raising ethical awareness among participants, supporting them in identifying ethical concerns within their data projects and facilitating the documentation of ethical decision making. Second, the DEDA is a research tool, through which we collect data for our research. It offers us a point of entry into Dutch governmental organisations and provides an opportunity to study data practices and the implications of datafication first-hand.

The workshop is requested by (local) government organisations, private companies or other organisations. In the case of municipalities, the request is most often motivated by a data project that is scheduled to start,

[5] Participants report their knowledge of data ethics through a brief questionnaire before and after a DEDA-workshop; this provides information about the learning impact of DEDA workshops.

[6] There is no foundational, guiding theoretical conception of values supporting DEDA, merely a common-sense understanding of values being fundamental beliefs that guide action.

[7] For an overview of AI and data ethics manifests and guidelines see the excellent inventory at Algorithm Watch: <https://algorithmwatch.org/en/project/ai-ethics-guidelines-global-inventory/>

[8] The impact also manifests in the adoption of DEDA in the field of public management in the Netherlands. The Association of Dutch Municipalities (VNG) has integrated DEDA into their Data Awareness Day hosted regularly for municipalities and hired a designated DEDA advisor; the consulting firm Verdonck Kloosters & Associates holds a license to use DEDA and carry out assessments, and DEDA has been covered frequently in professional publications in the sector of public management. See: VNG Magazine, DEDA geeft zicht op ethische kant van data, 22.10.2018, <https://www.vngrealisatie.nl/nieuws/deda-geeft-zicht-op-de-ethische-kant-van-data> or iBestuur, De vraag is: wat willen we met data?, 6.2.2019 <https://ibestuur.nl/praktijk/de-vraag-is-wat-willen-we-met-data> or Binnenlands Bestuur, Utrecht blij met ethisch beslissingsmodel data-projecten, 18.5.2017, <https://www.viag.nl/nieuws/2017/05/18/utrecht-blij-met-ethisch-beslissingsmodel-data-projecten>

Image 1 Assessing a data project with DEDA

which prompts organisations to call for a workshop to identify the ethical issues within their data project.[9] The participants consist of the various employees involved in the data project, often accompanied by the organisation's data protection officer. Policy advisors and domain experts related to the project also participate in the workshop. The role of UDS in these sessions as moderators is to guide the group in such a way so that all relevant topics will be discussed, and to ensure that everyone involved has a say. The moderators do not play a normative role in the process but a facilitating one, in which they merely ensure the process is carried out correctly and responsibly and in such a way that the participants document their process for later reflection and accountability. The moderators from UDS who are present during the workshop observe and take fieldnotes, as

[9]This workshop does not replace the legally required Data Protection Impact Assessment (DPIA): https://autoriteitpersoonsgegevens.nl/nl/zelf-doen/data-protection-impact-assessment-dpia

well as provide an expert opinion within the workshop process by helping the participants recognise ethical issues and values embedded in the project.

DEDA serves as an 'anthropological vehicle' to immerse ourselves into organisations not merely as researchers but as credible experts who gain privileged insight (Schäfer & van Es, 2017). This manner of doing research is informed by methodological approaches in communication and culture studies (Jahoda et al., 1975), anthropology (Malinowski, 2002/1922), and science and technology studies (Latour & Woolgar, 1979). With their groundbreaking study *Die Arbeitslosen von Marienthal*, Jahoda, Zeisel and Lazarsfeld set an example for the researcher's immersion into research object's domain, gaining trust and developing novel means of data collection, while simultaneously making an effort to improve the situation within the domain being studied. Malinovski is best known for shaping the anthropologists' imperative to follow the native point of view. In his inquiry into the social dimensions of trade in the Southern Pacific, he actually revealed—just as Woolgar and Latour have done with their participatory observation laboratories—how technology or artefacts affect and shape social relations (Malinowski, 2002/1922). Using ideas from actor-network theory (ANT) (*follow the actors!*) can be very powerful because it allows us to analyse the power relations between actors, both human and non-human.

Because a workshop using the DEDA is an educational exercise first and a research tool second, it can be approached as a kind of participatory observation. We involve ourselves in the practices of municipalities with the participants of our research.[10] The DEDA workshops can also be seen as focus groups for our research, giving us insight into the concerns of their organisation (Krueger & Casey, 2009). However, we did not carefully select the participants of the workshops like researchers usually do for focus groups, we ask the organisation to compose a group and to include

[10] Our research with DEDA relates to participatory observation rather than to participatory action research (PAR). Kemmis et al. (2013) name two core features of action research. First, the researcher recognises the capacity of people living and working in particular settings to participate actively in all aspects of the research process. Second, research conducted by participants is oriented towards making improvements in practices and their settings by the participants themselves (Kemmis et al., 2013, p. 4). The DEDA as a tool is designed to fit these criteria, the participants of our workshop set the agenda and are involved in all aspects of enhancing their data practices. However, in our research with DEDA, the participants of the workshops are not actively participating in the research. They are still research subjects, not participants.

participants from different departments to provide for a diversity of expertise during the workshop. We take the diversity of the group into account in our fieldnotes and in our analysis.

We borrow from these methods, but do not fully subscribe to any of these. Our method is distinctive from these in several aspects:

- Our role is not limited to being researcher-observers, but also being expert participants. The DEDA and the researchers in their role as experts have a direct impact on the organisation. It is not merely observing but also actively participating in the organisation's efforts to develop responsible data practices.
- DEDA is the 'vehicle' that grants us access to governmental organisations, and also funds our research, as we charge for our workshops. The latter emphasises our role as experts. Being experts, actively shaping and affecting data processes at societal organisations, makes us complicit but also provides more insight than being merely an observing researcher. It also allows for more effective impact within the organisations.
- By contrast to other forms of field research, we do not observe one site or one specific group for a longer period. With our workshops we see many different projects and organisations at a similar point in a process, during the start or early development of a data project. Because of this we are able to observe similar moments of reflection within many different organisations and relate those moments to many different projects. We do not know how the organisation carried out the findings and the decisions made during the workshop. However, we do get to distinguish different trends because of the variety and quantity of projects and organisations where these workshops take place. Furthermore, this method allows us to discern similarities and differences across a range of governmental organisations.

Insights from workshops are collected by taking fieldnotes. We have been collecting observations using the DEDA since 2016 and the structuring of these observations has seen many revisions. We have carried out workshops with over sixty organisations. For the purpose of this chapter, we have used our analysis of our fieldnotes from eleven of these workshops. There are always at least two researchers from UDS present, so the workshop is well-moderated while the researchers also have time to make fieldnotes. Though we do not have any explicit topics that guide these fieldnotes and try to make notes in a very 'open' way, we are guided by our prior experiences and informed by theory. During the taking of fieldnotes, we try to make note of what participants say, how they justify their actions, what explicit and implicit moral statements are made and what kind of project they are working on. We try to take into account the nature of the organisation (when looking at their explicit and implicit values), the backgrounds of the participants (role in the organisation, skills, and the views they express towards the project, the organisation, and data issues in general) and the group dynamic (how participants interact with each other). After the workshop, the researchers discuss the fieldnotes they made. During this discussion, everything that was written down is documented and provided with the necessary information on the context of the fieldnotes. After our discussion we document our notes in Nvivo, where we do qualitative, open coding to organise our findings into different subjects. All fieldnotes have been coded by more than one person. Through extensive coding, in multiple sessions over the course of a few years we have created a 'short list' of recurrent themes. Next to open coding, our coding process has been informed by the three angles of Moore's strategic triangle: operational capacities, authorising environment and public value outcomes.

Analysis: Moore's Triangle Made Tangible

Mark Moore designed the strategic triangle in his book *Creating Public Value* (Moore, 1995). The triangle is a way for those who govern to think about how public value can be created. Moore has a broad conception of public value. He argues that "the aim of managerial work in the public sector is to create *public* value just as the aim of managerial work in the private sector is to create *private* value" (Moore, 1995: 28, italics in original). He conceptualises public value as value for society, produced by public resources, which can both be "collective things that are individually desired" as well as "political aspirations that attach to aggregate social conditions" (Moore, 1995: 52). The first would concern products that cannot be provided through market mechanisms, the second would concern the proper distribution of wealth, rights and so on (Image 2).

Moore argues that in order to create public value, public managers need to consider which values need to be created, but at the same time consider their operational capacities, which involve finance as well as the

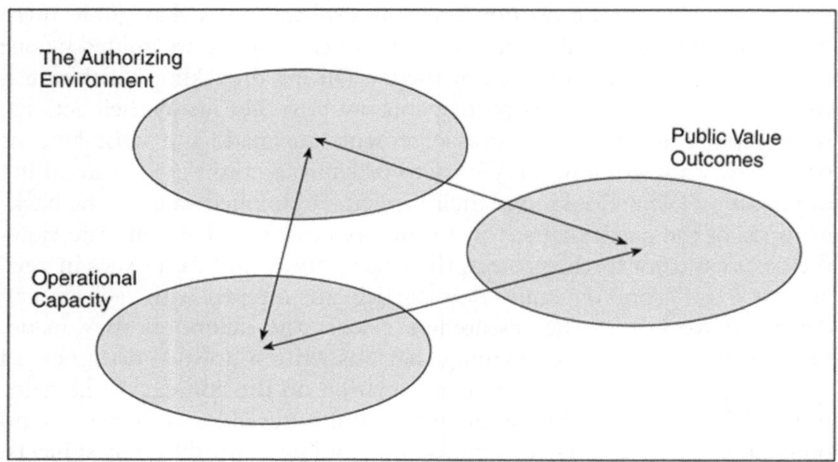

Image 2 the strategic triangle of public value, in: John Benington and Mark H. Moore, "Public Value in Complex and Changing Times", Public Value: Theory and Practice 1 (2011), 5

knowledge and expertise present in the organisation and the authorising environment, where the mandate for creating public value emerges. Public value, operational capacities and the authorising environment are three angles of a triangle that describe the strategy of creating public value. The three angles are seen as three processes that need to be in alignment in order to create public value, that is, these three angles of creating public value are interrelated and should be considered in an iterative process. Which value can be created depends on the operational capacities, but a public manager should also consider how capable her organisation is in the creation of public values and change their capacities accordingly. Operational capacities can be enhanced or developed. This can, however, depend on the other two angles. Creating public values will involve moving back and forth between the three angles while making trade-offs and renegotiations along the way (Benington & Moore, 2011; Moore, 1995).

The DEDA relates to Moore's triangle in two ways. First, it borrows from Moore's insights, and is designed in such a way that it makes participants think about their personal and organisational values, the goals they want to achieve with a project while asking questions about the means they have to achieve the project and how they can authorise it. Second, our research with the DEDA helps us gain insight into the practice of the three angles. We see how civil servants think about public value and which outcomes they want to achieve. We also see what they are capable of, the means they have and the means they lack. Finally, we see how they approach the authorising environment, how they try to gain legitimacy, communicate about their project to politicians as well as the broader public and their conceptions of transparency. Most importantly, we see how these three aspects influence one another: how the iterative process of the triangle is performed in practice, not only within a single organisation, but throughout various governmental organisations.

Operational Capacities

Moore describes the operational resources that the organisation has at their disposal as "finance, staff, skills, technology" (Benington & Moore, 2011: 4). Of the three angles of the strategic triangle, the operational capacity is the most straightforward one. Moore mostly uses it as a way of holding managers accountable for their use of resources (mostly funds), as well as for the creation of "adaptable and flexible organizations as well as controllable and efficient ones" (Moore, 1995). A public manager is responsible not just for taking their operational capacities into account, but also for managing their organisation in such a way that they have the operational capacity to create public value outcomes. Moving between the authorising environment to gain support and, as a consequence, possibly better operational resources is an important element of creating public value. During the DEDA workshops, we have gained insight into the operational capacities of municipalities and how this relates to their data practices.

Data literacy is one kind of operational capacity that we could study with DEDA. During our research, we take note of the organisation's commitment to data-driven practices and evaluate the data literacy of the project's participants. This differs very much across different organisations as well as within organisations. Some have more experienced civil servants than others. Some have hired data scientists, others have trained their own personnel in data skills. We have also encountered many government organisations who lack staff with experience in data research. Data literacy was easy to gain insight into, because the data literacy of the participants directly influenced the quality of the ethical discussion. One workshop at a large municipality showed that discussion flowed easily due to the presence of people with both data skills and domain knowledge [FN13/02/2019],[11] while a workshop at a different large municipality showed how participants were clearly limited by their lack of data skills, for instance, by not understanding the concepts of anonymisation, pseudonymisation and generalisation [FN15/10/2018].

Shortage of expertise also becomes evident when participants discuss who is responsible for the data project. In a lot of municipalities, it is clear that an alderman—or woman—is ultimately responsible for the data project, but there are concerns that they do not have the necessary expertise about the project to be aware of all of its complexities. In one municipality a participant said: "*I believe the alderwoman had never been sufficiently informed when she had to make her decision* [about the project]"[12] [FN01/04/2019]. Different versions of this statement were repeated during many of the DEDA workshops we moderated, in different organisations.

It can also be the case that there are experts, but that they are not involved in the project. In one workshop with a large municipality it seemed to be the case that the person with the necessary expertise was excluded from the project design process. This was the FG (*functionaris gegevensbescherming*), who should be responsible for thinking about issues of privacy on any data project. When we talked to this person after the workshop, it turned out that he had in no way been included in the project. His attendance at the workshop was the first time he heard about it. He told us that he thought the design of the workshop was unacceptable and he would never approve of it [FN13/12/2018]. This is an example where the person with the required expertise had no responsibility whatsoever for the project. Even when expertise is present in an organisation, it can be the case that it is not harnessed.

Data literacy can be low among civil servants and politicians and when the necessary expertise is present, a municipality may lack the structures necessary to make use of the expertise. As a consequence of this lack of data literacy, many municipalities lean on the expertise of external parties,

[11] In this chapter we refer to illustrating examples with a mix of observations and English translations of quotations using [FN#DATEOFWORKSHOP] as a reference. When English translations of quotations are used, the original Dutch quotations can be found in a footnote. For anonymisation reasons we refer to the relevant parties as small, medium or large governmental organs

[12] Original in Dutch: "Volgens mij is de wethouder nooit goed geïnformeerd geweest toen zij daar een besluit over moest nemen."

like consultants or developers. For this reason, consultants find themselves in positions of power. They can often negotiate to be part or sole owner of the data being used. A participant of a large municipality observed: "*We do not have the required expertise, so at the moment we are dependent on external parties.*"[13] The participant indicated that being dependent on external parties was something undesirable and something that they wanted to change in the future [FN15/10/2018]. During one workshop with a large municipality, we noticed that participants themselves could not explain why specific techniques had to be used for that data project. The use of these techniques was proposed by an external party hired by the municipality. This external party had convinced the municipality that these methods would be the best for achieving a specific goal. However, no one from the municipality could explain exactly why these methods were necessary or proportional. The justification for this was just that the external party thought these methods were the best [FN13/12/2018].

In this last case, the civil servants' lack of expertise and the involvement of an external party made ethical deliberation about the project almost impossible, precluding any possibility of transparency and justifying the project to the public. Here we see how the operational capacities of local government influence the authorising environment and public value creation. It can become harder to be transparent about a project when an external party is involved, or when those communicating about the project to the public are not familiar with its technical details. It also makes it impossible to think about the public values that are at stake in the project.

Authorising Environment

Moore's triangle involves an angle called 'authorising environment', where creators of public value have to find support and legitimacy for executing their plans. Building legitimacy and support from the public is essential for creating public value outcomes. It is achieved by "building and sustaining a coalition of stakeholders from the public, private and third sectors" (Benington & Moore, 2011: 4). This involves the support and mandate of elected politicians, but may also include authority from other parties, be it individuals, stakeholders or other organisations.

[13] All quotes have been translated from Dutch. We will provide the original Dutch quotes in the footnotes. Original: "We hebben de expertise niet in huis dus op dit moment zijn we afhankelijk van externe partijen."

Using the DEDA gives us insight into some aspects of the authorising environment of local government in the Netherlands. Data projects often involve issues of mandate. We mostly work with civil servants who are not responsible for the political aspects of the projects. They further develop policy as decided upon by the political institutions of the municipality. However, it is not the case that civil servants are not involved in the process of authorisation. During our research, we have seen several ways in which this involvement takes shape. With data projects in particular, it can be hard to distinguish the ethical and political aspects of projects from mere practical issues. It seems like civil servants only have to make some technical choices: which data to use, where to store the data, which data to store, how to design the algorithm and so on. It is only when discussing these questions explicitly and elaborately, that most civil servants realise that each of these questions is political.

One illustrative example comes from a workshop we moderated in September 2018 at a large government organisation. The data project was about modernising a government website. The project was initially not regarded as controversial or ethically complex. There were two options for the project: either personalising the website so that every citizen would see a different version, tailored to his or her needs, or keeping the website non-personalised, so that every visitor sees the same page and the same government information. The participants quickly related this discussion to the public value of equality and equal access to government information. Personalising the government website would mean that citizens would no longer have the exact same access to government information. And who decides what an individual sees and what he/she does not get to see? How can this process be made transparent and accountable? The participants of the workshop concluded that they did not have the mandate to make such decisions and that this discussion should be held within the political sphere by those with a political mandate [FN26/09/2018]. Because data projects often seem straightforward and 'value-free', civil servants can overlook the politically sensitive aspects of the project and make decisions that should be discussed in the political sphere.

After this realisation of how many political aspects are involved, the civil servants need to act on it. They may decide that they can estimate the political intentions of their assignment and translate that into the shaping of the project. They may decide that they do not have the authority to make these decisions and make sure the political council discusses them. Civil servants often have to decide in this way what is discussed as a political consideration and, therefore, what gets included in the authorising environment and what does not. Recognising when civil servants have a mandate and when they do not, is related to their data literacy and ethical awareness. It can be only after extensive discussion during a workshop that participants realise that they do not have the necessary mandate to make this decision.

Another thing we have noticed concerning the authorising environment is that for many data projects, responsibilities are dispersed and unclear. Somewhere along the way, from the commencement of the data project and the presentation of the final results or visualisation, responsibility gaps emerge. It is often unclear who is responsible for the different stages of the data project. For example, in a large municipality a participant said: "*We have not distributed the roles well yet within our organisation, so often the different responsibilities within a data project end up with the same person.*"[14] The speaker indicated that this happened because only one person took on these responsibilities [FN03/04/2019]. Questions concerning responsibility include issues regarding the responsibilities of the Data Privacy Officer (DPO) or *Functionaris Gegevensbescherming* (FG). In a workshop for a large municipality they noted that the DPO had only approved of a single project using a Data Privacy Impact Assessment (DPIA) because the DPO was not included in every project, plus it was not clear to them how to conduct a DPIA [FN01/04/2019]. In these cases, there is an awareness of the ethical aspects, but it is unclear how and where and with whom these should be discussed. Participants do not know what the authorising environment should include, and how they should give it shape.

Both examples show a relation between the operational capacities and the authorising environment. If data literacy and ethical awareness are absent in the organisation, it is impossible for the civil servants to recognise ethical and politically sensitive issues and make sure they are discussed in the political sphere. If the organisation is not well prepared to embed the responsibilities that come with conducting a data project in the organisation, responsibility gaps emerge and it is unclear who is accountable for the project. Because data projects are still relatively new and disruptive, it

[14] Originally in Dutch: "*Wij hebben de rollen binnen onze organisatie nog niet zo goed verdeeld, dus vaak komen de verantwoordelijkheden binnen dataprojecten bij dezelfde persoon terecht.*"

is unclear what gets included in the authorising environment, and when and how the project should be discussed by politicians and civil society.

Public Value Outcomes

Moore describes creating public value as the aim of managerial work in the public sector. We have seen how civil servants' perceptions of public value outcomes is dependent on the operational capacities of the organisation (data literacy) and its authorising environment (clear mandate and responsibilities). During our workshops, we have noticed two other interesting aspects of public value creation in local government. The first involves the civil servant's capacity to think about public value outcomes. In general, we see that civil servants tend to focus on the public good. This became especially apparent when moderating a workshop in October 2018 in a small municipality. A relatively large percentage of their population struggled with debt and financial issues. They wanted to start a data project, therefore, that helped them strengthen their poverty prevention policies. This data project would consist of an algorithm that could identify individual citizens with a high risk of finding themselves in financial trouble in the future. The municipality would then offer help to these people so that their situation would not worsen, ultimately preventing them from situations such as bankruptcy. During the entire session, which lasted three hours, participants would regularly ask questions like: "*When do we make the citizen feel happy and glad?*".[15] They were very aware of how the municipality could come across: showing up at someone's door unannounced and offering help can also be experienced as extremely patronising and invasive. The citizens' point of view was always at the forefront of the discussion. This was reflected in the goal of the project, which the participants agreed upon at the very start of the session: enhancing citizens' wellbeing. We observed (among other things) implicit value expressions when they expressed the need for being non-discriminatory towards citizens: "You have to be able to connect with people without labels or judgement and make them a non-committal offer"[16] [FN02/10/2018].

We have seen that overall, civil servants are very well-equipped to ethically justify their decisions. We have used the DEDA mostly with

[15] Original Dutch quotation: "Wanneer wordt de burger gelukkig en blij?"

[16] Original Dutch quotation: "Je moet zonder label of oordeel bij mensen binnen komen en ze een vrijblijvend aanbod kunnen doen."

municipalities or other forms of local government, but have done a few workshops within commercial organisations as well. What we learned was that by comparison, civil servants have a mindset for thinking about public value. The public interest is the main driver of their activities. We never consciously noticed how well civil servants can deliberate on public value until we saw how different this was for employees of commercial organisations. In our (limited) experience, participants from commercial organisations have trouble thinking about the ethical implications of data projects and the consequences of their projects for others in general. The maximisation of profit, instead of the public good, was at the centre of most discussions with these commercial organisations. During a workshop with a large commercial company in spring 2018, this point was specifically illustrated by one participant who said: "*all this talk of clients, let's just pretend for the sake of the efficiency of this data project that the clients do not matter.*"[17] Another participant during the same workshop could only think of one value that mattered within their organisation besides profit: maintaining a good reputation [FN02/02/2018]. For civil servants, however, thinking about the citizen first was the norm during most workshops.

Though the ethical awareness of civil servants was generally very high, we have seen that ethical deliberation about data practices can be seen as an obnoxious box that needs to be ticked. This is the second interesting aspect of public value creation we noticed during the DEDA workshops. Participants sometimes seem to think that by doing a DEDA workshop, they have taken care of ethical considerations. DEDA itself is then used as a means to wash their hands of ethical concerns. According to Elletra Bietti, 'ethics washing' occurs when an organisation makes an effort to self-regulate their ethical choices, with no need to involve other societal or political influences. For her, the biggest problem with ethics washing is that it can narrow the scope of the debate and, though it can have a good outcome in some questions of procedural fairness, distracts society from addressing structural problems with the technology (Bietti, 2020). Our workshops can be seen by organisations as a way for them to self-regulate the ethical choices involved in their projects. We try to make clear that DEDA can help point out where ethical problems occur, but cannot replace a political or societal discussion that needs to be had about these issues. Mostly this message hits home. However, we have seen that when

[17] Original Dutch quotation: "Al dit gepraat over klanten, laten we even omwille van de efficiëntie van dit data project even doen alsof the klanten er niet toe doen."

our workshop is seen as a '*moetje*',[18] it narrows the focus of the discussion. It takes more effort to get the participants to consider the broader questions about the project, most importantly whether or not the project should be launched in the first place. In these cases, the participants use the workshop to keep the ethical issues of the project out of the authorising environment.

In a few instances, the DEDA was seen as an 'ethical assessment' that could provide a green light for a data project. To give an example, the city council of a big municipality decreed that every data project has to undergo an ethical assessment of sorts, besides the (Data) Privacy Impact Assessment that was already mandatory by law. This led to a DEDA workshop for multiple projects, where project managers stated that it felt like a '*have-to*', a process that they themselves did not choose, but had to do because of decisions from higher up [FN08/01/2019] & [FN13/02/2019]. Some participants even expressed a wish for DEDA to be more like a checklist [FN15/10/2018]. In another workshop it became clear that some participants felt that as long as they walked through the DEDA poster and answered the questions, their projects would be ethically sound. They then tried to use ethical issues that arose during a workshop as indicators on how to best change the narrative of their project. For example, by reframing their project as a pilot, or wanting to break the ethical rules in order to find out where the limits are [FN08/01/2019]. In our role as advisors, we think this is problematic, and have always tried to prevent the DEDA from being used in such a way. We do this by making sure participants understand that we can help them to have the discussion about the ethics of their project but cannot tell them what decisions to make nor tell them or others that what they are doing is right. Referring to the concept of the honest broker introduced by Roger Pielke, we understand our role not as activists, consultants, or advocates, but as researchers/experts who merely point to the range of options available to policy makers (Pielke Jr, 2007). However, some workshops still ended with municipalities asking us questions concerning the further implementations of their results [FN08/01/2019], [FN13/02/2019] and [FN15/10/2018]. This is a moment where we have to emphasise that we are not responsible for decisions about the implementation of the results of their discussion. In all of these examples, ethical deliberation was not used to think about how to

[18] A 'have-to', named as such by one of our participants [FN08/01/2019]& [FN13/02/2019].

safeguard public values in data projects, but as a way of preventing further discussion by 'checking the ethics box'.

We have seen that ethical aspects of data projects are getting more attention than some years ago. There is a growing demand for tools that help to take ethics into consideration and there is more attention being paid to ethical issues concerning data projects in the public debate. However, we have also seen that along with this development, ethical assessments have become a part of the authorising environment, that can be mobilised to legitimise a project and create a narrative to gain public support for the project. We are not the first to point out that 'ethics' and 'public good' are in themselves not neutral terms but terms that are mobilised in the discussion around emerging technologies (Washington & Kuo, 2020). We need to take care and be critical of ways in which the DEDA itself can be mobilised. However, our insight into the practices of local government also gives us special insight into how this mobilisation of ethics works. Ethical deliberation is in these cases not integrated into the entire process of developing a project but is added on at quite a late stage during the development of the project [FN08/01/2019] and [FN13/02/2019]. At this stage it is very difficult to change the design of the project, the only thing that can be changed is the narrative in which the project will be presented. This kind of thinking precludes ethical deliberation about public value creation.

Conclusion

Working with the DEDA provides unique access to organisations and privileged insights into the ways they use data practices to meet their policy objectives. Our research makes explicit how organisations are challenged in applying new technologies while constituting legitimacy and safeguarding public values. It also highlights the dynamics between the three pillars of Moore's triangle: operational capacity, authorising environment and public value outcomes. As researchers we have a front row seat to the inner-organisational dynamics that unfold with the application of novel data practices. We also learn how they affect our understanding of citizenship and democracy as they transform public management processes, and capture citizens as data subjects. As such, datafication and data projects can carry or transform public values.

In this chapter, we have shown how DEDA makes Moore's triangle tangible. What we have tried to show is that public values are deeply affected by data projects and thoroughly interwoven with the operational capacities of an organisation and its authorising environments. First, in regard to operational capacity, we see that there can be significant limitations to data literacy and ethical awareness in Dutch municipalities. There also seems to be a strong correlation between these two. When public servants lack data literacy they are unable to recognise the ethical issues they will invariably exist in their data projects. This lack of expertise also causes organisations to rely on external partners. The need to rely on an external partner can affect the ability of an organisation to be transparent about their data project, which in turn affects the authorising environment. This relationship between governments and external partners raises further questions on how the former's values are affected by the collaboration with the latter. This highlights the tension between the (lack of) operational capacity and the expression of public values.

Second, the authorising environments in which data projects are situated are the same politically charged environments in which any governmental project is situated. The DEDA workshops show that data is also political, which in itself is not a novel conclusion. What *is* a novel insight, and one that also illustrates a tension between operational capacity and authorising environments, is that a lack of expertise can cause actors in this field to overlook the political aspects of their data projects, which can result in responsibility gaps. With the development of data projects, it is often unclear what the political mandate covers and it is, therefore also unclear how civil servants should approach the authorising environment and shape it. In our observations we saw that the aldermen can be poorly informed, sometimes even questioning the *raison d'être* of the entire data project while not understanding the ins and outs of it. This forces some of the political and ethical decisions into the hands of (back-end) data scientists with no political mandate.

Third, we have seen that among the civil servants who took part in our workshops, there can be a tendency to see ethical deliberation as a '*moetje*', which undermines the ethical discussion among civil servants, especially when the ethical deliberation is involved at a very late stage in the project. This mindset also prevents the ethical discussion being held in the authorising environment, or can even lead to the ethical assessment being mobilised to argue that the discussion about a data project does not need to be held with politicians or stakeholders.

Our purpose in this chapter was twofold. First, we hope to have shown that this kind of research can be a very fruitful way to gain new perspectives in both data practices and the practice of public value creation. Second, we have shown some instances of how civil servants relate to their own role as public value creators and how data practices complicate this role. Further research into how DEDA functions as a tool is needed. Due to the snapshot nature of a DEDA workshop, our experiences with and the results above are based on this one moment in time. We hope to be able to carry out further research into the long-term effects of ethical deliberation through DEDA for government organisations.

By investigating the data practices of Dutch local governments with DEDA, we have been able to gain insight into the practical context in which ethical problems of data practices arise. We can tentatively make some suggestions on how to improve ethical decision making in local government. Higher data literacy can likely increase ethical awareness and deliberation about data projects. Both because it makes ethical deliberation within the organisation possible, and because it pre-empts the need for external partners, who make open ethical deliberations more difficult. Ethical awareness would also benefit from a better internal structure in organisations such as municipalities, so that they know how to divide responsibilities and apportion accountabilty for data projects.

Acknowledgements The authors are indebted to the participants of sixty-three DEDA workshops, and the municipalities who trusted our approach and allowed us to take a detailed look at their projects and their data practices. We are also grateful to our students and interns who supported our research effort. Credit is due to Martin Jansen, manager for data-driven management at City of Utrecht, who initially commissioned the research project that led to the development of DEDA, and to Aline Franzke, who was essential in mapping the data practices at City of Utrecht and utilising it for developing DEDA. We would like to thank Marjolein Krijgsman for helpful comments on an early version of this chapter. We would also like to thank Andreas Hepp and Juliane Jarke for their valuable review of the chapter.

This research project was made possible with the support of the Utrecht Data School at Utrecht University.

References

Bannister, F., & Connolly, R. (2014). ICT, public values and transformative government: A framework and programme for research. *Government Information Quarterly, 31*(1), 119–128. https://doi.org/10.1016/j.giq.2013.06.002

Benington, J., & Moore, M. H. (2011). Public value in complex and changing times. *Public value: Theory and practice, 1.*

Bertot, J. C., Jaeger, P. T., & Grimes, J. M. (2010). Using ICTs to create a culture of transparency: E-government and social media as openness and anti-corruption tools for societies. *Government Information Quarterly, 27*(3), 264–271. https://doi.org/10.1016/j.giq.2010.03.001

Bietti, E. (2020). From ethics washing to ethics bashing: A view on tech ethics from within moral philosophy. *Proceedings of the 2020 Conference on Fairness, Accountability, and Transparency*, 210–219. https://doi.org/10.1145/3351095.3372860.

Eubanks, V. (2018). *Automating inequality: How high-tech tools profile, police, and punish the poor.* St. Martin's Publishing Group.

Floridi, L., & Taddeo, M. (2016). What is data ethics? *Philosophical Transactions of the Royal Society A: Mathematical, Physical and Engineering Sciences, 374*(2083), 1. https://doi.org/10.1098/rsta.2016.0360

Hintz, A., Dencik, L., & Wahl-Jorgensen, K. (2019). *Digital citizenship in a Datafied society.* John Wiley & Sons.

Jahoda, M., Lazarsfeld, P. F., & Zeisel, H. (1975). *Die Arbeitslosen von Marienthal. Ein soziographischer Versuch über die Wirkungen langandauernder Arbeitslosigkeit.* Suhrkamp.

Kemmis, S., McTaggart, R., & Nixon, R. (2013). *The action research planner: Doing critical participatory action research.* Springer Science & Business Media.

Krueger, R. A., & Casey, M. A. (2009). *Focus groups: A Practical Guide for Applied Research.* SAGE.

Latour, B., & Woolgar, S. (1979). *Laboratory life: The construction of scientific facts.* Princeton University Press.

Maarse, J. A. M. H., & Jeurissen, P. P. P. (2016). The policy and politics of the 2015 long-term care reform in the Netherlands. *Health Policy (Amsterdam, Netherlands), 120*(3), 241–245. https://doi.org/10.1016/j.healthpol.2016.01.014

Malinowski, B. (2002/1922). *Argonauts of the Western Pacific: An account of native Enterprise and adventure in the Archipelagoes of Melanesian New Guinea.* Routledge.

Moore, M. H. (1995). *Creating public value: Strategic Management in Government.* Harvard University Press.

O'Neil, C. (2016). *Weapons of math destruction: How big data increases inequality and threatens democracy.* Crown.

Pielke, R. A., Jr. (2007). *The honest broker: Making sense of science in policy and politics.* Cambridge University Press.

Redactie Gemeente.nu. (2016, april 28). Gemeente mag mensen volgen via afvalpas. *Gemeente.nu.* https://www.gemeente.nu/bedrijfsvoering/privacy/gemeente-mag-mensen-volgen-via-afvalpas/

Redactie Nieuws Digitale Overheid.nl. (2020, februari 6). Rechtbank: SyRI-wetgeving in strijd met recht op privacy [Nieuwsbericht]. *Digitaleoverheid.nl.* https://www.digitaleoverheid...nl/nieuws/rechtbank-syri-wetgeving-in-strijd-met-recht-op-privacy/

Schäfer, M. T., & van Es, K. V. (2017). Introduction: New Brave World. In M. T. Schäfer & K. V. van Es, *The datafied society: Studying culture through data*. Amsterdam University Press.

van Ark, T. (2018, juni 8). *Kamerbrief motie met betrekking tot Systeem Risico Indicatie (SyRI)—Kamerstuk—Rijksoverheid.nl* [Kamerstuk]. Ministerie van Algemene Zaken. https://www.rijksoverheid.nl/documenten/kamerstukken/2018/06/08/kamerbrief-motie-met-betrekking-tot-systeem-risico-indicatie-syri

Vermeulen, W. (2015). *Decentralisation of social policy in the Netherlands.* Decentralization of Education, Health and Social Protection. https://www.cpb.nl/en/publication/decentralization-of-social-policy-in-the-netherlands

Washington, A. L., & Kuo, R. S. (2020). Whose Side are Ethics Codes On? Power, Responsibility and the Social Good. Proceedings of the 2020 Conference on Fairness, Accountability, and Transparency, 230–240. https://doi.org/10.1145/3351095.3372844.

Open Access This chapter is licensed under the terms of the Creative Commons Attribution 4.0 International License (http://creativecommons.org/licenses/by/4.0/), which permits use, sharing, adaptation, distribution and reproduction in any medium or format, as long as you give appropriate credit to the original author(s) and the source, provide a link to the Creative Commons licence and indicate if changes were made.

The images or other third party material in this chapter are included in the chapter's Creative Commons licence, unless indicated otherwise in a credit line to the material. If material is not included in the chapter's Creative Commons licence and your intended use is not permitted by statutory regulation or exceeds the permitted use, you will need to obtain permission directly from the copyright holder.

Welfare Data Society? Critical Evaluation of the Possibilities of Developing Data Infrastructure Literacy from User Data Workshops to Public Service Media

Jenni Hokka

Introduction

Datafication has changed society and the economy in fundamental ways, blurring long-established social and institutional divisions (Constantiou & Kallinikos, 2015). The whole of human life is transmuted into data streams and is in danger of being exposed to continuous tracking either by profit-seeking companies (Couldry & Mejias, 2018) or government agencies (Dencik et al., 2016). Datafication allows companies to predict and even modify human behaviour as a means of producing revenue and gaining market control leading to what Shoshana Zuboff refers to as "surveillance capitalism" (Zuboff, 2015). Numerous scholars have raised concerns in regard to how this situation leads to surveillance and violations of the right

J. Hokka (✉)
Aalto University, Espoo, Finland
e-mail: jenni.hokka@aalto.fi

© The Author(s) 2022

A. Hepp et al. (eds.), *New Perspectives in Critical Data Studies*, Transforming Communications – Studies in Cross-Media Research,
https://doi.org/10.1007/978-3-030-96180-0_12

to privacy and how this may represent a serious threat to democracy and equality (Couldry & Mejias, 2018; van Dijk, 2014; Gangadharan, 2017; Gurumyrthy & Bharthur, 2018; Kennedy & Moss, 2015). It is very difficult, even impossible, for user-citizens to gain full knowledge of the personal data that corporations keep on them. As Zuboff notes, "Surveillance capitalism thrives on the public's ignorance" (2015: 83). However, in a datafied society in which data-intensive logics and practices have penetrated every aspect of human life (Mayer-Schonberger & Cukier, 2013), people have become accustomed to applications that make everyday life more convenient or even offer ways of earning a living. As Mai suggests, it is now virtually impossible to perform daily activities without giving away personal information which is then capitalised upon by either private enterprises, such as data brokers, or used by public organisations (Mai, 2006). The European Union has tried to protect its citizens by establishing the General Data Protection Regulation (GDPR), which went into effect in 2018. A number of earlier research projects (Selwyn & Pangrazio, 2018; Büchi et al., 2017; Park, 2013) have shown how people, on average, have a very weak understanding of exactly how their personal data is collected, linked, used, sold, and re-sold. Furthermore, as personal data is combined into data packages and sold to different parties by data brokers, it is nearly impossible for even a knowledgeable user to comprehend which parties have access to his or her data. As Micheli et al. (2018) suggest, the ability to protect his or her private data and minimise their 'digital footprints' should now be understood as an essential part of digital equality along with digital skills and online access. At present, only people with high levels of computational skills and expertise in data mining have access to data and data analytics tools which means that data power is concentrated within just a few, elite commercial companies such as Google, Facebook, and Amazon (Kennedy & Moss, 2015). Consequently, the GDPR regulation does not help if users do not understand that their data is being tracked and re-sold or how exactly this is done and give their informed consent without knowing what that actually means.

Because of the non-transparent nature of practices such as data mining and user tracking by online applications and platforms, there is, as a number of researchers have suggested, an urgent need to increase the level of digital literacy (Gray et al., 2018; Pybus et al., 2015; Park, 2013). The definition of digital literacy varies a great deal. According to Iordache et al. (2017), the concept of digital literacy most often includes three facets: knowledge, skills, and competence, in which knowledge refers to an understanding of the available digital tools, skills to the practical capabilities to use them and competence to the ability to use the knowledge and skills in different situations. Digital literacy is now seen as a crucial citizenship capability and most European countries have made digital competence a part of basic education, though, not so often as an independent subject but as a cross-curricular theme. Nearly thirty European education systems also mention data and privacy as one aspect of digital competence. For this reason, data privacy and the protection of personal data have recently become an ever more present and valued part of digital literacy (Iordache et al., 2017: 20). Yet, how data and privacy issues are taught in schools varies a great deal between European countries, districts, and even among individual schools depending both on the interpretations of the concept and on teachers' own digital abilities. The practical examples from the European Commission report on digital competence teaching in European education vary from strong passwords to legal issues in sharing information (European Commission/EACEA/Eurydice, 2019: 41–42).

Alternatively, there have been a number of public pleas from both the public and private sectors to develop not just digital literacy but data literacy. Nowadays, many European curricula also mention data literacy, but in basic education data literacy is understood simply as skills "to analyse, compare and critically evaluate the credibility and reliability of sources of data, information and digital content" (European Commission/EACEA/Eurydice, 2019: 38). According to this simplified definition, data literacy could be conceived as part of digital literacy. Yet, to be precise, data literacy overlaps digital literacy to some extent but comprises a different combination of skills and knowledge. A certain level of digital skills is inherently necessary to improve data literacy. However, while digital literacy emphasises general digital skills that are needed for creating, finding, and analysing digital content by using different kinds of software (Iordache et al., 2017), data literacy refers to the technical skills and statistical and informational literacy needed to produce, use, and interpret computational data (Gray et al., 2018). Data literacy is not a skill that can be learned overnight: it is a complicated knowledge framework that also requires statistical literacy, an understanding of the ethics of using data, and the ability to change the tools used according to purpose or discipline (Wolff et al., 2016). Because of this complexity it does not seem probable that data literacy would become a general skillset in the near future

It is certainly reasonable to try to improve a population's digital literacy. Yet, digital literacy skills alone do not contribute to people's understanding of datafication's political, economic, and legislative conditions. As proposed by Gray et al., there is a clear need for data infrastructure literacy that would "not only equip people with data skills and data science but also to cultivate sensibilities for data sociology, data culture, and data politics" (2018: 1). This is a vastly different skill to data literacy, as it requires more political and economic knowledge rather than more technical knowledge such as statistical literacy or coding skills. Data infrastructure literacy is essential in light of data justice (Dencik et al., 2016; Taylor, 2017), as it would help citizens of all ages, genders, and educational backgrounds grasp the full societal effects of datafication and then take a stand on fair conditions in data practices. Only through a more widespread understanding of datafication among the general public do awareness of the significance and vastness of data-gathering in our present societies, political discussion, and democratic decision-making about the conditions and regulation of datafication become possible. In a Nordic welfare society, the state has traditionally played a significant role in promoting equality and democracy among its citizens. By providing universal public services such as low-cost childcare and free education, the Nordic welfare society aims to help people gain skills and abilities that enable them to become full members of society (Holmwood, 2000; Kangas & Kvist, 2013). According to Hänninen et al., the Nordic welfare state has four dimensions: personal autonomy, participation, inclusion, and sustainability. "Autonomy refers to human condition [...] in which she is able to master and manage her own life and decisions. [...] Participation refers to a mode of action which influences people in a common endeavour to change their circumstances. [...] Inclusion refers to a state of circumstances in which all involved are so related that they belong together in such a fashion that they contribute to according to their own capacities. Sustainability refers to complex processes which relate people to each other and balance their relations with the environment helping them face (with precaution) uncertainty and contingency" (2019: 5).

How can welfare state thinking be applied to a datafied society, then? In a welfare data society the citizen should be free to master and manage her/his personal data. S/he should be capable of taking part in decisions on how data gathering practices are organised, regulated, and supervised as they now form one of society's key functions. All citizens' capability to participate in decision-making about the rules of datafication should be ensured through universal education provided by the public sector. Through digital literacy and data infrastructure literacy education, citizens would be more capable of taking precautions to protect their privacy and control the use of their personal data.

Well before the more intense datafication of everyday life Nordic countries have constituted a special model of the "media welfare state" (Syvertsen et al., 2014). In practice, the media welfare state has meant a policy in which all citizens have been granted universal access to education and information so that there exists equal opportunity for the understanding of the society in which they live. The policy has been successful in the sense that even in the present, platformised media environment, Nordic public service companies, such as NRK (Norway), DR (Denmark), and YLE (Finland) all reach a clear majority (roughly 60–90 per cent depending on the way of counting) of the population on a daily basis (Enerhaug, 2019; DR's public-service redegorelse, 2019; Nokela et al., 2019). Public service media have been an essential part of the (media) welfare state, as they should encourage participation and the inclusion of all citizens in the political and cultural public spheres (Syvertsen et al., 2014: 7). Furthermore, public service media have been an essential part of cultural policy that has aimed to diminish the influence of global market forces (ibid.: 25–28). In the previous period of mass media, global market forces have mostly referred to international cable channels and production companies. Now, in the present datafied and platformised media environment, the strongest global market forces are undoubtedly the so-called 'Big Tech' firms such as Google, Amazon, Facebook, Apple, and Microsoft (collectively ***known*** under the acronym *GAFAM*).

In this chapter, I claim that European public service media should take on new responsibilities in light of the datafied and platformised society. It should fulfil its mission by educating people of all ages to increase their general levels of digital literacy and data infrastructure literacy, and in this way, empower citizens to take part in determining how datafication could operate equitably. By ensuring that citizens understand the social and political effects of datafication, PSM could enhance citizens' capability to make informed decisions and protect themselves as users, and more importantly, to form an informed opinion on how data tracking and sharing should be regulated. In the long term, citizens should be able take part in discussing new options that pertain to the present situation in which a user is quite powerless in relation to data mining. In this chapter, I discuss the opportunities for PSM to raise the general level of digital and data infrastructure literacy. Although most other European public service broadcasters (such as ARD/ZDF or France TV) apart from the BBC play a somewhat minor role in their media market than their Nordic counterparts, they too could adopt a new role in providing adult populations both practical, digital literacy skills and a nuanced understanding of the political, societal, and cultural conditions and outcomes of datafication.

I first present the results from our research workshops organised in cooperation with the Finnish Broadcasting Company, YLE, in which participants were both educated on datafication and interviewed about their thoughts and experiences regarding datafication. The results of the workshops reveal the wealth of challenges faced by digital literacy education. In the second part, I describe what kind of educational content YLE already provides for adult citizens related to datafication and digital literacy. I then discuss whether using public service media to increase awareness of datafication and to develop data infrastructure literacy could be one of the essential large-scale practical solutions needed to tackle the imbalanced power structure between social media platforms and users.

Notions of Digital Literacy Based on User Data Workshops

The first part of the collaboration with YLE was to organise workshops with 'average' users who had no special education related to 'big data' such as programming or data-analytics. Methodologically, the research followed an emancipatory and educational action research approach (Carr & Kemmis, 2009). The workshops had several, overlapping features. First, the aim was to discuss with the participants their worries, thoughts, and hopes about datafication, especially in regard to the use of their personal data. Second, we wanted to examine how well participants protected their privacy and how willing they were or capable of doing that. In this way, we wanted to increase qualitative understanding, building on previous research (Büchi et al., 2017; Park, 2013; Kennedy et al., 2017; Selwyn & Pangrazio, 2018; Ruckenstein & Granroth, 2019), on users' perceptions of datafication and their digital capabilities. Third, we provided education on data collection practices and instructions on how to protect privacy online if participants were interested in learning those skills. Fourth, YLE examined what topics and perspectives of datafication interested different audiences. This had the further intention of producing educational media content about datafication to develop digital literacy, and I will discuss this aspect further in the second section. Fifth, we wanted to raise users' general awareness of datafication and data collection practices to increase their level of data infrastructure literacy. Sixth, we wanted to discuss and develop with the participants the possibilities of alternative data regimes and practices by offering them three alternative visions.

There was a total of six workshops which were partly organised in cooperation with the YLE Creative Content Unit and, more specifically, the Head of Development at YLE, Raimo Lång. The first two workshops were more experimental and concentrated more on acquiring knowledge on users' interest in YLE's potential datafication content, but they also included some of the same questions as the last four workshops. In the last four workshops, which were led by our researcher group and constitute the main research material in this section, we had an identical pattern of action. Each of the four workshops had four to seven participants, both female and male, adding to a total of twenty-five participants. Most were young adults of varying educational backgrounds, but one workshop was arranged for people in their seventies. Even though the participants differed in gender, age, and education, their answers and reactions exhibited similar patterns.

During these workshops, users were asked to familiarise themselves step by step with their Google account's privacy settings and the data that Google had collected on them. In addition, the participants were asked to try the Disconnect application, which informs users on how many third parties were gathering data on them through the websites they visited. After each section, the participants were asked to answer related questions on a Google Forms questionnaire. The reason for asking them to answer in a literal form first was to diminish the effect of participants' views on each other. After participants had sent their answers online to us after each section, they were also asked to discuss their answers with us and other participants. The questions concerned their thoughts and feelings regarding the data that Google collected on them and if they now wanted (or did not want) to change their privacy settings and why. There were also questions about their thoughts on the results of the Lightbeam and Disconnect applications that show the number of third-party requests on each site. In the end, the participants were given a more complex question on their vision for data collection practices of platforms, online applications, and websites in the future.

At the beginning of each workshop, we asked participants if they were slightly, very, or not at all concerned about the gathering of private data on the web. Most people were slightly concerned, except for the group of women in their seventies, who were in the main very concerned. During the workshops in which participants were introduced to how their online behaviour was tracked, the participants described a range of feelings, including contradictory ones. Nearly half the participants (eleven) described feeling frightened, concerned, confused, shocked, startled, and even angry upon learning the amount of data that Google or third parties were collecting online. In particular, the data from Google Maps, their Google browsing history and/or the results of the Lightbeam and Disconnect applications seemed to be unpleasant for many participants.

After looking at her browsing history in Google's My Activity section, one participant commented:

> I am slightly frightened, as all the search words that I have used tell something about me and my life situation, and I am concerned where this data can be further transmitted.

A participant in her twenties expressed her concerns on Google Maps:

> It is awful how well Google knows where I have been. Somebody could easily follow my movements through Google. I started feeling insecure.

Another participant in her twenties commented on her Disconnect results:

> Confusingly, many parties follow my every step on the web. There are some parties that, luckily, I am able to prevent from gathering data, but there are way too many parties that I can't prevent from doing that. I would like for the Internet to be the anonymous world that people still describe it as being. This feels like a rough coming back to reality.

A participant in his thirties responded to his Disconnect results by saying:

> This is very confusing! It was not surprising that advertising or data analysis companies were tracking data on popular web sites, but I was really confused to notice that Imgur was in contact with a Russian news site. What should I think about this?

In two answers, the participants stated that they were not concerned on behalf of themselves and their own data but saw the vastness of data-gathering practices as worrying or interesting on a more general, sociopolitical level, especially regarding attempts to influence and/or interfere with elections. One respondent stated a feeling that "there is a risk that at some point, delivering private data may go too far" but felt that the limit had not yet been surpassed. One participant stated that she would have been devastated by this kind of data collection if she had been asked before the present situation, but she was now used to it. In addition, ten respondents had already changed their My Activity settings in some respect to protect their privacy. Two participants stated that they used the incognito setting when browsing, and one had already installed an ad-blocking application.

However, concern and upset were not the only feelings people experienced when looking at their Google data. Nine respondents felt indifferent about at least part of their results, stating that they did not consider this kind of information dangerous, or alternatively, that they were already aware of these data collection practices. It was mainly browsing and location history that these respondents felt were safe to give to Google. Yet it is noteworthy that feelings related to privacy could vary according to the type of information. For example, one participant did not consider location history to be dangerous, but was startled to find that Google had a recording of her voice and that Google was still exchanging data with some third-party applications, such as games, that she had removed from her mobile phone long before: "Why does a mobile game have access to my Google Drive?"

A few respondents underlined the convenience to the user that data collection practices enabled and wanted to continue providing this data in the future as well. Google Maps was seen as especially helpful, for example, when jogging or driving. Many participants considered the location history aspect of Google Maps beneficial because they could see which places or restaurants they had been to, and for many of them the location history acted as a kind of personal diary that stirred pleasant memories. As Mark Andrejevic already stated in 2011, Google has come to be treated much like a public service both by users and institutions. The participants' responses demonstrate how Google Maps especially, along with Google Search, has become a necessary and unquestionable part of everyday life.

The Ads personalisation section in particular generated mixed feelings from the respondents. The results from our workshops are in line with a previous study by Ruckenstein and Granroth (2019). Similar to their interviewees, targeted advertising was, in the first place, the main or even the only feature from which people could observe that their actions online were being tracked. In our workshops, a number of respondents were amused by their list of interests for Ads personalisation either because of its accuracy or non-accuracy, but it could cause strong negative emotions too. Over half (fourteen) of the respondents, even those participants who were concerned about their privacy, wanted to have personalised rather than non-personalised advertising. When seeing their lists, many revised them to better correspond with their interests. Participants were, therefore, voluntarily providing more data to Google and in a more targeted way. It appeared that the users responded emotionally to their list of (commercial) interests as markers of their identity and, because of that, felt a need to make it correspond to their 'real' interests. One participant was very irritated when seeing her presumed list of interests because she had thought that she had changed her settings to prevent personalisation but also because her profile was "generic and erroneous".

Several times during the workshops, participants expressed their surprise about their own Google settings that they thought they had set otherwise or for which they did not recall taking any action. Several respondents were also surprised that some third-party applications they had used still had access to their Google account, and they either did not remember or had not realised they were giving this permission when using their Google account for some third-party applications. In everyday use, people easily forget or are not capable of protecting their privacy even if they would consider that important on a more general level. As Park et al. (2018) have noted, privacy regulations are based on the assumption of a rational user—but for the most part, people do not actively and rationally weigh the consequences of their every step online from a privacy point of view, especially when the online environment is structured to provide all kinds of pleasurable feelings from sharing one's personal data.

Furthermore, especially in the group in which the participants were in their seventies, the practical skills to protect one's privacy were quite low. For a few, basic skills such as using several different browser windows at the same time were difficult, and some had not realised how a simple online toggle button works—even though they actively used many kinds of online applications and services. Many of them had difficulties managing junk mail, and they asked us, the organisers, what they should do to block it. To be able to protect one's privacy, therefore, a user should have fairly good digital skills (Büchi et al., 2017). The participants in this group felt great unease with these practical problems the logic of which they did not understand. As a previous study shows (Schreus et al., 2017), the socio-emotional aspects and the level of self-efficacy are crucial factors to bear in mind when thinking of digital literacy education among older users. This is challenging but still very important, as, especially when thinking of users with low digital skills, the idea of users' 'informed consent' seems very unrealistic.

In general, participants were suspicious about the data-gathering practices of social media platforms such as Facebook or Instagram but had little knowledge of the data collection practices carried out by third parties on ordinary websites. This is probably related to the fact that news media have reported many of the privacy scandals related to social media applications; a few participants even mentioned that they had become more cautious after the scandal related to Facebook and Cambridge Analytica. However, the news media, which itself takes part in data collection through its own sites and applications (Helberger, 2016; Turow, 2011; Ruohonen & Leppänen, 2017), have not been so eager to inform people of the regular practices of data-selling to third parties that they also use.

Even for those who had adjusted their Google settings in efforts to protect their privacy in some ways, the vast nature of data collection by third parties through ordinary websites came as a surprise. This particularly held true for the youngest and the oldest participants. Even in the group of the most educated participants (who had all completed their master's studies at university), many had no prior knowledge of how much personal data Google had about them and had not checked their My Activity information before. Research from the last decade shows that, on a general level, most users do not fully comprehend how cookies work (Ha et al., 2006; Jensen et al., 2005), and users grossly overestimate their knowledge of cookies on a research survey (Jensen et al., 2005). In addition, many people tend to think that if one wants to use a certain site, one has no other option than to accept all the cookies (Selwyn & Pangrazio, 2018). Our preliminary findings based on workshops suggest that even though people are capable of linking privacy notices and cookies, they mostly do not realise that cookies do not only share data with the provider of the site but also with a number of other data analysis and advertising companies. As noted (Luzak, 2014), there is no point in asking users for their 'informed consent' to share their data if a majority of users are not aware of cookies or do not understand how they work.

At the end of the two-hour workshops, the participants were introduced to three options for organising their online world if the internet were invented now and there were no personal data already shared through applications and services. The reason for setting the imaginary frame was to create free space for the participants to consider an ideal scenario without cynically figuring out how much of their data is already available to outside interests. The idea was that this kind of ideal scenario would offer guidelines for future discussions on how to make the present situation better, both for present and future user generations. The three options were:

1. As the user, I share my data so that I can use the applications and sites I want. My personal data can also be shared with third parties. The service provider may come from any country, and it acts upon its information security legislation regarding my data.
2. Service providers have permission to sell my personal data to third parties, but I am able to see what kind of data each service provider and company have about me through my personal *data bank*. Through my data bank, I am able to remove my data from these companies and service providers when I stop using their services or applications.
3. All the online services and applications work through subscriptions. I pay a monthly fee for each service and application, and my data is not sold to third parties. However, I understand that this would make the innovation of new applications and services more difficult.

The first option responds to the situation before the GDPR. The second option was developed by the research group from the ideas of the My Data movement (Lehtiniemi, 2017; Lehtiniemi & Ruckenstein, 2019). The third offers a realistic option in which no data would be exchanged for services and applications, but these would be financed through user payments.

With the exception of four respondents, every participant chose the second option. The second option was often considered as a reasonable compromise. Most participants commented that they still wanted to use free services, but at the same time they wanted to know about the use of their data. Many comments underlined that users should have the right to control the use of their data:

> I think that the user whose data forms part of service providers' increase in value, should have a right to know what data is collected and to have a right to control its use. Overall, it is important that data collection is done according to lawful principles. Preventing the irresponsible data gathering by ser-

> vice providers should not only be users' responsibility. Even though data collection would be performed according to the rules, the user should have the right to control their data.
>
> This option would secure that service providers would compete with each other to offer better services surrounding shared data. So, users could directly influence service providers.

Even those participants who, in general, did not consider sharing their personal data to be harmful stated that they felt uneasy about the option that service providers would act according to the information security legislation of the provider's origin. Apparently, many of these participants had not realised that this was the situation before the GDPR.

Younger participants in particular noted that they would not have enough money to pay for every application, and a few respondents also thought that subscribing and paying for every application would be very inconvenient. Two participants said that they did not choose the third option but the second because they regarded it as important that there still exists the possibility of developing new applications through data collection; one mentioned that she would happily share her data to be used by health companies. Projects in which organisations or companies have donated their data for the public good have been implemented (Susha et al., 2019; Petersen, 2019; Taylor, 2016), and the participant might have been aware of the idea, as in Finland The Finnish Institute for Health and Welfare already openly shares their anonymous statistical health data concerning, for example, the number of visits and different procedures in each county (THL Open Data). However, when private companies seeking profit develop innovations from open data, the definition of 'public good' might become tenuous, and again, it is questionable as to whether or not most users really understand how revealing their data might be when giving their consent (Taylor, 2016; Lindman & Kuk, 2015).

Yet, some respondents also criticised the second option, even though they had chosen it. Since they thought the data bank option would also be very risky, they offered a new, more developed version of this option:

> I would choose the second option, if I could choose a data bank that agrees with third parties that they would have access to my personal data but would not own it. This way data could really be removed so that I would vanish and cease to exist from everybody. I would also like to limit in advance the selling of my data to third parties. I would also like to have the possibility of

making several agreements with different data banks so that the data in different banks could not be linked with one another.

Another participant would have added an obligation for all service providers to report to the user the data they had collected on them. He would have also included all the health companies to service providers that a user could check through a data bank.

One participant considered all the options confusing, including the prevailing situation with present data tracking practices, saying that she "would regard them as absurd if she hadn't got used to them". She doubted that the second option could be technically feasible. She also criticised all the options for being commercially orientated and reminded us that originally the internet was not a commercial space. This comment also reminded us researchers how adjusting to the present neoliberal online environment can narrow the ability to imagine other kinds of systems. Furthermore, the idea of a 'non-commercial internet' is not only utopian; the BBC, for example, as well as other public service actors, have already taken initiatives to build a "public service internet" (Building a Public Service Internet, BBC Research & Development; Nikunen & Hokka, 2020, see also Fuchs, 2018).

In general, there was a strong tendency among our respondents, corresponding with findings by Selwyn and Pangrazio (2018), that the burden of protecting online privacy should not be left to the user alone. When choosing the second option, many participants explained that that they would need some reliable party to take care of privacy protections on their behalf. Some of the youngest and oldest participants considered it particularly unfair that they were left to personally take care of protecting their personal data against parties they had not even realised were tracking their actions. This is noteworthy, as the traditional understanding of digital literacy places pressure on the individual and underlines the skills that the user should learn to protect herself.

Our workshops also showed that when people with average ICT-knowledge were shown in practice how data-gathering practices work and taught how and why their data is sold by third parties, they were perfectly capable of forming an opinion, and few of them even developed new ideas based on the three options about how they would like data gathering to be organised and regulated. Similar to the results of a study by Kennedy et al. (2017), many respondents thought that they would need more information on how this system works and that there should be more public discussion on data collection. Offering practical knowledge on data-gathering practices is a good starting point for increased data infrastructure literacy that, in turn, could help average users/citizens better participate in the discussion about the conditions of datafication. As Gray et al. (2018: 9) suggest:

> Drawing attention to the politics and making of data and data infrastructures could open up new sites of contestation and controversy as well as creating opportunities for new forms of mobilization, intervention and activism around what they account for. [...] Gaining a sense of diversity of actors involved in the production of digital data (and their interests, which may not align with the providers of infrastructures that they use) is crucial when assessing not only the representational capacities of digital data but also its performative character and role in shaping collective life.

The results mentioned above reassert the ideas of a 'welfare data society'. Users feel insecure and burdened by the requirement that is built into online environments requiring all users to be solely responsible for her/his own safety against the large platforms or the third parties whose actions were mostly hidden from a user. They long for help from some kind of organisation or institution that they could trust. Fairer user environments can be achieved along two paths that a 'welfare data society' should offer: (1) practical digital literacy education to help people grasp the ways in which data gathering works and (2) more analytical data infrastructure education that would help people understand, discuss, and even demand new options to the present situation in which global giants monopolise the online user environment.

In sum, there is a clear need for better data infrastructure literacy so that average users and citizens can be capable of having a political discussion on the ethical aspects of datafication and appropriate regulation. Raising awareness through workshops is effective for small groups, but they are very time-consuming. At the same time, it is urgent to raise general awareness of datafication practices, as a growing number applications that use personal data are continually developed. As Selwyn and Pangrazio (2018) have proposed, there is a strong need for more structural, large-scale solutions to raise the level of digital and data infrastructure literacy. In Finland, one of the major actors in digital literacy education is the Finnish Broadcasting Company YLE. In the next section I analyse the YLE Learning's content and the actions they have taken in pursuit of raising data infrastructure literacy and discuss whether European public service media could play a major role in achieving this goal.

YLE Learning as a Content Provider for Digital and Data Infrastructure Literacy

The YLE Learning (Oppiminen)[1] editorial staff includes an executive editor, a producer, a subeditor, a community manager, and four journalists. As part of the general organisational structure of YLE, it is part of YLE's Creative Content Unit. For this chapter I have interviewed YLE Learning's producer Anna-Leena Lappalainen, with whom we cooperated during the project. According to Lappalainen,

> YLE Learning's main task is to promote lifelong learning. It covers categories ranging from digital and media skills, learning skills, school environment, well-being and human relationships, to how society and economics works, and how to develop oneself as a citizen. We approach our topics in an experimental and exploratory spirit.

Unlike many of its European public service counterparts such as BBC Learning[2] or NRK Skole,[3] YLE Learning is not focused on providing content for schoolchildren but to citizens of every age in the spirit of lifelong learning. YLE Learning provides feature articles, educational videos, and quizzes. It has provided digital skills education for several years already and has been producing practical 'digital skills training' content since 2016. In this way, YLE fulfils the traditional public service mission: it provides universal access to education on digital literacy and in this way seeks to empower all kinds of citizens so that they might better cope in a digitalised environment, even if their background education had been left wanting in this respect.

During our cooperation, YLE Learning took datafication as one of their major topics. The decision was grounded in our joint preliminary workshops and YLE's own user workshops in which participants expressed a strong interest in datafication as a journalistic topic. From August 2019

[1] https://yle.fi/aihe/oppiminen.

[2] https://www.bbcstudioslearning.com/index.html, https://www.bbc.co.uk/programmes/articles/37gYmkZ17J23P5cxFSL7Q9W/about-the-learning-zone.

[3] https://www.nrk.no/skole/.

to May 2020, YLE Learning produced seven exploratory pieces that shed light on datafication from different perspectives.

When looking at the seven pieces by YLE Learning analytically, they can be divided into those that support digital literacy and those that could increase data infrastructure literacy. The first group mainly comprises quizzes or short informational packages. In terms of digital literacy (Iordache et al., 2017: 23), they teach users the operational, technical, and formal skills related to digital use and provide guidance in the analysis and evaluation of digital content—a skill that is considered central to digital literacy but also a necessary step in gaining data infrastructure literacy (Büchi et al., 2017; Gray et al., 2018). The second group comprises generally lengthy articles that provide detailed analysis of their topics. Those articles are relevant content for increasing data infrastructure literacy: they help in providing understanding of the present political, social, and economic situation, the "actors involved in the production of data" and their interests, and how digital data now shape everyday life (Gray et al., 2018).

Most of YLE Learning's datafication pieces are published as pairs, so one piece gives practical advice while the other offers deeper insight into the matter. For example, the first and, so far, the most popular piece is a feature article of an ordinary young woman who tests the GDPR for her own online data. The article explains how she requested that fifteen companies and organisations, from Airbnb to the city of Lahti, provide her with the personal data that they have on her, and explains, in a thriller-like narrative, how each company responded. The article also includes a fact box on how and why companies and organisations gather personal data and what kind of rights the GDPR grants to a private citizen regarding their data. The first story is linked with a second, more practical piece that explains how one can make a request for his or her personal data based on GDPR regulations.

Another pair of YLE Learning's datafication pieces is a quiz and an educational article related to digital footprints. The quiz, named "What kind of a trace do you leave online?" is made up of questions like, "Do you switch off location data from your mobile phone if you are not using an application that needs it?" or "Have you changed your privacy settings to correspond with your needs in the applications you use?" The quiz has a somewhat similar approach to the user workshops in our research in that it asks the user about her privacy settings. After each answer, YLE's digital footprint quiz provides a short explanation of why this is important and what option would be useful in light of privacy protections. The quiz is linked with an educational article that provides nine grounded tips on how to minimise the amount of personal data one shares online.

When we believe that the ability to manage digital footprints is an essential part of digital equality (see Micheli et al., 2018), this kind of accessible content may be quite valuable when trying to raise broad awareness of how to protect personal data. In particular, quizzes may act as effective routes taken to raising awareness of data-gathering practices—such as checking one's own privacy settings and data, as in our user workshops—though, in face-to-face workshops, users are provided with the opportunity to ask more questions. In addition, the practical tips offered in educational pieces will certainly provide a few more digital skills that will help individuals protect their privacy online.

The fifth article has a slightly different approach as it sheds light on YLE's own data-gathering practices. The article begins by describing what technically happens when the user opens this webpage and how cookies start to gather data about his or her movements on the site. The article explains which data YLE gathers, why it gathers data, and for what purposes. It also explains that if the user reads the article through Facebook, then Facebook will obtain data about that visit to YLE's site and use it according to Facebook's own privacy rules. Again, it expounds the idea that if a YLE news story includes a tweet, Twitter also obtains some user data. The article reveals, on YLE's behalf, how and why media companies gather user data. This helps the user to understand the now prevalent practices of datafied media with the aim of improving their data infrastructure literacy.

The sixth and seventh articles are again connected to one another. First, YLE Learning has published a nine-minute-long video on the topic "The Internet wants to know everything about you—why should you bother to take interest?" The video features Laura Kankaala, an information security expert who is also known from YLE's TV series *Team Whack*, in which three 'white hat hackers' demonstrate through different case studies how easy it is to hack someone's personal data. In the video, which also illustrates its points using actors and storytelling, Kankaala explains in detail the many aspects of datafication: why personal data is valuable, how algorithms work through profiling, how algorithms try to make users addicted to social media content, how they may even expose users to political propaganda, and so on. In the end, she urges everyone to take a critical stance towards online incitements and take care of their personal privacy. The seventh article is a profile of Laura Kankaala, in which she also describes what everyone should take into account when using social media and other online applications. The video sheds light on data-gathering practices but also on the underlying models and the thinking behind data gathering, offering insight into the "politics and making of data" (Gray et al., 2018: 9) and underlying data infrastructures. Content-wise, YLE Learning has much to offer in its attempt to raise the general level of data infrastructure literacy.

Like the first story of the young woman tracing her personal data, the articles on datafication explained by Laura Kankaala have been widely shared on Facebook, and both have managed to reach many readers. In addition, the article providing information on how to make a request for one's own data based on the GDPR was also very widespread. YLE's article on data-gathering practices was not so popular among average users but, according to producer Anna-Leena Lappalainen, has gained a lot of positive attention on LinkedIn among IT professionals. Furthermore, Lappalainen noted that unlike regular news articles, YLE Learning's articles have a long lifespan and are typically found through search engines long after publication by people looking for information on digital skills, digital media, and datafication.

Lappalainen has admitted that datafication is a complicated issue and it is not easy to find story angles that would make datafication interesting to the average reader. However, along with the central PSM ideas of egalitarianism and universalism (Hokka, 2018; Brevini, 2013), they also try to reach those people who are not interested in datafication in the first place and/or have limited education on the subject that could help them understand what datafication and data gathering mean in practice. Yet, what YLE Learning has noticed through their work is that certain storytelling techniques help make datafication a more comprehensible topic. Datafication needs to be linked to everyday life in a very concrete way and the article has to offer something that seems useful, not just something interesting. It helps to explain datafication through an individual and personal perspective, such as in the story of the young woman who requested her own data or through the perspective of some fairly well-known character such as 'white hat hacker' Laura Kankaala. Naturally, quizzes that reveal something about the user also pique readers' interest. However, Lappalainen noted that even though many people, such as the respondents in the user workshops, claim that mere information is enough to get their attention, reader statistics from YLE Learning show that in real terms, sharing pure facts does not induce average readers to get to know more about datafication; most need some kind of journalistic "kicker" to get started. The insights of YLE Learning's journalists correspond with previous journalism audience research (Costera Meijer, 2012) in which news readers not only valued journalism that improved the 'quality' of their lives but also innovative narrative forms that would increase the pleasure of reading.

Still, when looking at reader statistics at the level of the whole population, the effect of YLE Learning is probably still fairly modest. YLE's online site,[4] where YLE Learning's content is published, reaches 37 per cent of Finns over fifteen-years-old weekly (Nokela et al., 2019), which is impressive when compared to many other European public service media (Schulz et al., 2019: 13). But the average number of readers of YLE Learning's content is lower, approximately 4 per cent of the Finnish population. The user workshops from our research project indicate that datafication as a topic interests mainly those people who are already somehow familiar with the subject. Despite the positive outcomes of YLE Learning, the question is, what measures could be taken to raise data infrastructure literacy to the level that would make possible a well-informed political discussion on datafication and the democratic decision-making that arises from that newly gained knowledge?

The answer partly lies in what YLE Learning already does. YLE is well-connected with adult education institutes, community high schools and libraries, and they also advertise their content to grammar and high school teachers. In this way, the content helps build digital literacy, and data infrastructure literacy gradually spreads beyond the regular readers of YLE's website. Furthermore, unlike commercial media, YLE, as a public service media organisation, is free to talk about present data-gathering practices, as their financial stability is not dependent on them unlike more profit-oriented media companies. However, the spread of educational and journalistic content related to datafication currently reaches only those who are seeking out such information. The question remains of how to reach the majority who have trouble allocating the time and/or effort to understanding datafication as a phenomenon with the tremendous effects it has on the development of societies. While public service media are clearly able to produce material that could increase data infrastructure literacy and be an important part of the solution, there is still a need for more coordinated cooperation between different kinds of public organisations

[4] https://yle.fi/.

and educational institutions. We clearly need more structural, large-scale solutions for increasing the level of digital literacy (Selwyn & Pangrazio, 2018) and data infrastructure literacy, but the experiences from both user workshops and YLE Learning show that no single actor will manage that mission on their own.

Conclusion

Work in the field of critical data studies has recently made great strides in highlighting the many downsides of datafication: surveillance capitalism and dataveillance (Zuboff, 2015; Andrejevic, 2019, Lee, 2019), data colonialism (Couldry & Mejias, 2018; Ricaurte, 2019), data mining, digital footprints and digital traces (Kennedy et al., 2017; Breiter & Hepp, 2018; Micheli et al., 2018), anxieties caused by datafication (Ruckenstein & Granroth, 2019; Lupton, 2019) or the power that algorithms and automatic decision-making possess (Andrejevic, 2020). This remains, however, a work in progress as developing technologies and new products and systems will always force us to confront new ethical dilemmas. Yet, critical data studies should also seek solutions to those dilemmas. That work has already started as many researchers develop new methods and practices that aim to help users as citizens protect their privacy and to look for ways to handle data so that it would benefit users more than it does now (Selwyn & Pangrazio, 2018; Jarke, 2019; Pybus et al., 2015; Kennedy & Moss, 2015; Markham, 2020). In my article, I also attempt to take part in finding solutions for the problems of datafication. As datafication inevitability proceeds, it should be asked, what kind of data society takes care of all citizens' wellbeing and treats citizens in a fair way. How could a datafied society become a welfare data society?

In a welfare data society, the rights and wellbeing of citizens are strengthened through education: by increasing the level of digital and data infrastructure literacy the user/citizen knows her rights and is capable of using them. European countries have already made efforts to improve digital literacy in schooling, but education on digital literacy and especially data infrastructure literacy should also reach those who have not gained this kind of education or whose knowledge is outdated. In my chapter I propose that public service media should also reinforce citizenship through education in an age of datafication.

The EU has taken a leading role in regulating data gathering and granting European citizens the opportunity to manage and control the use of their personal data. However, for the regulation to be as effective as it should, European citizens need to be more aware of the practices and outcomes of data collection. The results from the user workshops in this study showed how even highly educated users often do not know the amount and type of personal online data about them that is gathered by many kinds of private companies and institutions. Instead, half of the participants were astonished, confused, shocked, and even angry when they realised how much personal data different online services have on them and how detailed it is. Taking care of online privacy requires a certain level of digital literacy that most people do not possess. This leads to digital inequality at the level of the individual when only those users with a fairly good technological education are able to control their digital footprints and protect their privacy online. But, more importantly, it results in an imbalanced and unfair power structure between the 'Big Tech' companies and users.

To strengthen the position of users and citizens and to transform the data infrastructures so that they may be fairer towards users, citizens should be informed about the present legal conditions of data gathering so that they might gain reasonable understanding of the political, societal, and cultural consequences of datafication. While this may sound ambitious, our workshops demonstrated that on average, people are capable of forming a thoroughly considered opinion on fair data-gathering practices. Furthermore, they were able to discuss and develop "alternative data regimes and practices" (Kennedy & Moss, 2015) after being introduced to data collection in practice.

In Finland, the public service broadcaster YLE has taken on an educational role as far as the subject of datafication is concerned. The results from our cooperation with YLE Learning show that public service media already possess inventive means through which different kinds of users can be reached—even those users who are not learning digital skills at school or university. Still, more controlled cooperation is needed among different public institutions to increase the number of people who can access the necessary information.

At the same time, work and solutions to increase the level of data infrastructure literacy cannot be left to the cooperation of national institutions only. The average user faces an online environment that is fundamentally global. If we really want to support citizens' right to be informed and, therefore, able to value the conditions of datafication, there needs to be European-wide cooperation. European public service media organisations could work together and bring to the fore topical issues related to datafication so that the social and political questions of datafication would not only be discussed by political and academic elites, but by the public at large as well. Possibly, through European-level discussion, new data regimes and practices, such as the data banks that participants preferred, this proposal could be taken forward.

It should also be discussed whether the EBU (European Broadcasting Union) should actively encourage the European PSM organisations to integrate education in digital literacy and data infrastructure into their regular content. At the very least, the EBU should actively support fair and transparent data collection and data use by European PSM companies—and not just to advise PSM organisations to benefit from their user data, as their AI and Data Initiative seems to do.[5] Only through raising awareness of current data mining practices and their threat to privacy and democracy will citizens be capable of imagining and insisting on new and possible options for the present online environment that GAFAM corporations dominate. The EBU could also take an active role in supporting the initiatives that some PSM organisations have already begun to put into practice, such as the BBC's public service internet model.

If the EU wants its citizens to appreciate and effectively use the GDPR and have more control over their rights to their personal data—or even judge companies and institutions by their ethical standards in data gathering—it should take an active role in increasing the level of data infrastructure literacy among citizens. As our experiences from the user workshops demonstrate, education cannot be left to schools and universities alone, as all kinds of users, and users of every age, need help in gaining the necessary digital skills and data infrastructure literacy. Public service media must also be considered as part of the solution.

References

Andrejevic, M. (2011). Public service media utilities: Rethinking search engines and social networking as public goods. *Media International Australia, 146*(1), 123–132. https://doi.org/10.1177/1329878x1314600116

Andrejevic, M. (2019). Automating surveillance. *Surveillance and Society, 17*(1–2), 7–13. https://doi.org/10.24908/ss.v17i1/2.12930

Andrejevic, M. (2020). *Automated media* (1st ed.). Routledge.

[5] https://www.ebu.ch/aidi.

Breiter, A., & Hepp, A. (2018). The complexity of datafication: Putting digital traces in context. In A. Hepp et al. (Eds.), *Communicative figurations, transforming communications: Studies in cross-media research* (pp. 387–405). Palgrave Macmillan. https://doi.org/10.1007/978-3-319-65584-0

Brevini, B. (2013). *Public service broadcasting online. A comparative European policy study of PSB 2.0*. Palgrave Macmillan.

Büchi, M., Just, N., & Latzer, M. (2017). Caring is not enough: The importance of Internet skills for online privacy protection. *Information, Communication & Society, 20*(8), 1261–1278.

Carr, W., & Kemmis, S. (2009). Educational action research: A critical approach. In S. E. Noffke & B. Somekh (Eds.), *The SAGE handbook of educational action research* (pp. 74–84). Sage.

Constantiou, I. D., & Kallinikos, J. (2015). New Games New Rules: Big Data and the Changing Context of Strategy. *Journal of Information Technology, 30*(1), 44–57. https://doi.org/10.1057/jit.2014.17

Costera Meijer, I. (2012). Valuable journalism: A search for quality from the vantage point of the user. *Journalism, 14*(6), 754–770.

Couldry, N., & Mejias, U. A. (2018). Data colonialism: Rethinking Big Data's relation to the contemporary subject. *Television & New Media, 20*, 1–14. https://doi.org/10.1177/1527476418796632

Dencik, L., Hintz, A., & Cable, J. (2016). Towards data justice? The ambiguity of anti-surveillance resistance in political activism. *Big Data & Society, 3*, 1–12. https://doi.org/10.1177/2053951716679678

DR's public-service redegorelse. (2019). https://www.dr.dk/om-dr/fakta-om-dr/dr-i-2019/dr-i-2019-ny-strategi-massiv-valgdaekning-og-et-meget-mere-digitalt-dr-0

Enerhaug, R. F. (2019). Allmennkringkasterregnskapet. Statistikk 2019. https://www.nrk.no/informasjon/her-er-nrks-rapporter-for-2019-1.14977138

European Commission/EACEA/Eurydice. (2019). Digital education at school in Europe. *Eurydice report*. Publications Office of the European Union. https://eacea.ec.europa.eu/national-policies/eurydice/content/digital-education-school-europe_en

Fuchs, C. (2018). The online advertising tax as foundation of public service internet. *A CAMRI extended policy report*. University Westminster Press. https://doi.org/10.16997/book23

Gangadharan, S. P. (2017). The downside of digital inclusion: Expectations and experiences of privacy and surveillance among marginal Internet users. *New Media & Society, 19*(4), 597–615.

Gray, J., Gerlitz, C., & Bounegru, L. (2018). Data infrastructure literacy. *Big Data & Society, 5*, 1–13. https://doi.org/10.1177/2053951718786316

Gurumyrthy, A., & Bharthur, D. (2018). Democracy and the algorithmic turn. *Sur International Journal on Human Rights, 15*(27), 39. https://libproxy.

tuni.fi/login?url=https://search-proquest-com.libproxy.tuni.fi/docview/2161599967?accountid=14242

Ha, V., Al Shaar, F., Inkpen K., & Hdeib, I. (2006). An examination of user perception and misconception of internet cookies. *CHI 2006 extended abstracts on human factors in computing systems*. Association for Computer Machinery, 833–838. https://doi.org/10.1145/1125451.1125615

Hänninen, S., Lehtelä, K.-M., & Saikkonen, P. (2019). Introduction: The Nordic welfare state as a state of civilization. In *The relational Nordic welfare state* (pp. 1–13). Edward Elgar Publishing.

Helberger, N. (2016). Policy implications from algorithmic profiling and the changing relationship between newsreaders and the media. *Javnost—The Public, 23*(2), 188–203.

Hokka, J. (2018). Towards nuanced universality: Developing a concept bible for public service online news production. *European Journal of Communication, 34*(1), 74–87. https://doi.org/10.1177/0267323118810862

Holmwood, J. (2000). Three pillars of welfare state theory. T.H. Marshall, Karl Polanyi and Alva Myrdal in defence of the national welfare state. *European Journal of Social Theory, 3*(1), 23–50.

Iordache, C., Mariën, I., & Baelden, D. (2017). Developing digital skills and competences: A quick-scan. Analysis of 13 digital literacy models. *Italian Journal of Sociology of Education, 9*(1), 6–30.

Jarke, J. (2019). Open government for all? Co-creating digital public services for older adults through data walks. *Online Information Review, 43*(6), 1003–1020.

Jensen, C., Potts, C., & Jensen, C. (2005). Privacy practices of internet users: Self-reports versus observed behavior. *International Journal of Human-Computer Studies, 63*(1–2), 203–227. https://doi.org/10.1016/j.ijhcs.2005.04.019

Kangas, O., & Kvist, J. (2013). Nordic welfare states. In *Routledge Handbook of the Welfare State* (pp. 148–161). Routledge.

Kennedy, H., Elgesem, D., & Miguel, C. (2017). On fairness: User perspectives on social media data mining. *Convergence, 23*(3), 270–288.

Kennedy, H., & Moss, G. (2015). Known or knowing publics? Social media data mining and the question of public agency. *Big Data & Society, 2*, 1–11. https://doi.org/10.1177/2053951715611145

Lee, C. S. (2019). Datafication, dataveillance, and the social credit system as China's new normal. *Online Information Review, 43*(6), 952–970.

Lehtiniemi, T. (2017). Personal data spaces: An intervention in surveillance capitalism? *Surveillance & Society, 15*(5), 626–639.

Lehtiniemi, T., & Ruckenstein, M. (2019). The social imaginaries of data activism. *Big Data & Society, 6*, 1–12. https://doi.org/10.1177/2053951718821146

Lindman, J., & Kuk, G. (2015). From open access to open data markets: Increasing the subtractability of open data. *48th Hawaii international conference on system sciences*. VIELÄ OSOITE

Lupton, D. (2019). 'It's made me a lot more aware': A new materialist analysis of health self-tracking. *Media International Australia, 171*, 66–79. https://doi.org/10.1177/1329878X19844042

Luzak, J. A. (2014). Privacy notice for Dummie? Towards European guidelines on how to give "clear and comprehensive information" on the cookies' use in order to protect the internet users' right to online privacy. *Journal of Consumer Policy, 37*, 547–559.

Mai, J.-E. (2006). Big data privacy: The datafication of personal information. *The Information Society, 32*(3), 192–199.

Markham, A. N. (2020). Taking data literacy to the streets: Critical pedagogy in the public sphere. *Qualitative Inquiry, 26*(2), 227–237.

Mayer-Schonberger, V., & Cukier, K. (2013). *Big data: A revolution that will transform how we live, work, and think.* John Murray.

Micheli, M., Lutz, C., & Büchi, M. (2018). Digital footprints: An emerging dimension of digital inequality. *Journal of Information, Communication and Ethics in Society, 16*(3), 242–251.

Nikunen, K., & Hokka, J. (2020). Welfare state values and public service media in the era of datafication. *Global Perspectives, 1*(1), 12906. https://doi.org/10.1525/gp.2020.12906

Nokela, S., Rouvainen, O., & Hyvärilä, A. (2019). *YLE's customer relationship. YLE's smart data and audience insight.* https://drive.google.com/file/d/1gGZUpAU5PzPNrJ9CR-pf9WA4nqku_VQP/view

Park, Y. J. (2013). Digital literacy and privacy behavior online. *Communication Research, 40*(2), 215–236.

Park, Y. J., Chung, J. E., & Shin, D. H. (2018). The Structuration of Digital Ecosystem Privacy and Big Data Intelligence. *American Behavioral Scientist, 62*(10), 1319–1337. https://doi.org/10.1177/0002764218787863

Petersen, C. (2019). User-focused data sharing agreements: A foundation for the genomic future. *JAMIA Open, 2*(4), 402–406. https://doi.org/10.1093/jamiaopen/ooz043

Pybus, J., Coté, M., & Blanke, T. (2015). Hacking the social life of Big Data. *Big Data & Society, 2*, 1–10. https://doi.org/10.1177/2053951715616649

Ricaurte, P. (2019). Data epistemologies, the coloniality of power, and resistance. *Television & New Media, 20*(4), 350–365.

Ruckenstein, M., & Granroth, J. (2019). Algorithms, advertising and the intimacy of surveillance. *Journal of Cultural Economy, 13*(1), 1–13. https://doi.org/10.1080/17530350.2019.1574866

Ruohonen, J., & Leppänen, V. (2017, 26 December). Whose hands are in the finish cookie jar? *Proceedings—2017 European intelligence and security informatics conference*, 2017, 127–130. https://doi.org/10.1109/EISIC.2017.25

Schreurs, K., Quan-Haase A., & Martin, K. (2017). Problematizing the Digital Literacy Paradox in the Context of Older Adults' ICT Use: Aging Media

Discourse and Self-Determination. *Canadian Journal of Communication, 42*(2). https://doi.org/10.22230/cjc.2017v42n2a3130

Schulz, A., Levy, D. A. L., & Nielsen, R. K. (2019). Old, Educated, and Politically Diverse: The Audience of Public Service News. Reuters Institute. https://www.mediaguru.cz/media/9835/the_audience_of_public_service_news_final.pdf

Selwyn, N., & Pangrazio, L. (2018). Doing data differently? Developing personal data tactics and strategies amongst young mobile media users. *Big Data & Society, 5*, 1–12. https://doi.org/10.1177/2053951718765021

Susha, I., Rukanova, B., Gil-Garcia, J. R., Tan, Y.-H., & Hernandez, M. G. (2019). Identifying mechanisms for achieving voluntary data sharing in crosssector partnerships for public good. *Proceedings of dg.o 2019, 20th annual international conference on Digital Government Research.* https://doi.org/10.1145/3325112.3325265

Syvertsen, T., Enli, G., Mjøs, O. J., & Moe, H. (2014). The media welfare state. In *Nordic media in the digital era.* University of Michigan Press.

Taylor, L. (2016). The ethics of big data as a public good: Which public? Whose good? *Philosophical Transactions of the Royal Society A, 374*, 20160126. https://doi.org/10.1098/rsta.2016.0126

Taylor, L. (2017). What is data justice? The case for connecting digital rights and freedoms globally. *Big Data & Society, 4*, 1–14. https://doi.org/10.1177/2053951717736335

THL Open Data. (2022). https://thl.fi/en/web/thlfi-en/statistics/statistical-databases/open-data

Turow, J. (2011). *Daily you. How the new advertising industry is defining your identity and your worth.* Yale University Press.

van Dijk, J. (2014). Datafication, dataism and dataveillance: Big Data between scientific paradigm and ideology. *Surveillance & Society, 12*(2), 197–208.

Wolff, A., Gooch, D., Montaner, J. J. C., Rashid, U., & Kortuem, G. (2016). Creating an understanding of data literacy for a data-driven society. *The Journal of Community Informatics, 12*(3), 9–26.

Zuboff, S. (2015). Big other: Surveillance capitalism and the prospects of an information civilization. *Journal of Information Technology, 30*, 75–89.

Open Access This chapter is licensed under the terms of the Creative Commons Attribution 4.0 International License (http://creativecommons.org/licenses/by/4.0/), which permits use, sharing, adaptation, distribution and reproduction in any medium or format, as long as you give appropriate credit to the original author(s) and the source, provide a link to the Creative Commons licence and indicate if changes were made.

The images or other third party material in this chapter are included in the chapter's Creative Commons licence, unless indicated otherwise in a credit line to the material. If material is not included in the chapter's Creative Commons licence and your intended use is not permitted by statutory regulation or exceeds the permitted use, you will need to obtain permission directly from the copyright holder.

PART III

Everyday Practices and Collective Action

(Not) Safe to Use: Insecurities in Everyday Data Practices with Period-Tracking Apps

Katrin Amelang

Introduction

Application software for mobile computing devices forms part of the numerous digital technologies that pervade and mediate everyday life. Given the normalisation of the use of mobile apps to support, extend, delegate, or reorganise all kinds of mundane tasks and routines over the last ten years, daily life has been increasingly "appified" (Morris & Murray, 2018). Out of the multitude of available mobile apps, period trackers provide the exemplary case considered in this chapter on everyday data practices around the use of apps. Besides being an integral part of everyday communication and everyday life, mobile apps, in conjunction with smartphones take part in the ongoing creation of data (sets) by and about people, their devices, and interactions. Voluntarily or involuntarily, app or smartphone users leave data traces and participate in the generation of data or data profiles about themselves in different ways. Whereas providers of mobile apps rely on such data as a valuable resource and prized

K. Amelang (✉)
University of Bremen, Bremen, Germany
e-mail: amelang@uni-bremen.de

© The Author(s) 2022

A. Hepp et al. (eds.), *New Perspectives in Critical Data Studies*, Transforming Communications – Studies in Cross-Media Research,
https://doi.org/10.1007/978-3-030-96180-0_13

commodity, the unanticipated collection, distribution, and utilisation of user data by private corporations (and states) have become the subject of public debates on privacy issues and data ownership. Overall, mobile apps and devices can be understood as mundane tools of "datafication" (Cukier & Mayer-Schoenberger, 2013: 35) and are elements of the socio-technical systems or "data assemblages" (Kitchin, 2014: 24–26) that organise contemporary data practices and are a principal subject for critical data studies.

Within critical data studies' heterogeneous examination of the impact and challenges of data and its power in society (see Iliadis & Russo, 2016) calls have been made to pay more attention to the everyday experience of (big) data (see Michael & Lupton, 2016; Ruckenstein & Schüll, 2017; Kennedy, 2018). Scholars studying the wide array of digitised self-tracking practices underpin this focus empirically and complicate the picture of datafication, data power, and dataveillance on the level of the everyday by pointing to the ambivalent effects and appropriation of data technologies and dataflows (e.g. Weiner et al., 2020; Pantzar & Ruckenstein, 2017; Fiore-Gartland & Neff, 2015). In addition, Helen Kennedy links the importance of research on how people engage and live with data to the endeavours of data activism: "One of the main purposes of exploring how ordinary people experience datafication in their everyday lives is to develop understanding of their perspectives on how they might live *better* with data" (Kennedy, 2018: 21, emphasis in original). I join this literature's concern to take a closer look at datafication and "its agential possibilities" (Ruckenstein & Schüll, 2017: 268) from an everyday perspective, and thus broadening perspectives in critical data studies, by using a genre of mobile apps as an entry point for examining mundane data practices.

Drawing on empirical material, in particular interviews with app users, I will explore how people encounter menstruation with and through data. App-supported period tracking provides an interesting example because of its commonness, its non-digital precursors and its banality (not triviality) in terms of self-tracking. Further, it is insightful regarding typical data traces produced through mobile app usage and regarding its enmeshment with the sociocultural circumstances and the gendered politics of birth control—turning the question of *safe use* in a twofold matter. To illustrate, I will begin with situating the app-genre of period trackers, the motives of users, and the promises of app-providers by outlining the sociotechnical constellations in which app-mediated menstrual self-observation as an everyday engagement with data unfolds. Then, I will discuss two critiques that regularly come up in public discussions about period trackers, which address two forms of data insecurity.

The first kind of insecurity concerns the promises and failures of measuring and 'taming' bodily processes with self-tracked data, quantification, and predictive algorithms. It addresses the reliability of bodies as data repositories (Lupton, 2013), the accuracy of algorithms just as the confidence of app users interpreting data and aligning their embodied selves with their data(fied) bodies. The second kind of insecurity concerns the question of what data apps reveal to whom and who can use the data entered into the app for what purpose. It brings the ambiguities of dataflows, commercial data collection, and issues of privacy to the fore and addresses the ways app users deal with the accompanying "intimate surveillance" (Levy, 2015) or "dataveillance" (van Dijck, 2014). Both insecurities do not pertain to period trackers alone. The first applies in particular to mobile apps that allow the monitoring of bodily processes and support self-tracking practices but ties in with wider discussions of the calculability of the body and the self. The second is a central subject in considerations of mass data collection and consumer tracking for commercial purposes (be it via mobile apps or web cookies).

Exploring how interviewed app users in Germany make use and sense of period trackers and menstruation data while negotiating the two insecurities involved, this chapter will show the multi-faceted ways people with periods engage with data in everyday life. Their narratives point to ambivalences resulting from the tension that app-related data present both a source of insecurity and security. Moreover, their accounts of becoming entangled with, while pragmatically tackling, datafication and its challenges in the field of appified menstruation indicate a reflexive (data) practice. In this way, this chapter aims at contributing to new perspectives in critical data studies that pay more attention to the contingencies and contradictions of people's daily involvement with and understanding of data. Such an approach encourages the reframing of dominant narratives about the data(fied) worlds we live in and to envision alternative forms of living with data.

Material and Methods

The analysis presented here draws on an ongoing case study of digitised period tracking that is part of a broader ethnographic study, in which I explore the cultural dimension of software, data, and algorithms, especially with respect to body-technology relations. The empirical material has been created in the process of my "polymorphous engagement" (Gusterson, 1997: 116) with period trackers since 2017. It consists primarily of conversations with menstruating people in Germany, with a focus on those who track their period and use an app to do so. Further, it includes a mix of online app reviews and discussions of period trackers in different (social) media, my (self-)testing of apps,[1] and the participation in events of/with actors involved in the education and counselling of sexual and reproductive health and rights. For this chapter, I focus on the manifold conversations I had with app users. Of these, six were formal interviews of 1–2 hours that were audio recorded and, to a varying degree, included an engagement with the participant's respective app/smartphone. Yet, most of these ongoing conversations about experiences with menstruation and period trackers took place in a less formalised manner and occurred rather impromptu in diverse social settings and interactions as part of my daily (private and academic) life. These informal ethnographic interviews range from a vast number of brief exchanges of a few minutes to, by now, 23 more focused conversations of 15–30 minutes, which I recorded via memory logs or notes on the spot. In terms of demographics, participants were between 18 and 44 years old, identified themselves as women, mostly as heterosexuals, and are—due to the bias of recruiting through personal networks and snowball sampling—largely educated, middle class, and *white*. To take into account the diversity of forms and experiences of app-supported period tracking, I did not limit recruitment to a specific purpose—like the use of fertility tracking apps in trying to conceive (e.g. Hamper, 2020) or of period trackers supporting the symptom-thermal method in order to prevent pregnancy (e.g. Rotthaus, 2020). Despite the dissemination of period trackers, it is important to remember that not all people with periods track their cycles, nor do those who track their periods use an app to do so.

[1] In order to better understand, compare, and get a feeling for different period trackers, their features, functions, and ways of addressing or guiding users, I tested several popular period trackers [on an Android device] and those my interlocutors mentioned, each for about 3–5 months. In the course of the process, I came across the "walk-through method" proposed by Light et al. (2018), which provides a good systematic for what I was doing in the wild and helps with assessing the embedded meanings, norms, scripts, and ideal use(r)s of apps.

Situating Period Trackers and Period Tracking

Period trackers, also known as menstrual cycle apps, ovulation trackers or fertility tracking apps, help to record, monitor, and forecast menstrual cycles. They have names like Ava, Clue, Cycles, Eve., Flo, Glow, Kindara, Magic Girl, Maybe Baby, My Calendar, Natural Cycles, Ooops!, Ovy, Period Calendar, or Women's Log and are considered part of the booming field of health and wellness apps, which promise to coach healthy users in their self-care. Increasingly launched since 2012,[2] they also filled a biased gap: fitness and health trackers of the time allowed collecting all kinds of body data but left out menstruation. Period trackers range from simple calendar programmes to more advanced applications that support fertility awareness methods. They draw on long-established practices of monitoring menstruation and a familiar tool—the menstruation calendar, of which the apps represent a digitised and extended version. Based on a user's self-observation and logged data the apps calculate and predict the onset of period, estimated ovulation and fertile window, and (pre)menstrual symptoms. To a differing degree, these apps include user community features, additional information on menstrual health, and the possibility of sharing period data with ones' intimates or to connect with wearables. They all edit and visualise a user's menstrual data by means of colourful icons and charts and above all offer a broad variety of tracking categories that cover, besides bleeding, numerous symptoms and complaints associated (more or less) with menstruation as well as information on sexual behaviour, mood, fitness, health, or lifestyle. In terms of business-models, some period trackers are part of a larger company, others rely on venture capital (Rizk & Othman, 2016), and most of those that are for free are heavily supported by advertising however, subscriptions models seem to be on the rise.

Situating Period Trackers as Gendered Technology

When one takes a window shopping tour at one of the app stores, one can be overwhelmed as much by the number of available period trackers as by the quantities of pink, flowery-lovely-girly-cute design. The colour scheme

[2] Despite earlier examples, e.g. Period Tracker (launched 2009) or Maybe Baby (2010), there was a boom of period-tracking apps in 2012/2013. Moreover, several share a similar foundation narrative (Fluhrer, 2018: 48–53, 59–65).

of period-tracking apps has been a source of amusement or annoyance in some of my interviews, has been mocked in many reviews, news and social media pieces, and has even been taken up in the tagline "reliable, scientifically based and definitely not pink" with which the app *Clue* was introduced in 2013. Yet, the feminised design is only one aspect illustrating menstrual cycle apps as a gendered technology that reflects and reproduces gender stereotypes and social norms.[3] The "gender scripts" (Rommes et al., 1999) or inscribed designers' imaginations of (potential) uses and users entail the ways theses apps address and judge users as well as socio-cultural ideas of menstruating people, which might correspond with some users, their experiences, or aims of period-tracking but marginalise, exclude, or simply annoy others.[4] Built-in gender assumptions are reflected as much in the apps' focus on fertility, pregnancy, family planning and heterosexuality as in the vocabulary and symbols that are used to present tracking categories, menstrual information, or remind users via push notification. Obviously, there are more binary systems at work in these apps than binary code.

Although several research participants stated that they have decided for a specific period tracker "for aesthetic reasons" and app creators inevitably react to gender stereotypes by carelessly reproducing, playfully appropriating, or deliberately avoiding them (see also Klein, 2020), choosing a period-tracking app is about more than an individual consumer decision based on personal preferences of design. As one interviewee explains: "I don't mind the pink or sometimes silly icons, which certainly do not meet everyone's sense of humour, right [...] as long as the app does what it is supposed to do. [...] help me to keep an eye on my menstruation, to know where I am in my cycle and to be somewhat in control" [ptiv5]. Regardless of why research participants pay attention to their menstrual cycles, the subject of control or lack thereof came up in almost all conversations eventually. For this 27-year-old (and other interlocutors), controlling one's period is about being able to handle bodily processes, which are recurring

[3] See Faulkner (2001: 83–84) for an overview of how technological objects are gendered symbolically or materially to varying degrees.

[4] This applies not only but particularly to queer, non-binary, and Trans people with periods or to those who are infertile or not interested in procreation. In 2013, the crowdfunding campaign for mcalc (iOS), a "gender neutral menstruation calculator", which "can be used by almost everyone regardless of their gender identity" and "is inclusive with the LGBTQA community" failed. Apparently, the project and existing beta version for Android got shelved (https://www.indiegogo.com/projects/mcalc-for-iphone#/).

yet tend to be or are perceived to be unreliable and raise questions about what is considered normal. For her, control also means being able to deal responsibly with sex: "Of course, period tracking is first of all about understanding your cycle and body. But in the end it is quite often about the dread of potential pregnancy, isn't it?" [ptiv5]. In line with the dominant marketing of period trackers as an aid to control one's fertility the practice of period tracking becomes part of the interviewee's (contraceptive) "fertility work" (Bertotti, 2013; Kimport, 2018). The individual wish to keep an eye on one's cycle and the apps' offer to get to know one's body with the help of data must hence be placed in the wider socio-technical arrangements and gendered politics of period tracking and sexual reproduction—period trackers are enmeshed in both, body and data politics.

Resonances in Observing and Measuring Menstrual Cycles

In addition to offering fertility monitoring to achieve or avoid pregnancy, app providers invite users to get to know their bodies better. While the idea of learning about one's body (self) conjures up memories of the feminist legacies of the women's health movement of the 1970s,[5] the idea of perhaps not having numbers but data as a (superior) form of self-knowledge resonates with contemporary promises of quantified self-measurement. That the idea of measuring and knowing (and often improving) oneself is older than the public discourse around the recent Quantified Self movement suggests, as Crawford et al. demonstrate (2015), by juxtaposing the introduction of the domestic weight scale at the early twentieth century and current wrist-worn fitness trackers. Schmechel, who discusses self-measuring from a gender perspective, even views menstrual self-observation shaped by the development of the menstruation calendar as a precursor to today's quantified self-tracking (2016: 148–150). Although such comparisons are instructive, there are differences in terms of measuring menstrual cycles.

[5] The women's health movement also opposed the objectification of female bodies by patriarchal medical authority (we are our bodies) and considered self-knowledge as a way to gain control over "our bodies/ourselves" and medical technologies. Becoming knowledgeable rested upon the collective compilation of scientific information but also on sharing personal experiences (see Boston Women's Health Collective, 1970). Ford (2019) discusses in what way period-tracking apps take up this feminist legacy by offering self-knowledge via technology.

Two of my interlocutors firmly distinguished period tracking from fitness-related self-tracking, which would be aimed more at optimisation or self-improvement: "The wish for better apprehending and feeling more in control of one's body is perhaps similar. But you cannot improve your cycle" [ptiv4]; instead, period tracking can help you to "feel less at the mercy of menstruation" [ptst17]. Moreover, in contrast to sensor-equipped self-tracking devices, which make invisible bodily processes visible and render them into digital data (Lupton, 2018: 2), period trackers cannot rely on a continuous recording of physical conditions and body activities.[6] Rather, users estimate qualities (e.g. how strong is the bleeding, how is my skin, hair and mood) or check boxes (e.g. had [unprotected] sex, headaches, voracious appetite) in order to log their observations and activities before those can be processed and quantified by the app. While the apps' tracking categories surely preconfigure what users should observe, log, and how it should be classified, users have to enter metrics (values of self-measurement) manually. This extra effort creates uncertainty, but also more opportunities to tinker with (Mol et al., 2010) or "curate" (Weiner et al., 2020) the self-generated data records—in the spirit of self-care.

Motives, Benefits, and Promises of App-Based Period Tracking

The reasons research participants expressed about their wishes to track their menstrual cycles via app vary and can change over time. In the following, I outline some of the typical reasons they provided. I present what interviewees appreciate about period trackers and their app-mediated engagement with menstrual cycles. Finally, I discuss the role of tracking data and apps as hubs.

Reasons for Tracking Menstrual Cycles

The most frequent reasons given for period tracking were getting to know one's body and getting "a better picture" of one's cycle. Meaning to "recognise more clearly how the menstrual cycle works" [ptst5] and to

[6] There are period trackers that include sensors or measuring systems that record the basal body temperature or hormones in the urine. I do not consider these apps/systems in my research or this chapter.

understand how one's own body is responding to menstrual processes. Comparable to other forms of self-tracking, gaining body knowledge through data was for many research participants about identifying patterns and understanding co/relations[7]—for example, concerning their fertility, well-being, or troubles in specific phases of their menstrual cycles. Interestingly, they described their app-based learning about their body equally often as a purpose and a side benefit of using a period tracker. For several, this included realising how much they actually do not know about menstruation.

In addition to the motive of learning and the above-mentioned concerns of controlling fertility, interviewees emphasised the possibility that they could "check quickly the cycle in between". For example, one interlocutor used a period tracker "to have less random sex to get pregnant" [ptst8]; checking the app/cycle daily was essential for her to realise her goal. More common was a "glimpse in between", in order to not be surprised when the period starts, "whether tampons or condoms should be taken on a weekend-trip" [ptiv3], "when to better not schedule a visit to a sauna or the beach" [ptst11], to know "everything is okay, nice and regular" [ptst20] or that one is not pregnant. These reasons for tracking menstrual cycles concur with the outcomes of a study in the U.S. (Epstein et al., 2017). Also mentioned was the motive to answer the standard question of gynaecologists (When was the first day of your last period?) without having to think twice. Although this was a statement regularly made tongue-in-cheek, it points to period tracking as an established everyday practice that is entrenched in gendered medical regimes of knowing and controlling menstrual cycles and menstruating bodies/people.

Benefits of Using an App

Among my research participants were those for whom period tracking was a routine before they used an app and those who only began regular period tracking through one. Some started out with recording "the typical information" (first and last day of period, strength of bleeding), yet were encouraged to log further details by the app's diverse tracking categories and frequency of prompts to insert more data. Others depicted the first

[7] Differentiations between types of self-trackers, e.g. "pragmatic and enthusiastic self-trackers" (Gerhard & Hepp, 2018), provide an interesting template for comparison but they need to consider the socio-cultural specificities of documenting menstrual cycles.

few months with the app as playful exploration before deciding what to monitor—often after realising "the amount of work all the tracking means" [ptst14]. Over time, quite a few reduced their logging to a few aspects or "decided to restrict data entry to the bloody days and 2–3 PMS-days prior to that" [ptiv6]. Except for those who use the symptom-thermal method and, therefore, must be disciplined self-trackers, research participants described tracking routines of varying discipline when reflecting on their app-related data practices. In these descriptions, comparisons with other (non-digital) forms of menstrual observation were common.

According to interviewees, period trackers are "more fun but also more sophisticated than the good old paper version" [ptiv4]. An app allows them to "record all that actually happens" and offers information "with a few swipes", "in nice graphs", or "in the form of a neat overview". Besides increasing the calendar's availability through the smartphone, the app does the calculating and predicting, is perceived as "less schematic", often includes a glossary, encyclopaedia, explanatory articles or "even scientific reference", and, last but not least, "involves you differently" in dealing with the menstrual cycle. As to the different involvement through the app, two points regularly came up: the data visualisations and the push notifications. All interviewees welcomed (and were often fascinated by) the functions of processing tracking data, making it visible in charts or diagrams, and showing forecasts and connections. This would help them to become more aware of their menstrual cycle *as* a cycle and of patterns, which they may have only imagined or never considered. There was more disagreement regarding the push notifications. While some appreciated these reminders as an occasion for self-observation and for "getting more involved" with their cycle, others stated that notifications that especially prompted log entry can "turn into an obligation one is kind of dragged into" [ptiv4]. The "tone of the app" perceived as either supportive, intrusive, or disciplinary also appeared in various images or ambivalent roles interlocutors attributed to their period trackers—for example as a daily companion, advisor or backup, as fusspot or "clever calendar, which for better or worse speaks back" [ptst1].[8] Regardless of the app-user relationship, research participants highlighted their app-mediated engagement with data as a positive, notably "new way" of dealing with menstruation.

[8] A more detailed analysis of the app-user relationship captured in these images would take me in this chapter too far afield. Ambivalent feelings towards period trackers were common; see also Hamper (2020: 11).

The Promise of Data and Smart Algorithms: To Be on the Safe Side

The helpfulness of data is, according to taglines and self-representations, exactly what providers of period-tracking apps promise. To illustrate, I quote four popular period trackers:[9] The above-mentioned *Clue* invites you to "Find the unique pattern in your cycle" and offers "an algorithm that learns from the data that you enter. The more you use it, the more intelligent it gets" as well as period, PMS and fertile window "predictions, you can trust". *Glow* presents a data-driven ovulation calculator that "helps women take control of their fertility" and "predications that become smarter over time". It asks potential users if they "[W]ant a period tracker or a woman's health app you can rely on for detail and accuracy?" to confirm, "We've got you covered." *Natural Cycles* summons potential users to "take birth control in your own hands" by "track[ing] your cycle naturally and effectively" with "the only birth control app that is cleared by the FDA in the US and CE-marked in Europe as a medical device". Finally, *Flo* advertises that users can "log over 70 menstrual symptoms and activities to get the most precise AI-based period and ovulation predictions". Self-observation paired with personalised calculations and data-driven accuracy, self-knowledge through aggregated data and 'learning' algorithms—this sounds promising and fits in well with what research participants expect from or value in period trackers.

Overall, app users consider (their) period-tracking apps to be valuable as both a tool to find out more about themselves (thus creating or adding to self-knowledge) and a more graspable and formalised form of self-observation. Referring to the often felt waywardness of the human body in general and menstrual cycles in particular, some described the app and the insights gained as assuring support or as a backup of self-perception. Several felt better informed since using the app and stated that the calculations and visuals of data (i.e. displays of PMS clouds, typical cycles summarised in numbers or that "today is your calculated ovulation") would particularly help them to feel more certain and confident. As one of them put it, it is "like an insurance against failure" [ptst9]—failure understood in the sense of "getting it [her body or interpretation of hormonal processes] wrong". Very often, interviewees used in this context, the German

[9] The quotes are from the blurbs used on websites and app-store presentations of the respective apps.

phrase *sicherheitshalber* [for good measure, just to be on the safe side], a word regularly used in everyday life for all kinds of precaution that contains the German word for security in its root word. By contrast, insecurities were associated with menstruating bodies rather than data.

The idea that data provide a better access to bodies than other forms of self-knowledge or self-observation (Lupton, 2013; Berson, 2015) ties in with cultural beliefs of what counts as secure or accurate information. Participants' reliance on data points less to the alleged data fetishism commonly associated with self-trackers than to the historically established appeal of quantification in modern societies (Porter, 1995) or to the medical model of knowing and its "medical gaze" (Foucault, 2012). The notion of the body as something that is measurable, understandable, and manageable through data or assessable and judgeable in comparison with norms has a long history intertwined with biomedicine as well as its critique (Lock & Nguyen, 2010). It might not be surprising, therefore, that participants' narratives echo understandings of the body as a source or depot of data, which technology helps to read, visualise, and make sense of. In contrast to the "reading" of their bodies through self-perception, interviewees often considered data provided by their period tracker to be more "reliable", "objective", and "credible" (see also Ruckenstein, 2014: 10). Some used this approved credibility of (visualised) data for their own purpose, such as to support their subjective claims or actions towards a doctor or a sexual partner. The view that menstrual data is not only useful but also gives a sense of confidence and security is where the wishes of users and the advertising of app providers meet. The promise of period trackers is one of self-knowledge and security through accuracy and, as a result, self-empowerment by enabling users to take control over assumed chaotic bodily processes with the help of up-to-date science and technology. In other words, it is a story that once again tells of either the utopian character of or the mythically charged belief in (big) data and datafication (boyd & Crawford, 2012). So what is the problem, when apps provide what users ask for?

Critiques of Period Trackers' Smartness: Two Safety Concerns or Data Insecurities

With its rising popularity,[10] menstrual cycle apps have increasingly become the subject of warnings in the media. Both the apps' smartness and the privacy of the data entered gave rise to criticism. In Germany, the results of a test of 23 popular period trackers by the state-funded German Foundation of Product Testing received a lot of attention. They had rated most of the tested apps as "faulty" because of their predictive deficiencies and, in addition, had classified many of them as "critical" regarding privacy (Stiftung Warentest, 2017).

Unwise Algorithms and Old Methods Repackaged

Dealing with the question of how reliable period trackers are in predicting ovulation and fertility (and, therefore, how good they are for achieving conception or contraception), medical studies detected that the apps' algorithms are in most cases neither particularly smart nor precise (e.g. Duane et al., 2016; Moglia et al., 2016). Most apps not only lack independent, scientific studies proving their efficacy but also base their predictions primarily on the average data of previous cycles while ignoring information on the current one (Freis et al., 2018). These tests call into question the extent to which most period tackers actually present an improvement (ibid.). To understand what went wrong with the smart determination of fertility, one should go back to the pen-and-paper version of period tracking. Schlünder (2005) has convincingly shown that today's well-known menstruation calendar resulted from a medical dispute in the 1920s over the question of how to calculate "female natures" and eventuated in the calendar-based contraception method by Knauss and Ongino in the 1950s. The method demands the recording of at least 12 menstrual cycles and applies the retrospectively gained data of the longest and shortest cycle in order to estimate the length of the pre-ovulatory infertile phase, the fertile days, and the start of the post-ovulatory infertile phase. In other words, the method is about recognising a pattern in retrospect and draws on menstrual data from the past to forecast the next ovulation and fertile

[10] It is necessary to note that apart from the high download numbers displayed in app-stores, public figures on the actual user-base or regular uses of period-tracking apps are lacking.

window. A rough estimation, though, does not count as a safe method of birth control. Considering normal menstrual cycles varying between 21 and 35 days, (inter)individual variability of cycle length, normal fluctuations and diverse activities influencing ovulation (e.g. travelling, illness, sport, lack of sleep, alcohol—simply life as it is lived), predicting distinct fertile windows with retrospective data and mathematics or statistics is a rather risky business—regardless of whether a human or an app calculates.

Various professionals in the fields of gynaecological care, family planning, and sex education, who I met during my research, viewed period trackers as an old method repackaged in shiny new clothing and their use (especially among young people) with concern. Albeit controversial, the German Professional Association of Gynaecologists even attributed an increase in the number of abortions in 2017 to the use of menstrual cycle apps (BVF, 2018). In a similar vein, the app *Natural Cycles*, marketed as an effective hormone-free contraceptive method, hit the headlines in 2018: In Sweden, a large hospital reported that 37 of 668 women, who sought an abortion in the last quarter of 2017, claimed that they had been relying on the *Natural Cycles* app for birth control (Wong, 2018). The app includes the basal body temperature, measured and entered by users in the morning, and informs users whether it is safe to have unprotected sex. Yet, despite this extension of data, again the app only considers previous cycles for predictions (Freis et al., 2018: 5).[11] Not all period trackers employ calendar methods. In the test carried out by the German Foundation of Product Testing, the three apps judged best ("good") were those based on the symptom-thermal method. What is also known as Natural Family Planning (NFP) employs parameters of the current cycle (self-observation of bodily signs such as temperature and cervical mucus changes) to determine fertile days. If applied correctly, it is considered a safe (albeit marginalised) method of birth control—which works well for some people but requires daily, disciplined, and differentiated self-observation/-analysis and requires knowledge, instruction, effort put into learning, and regularity.[12] The key benefit of an NFP-app is that it analyses

[11] Later that year, Swedish authorities cleared the app because they could confirm the indicated failure rate of 7 per cent but asked the company to state the risk of unwanted pregnancy more clearly (Leonard, 2018). Since then, some other period trackers also notify users that they should not use the app as a contraceptive method.

[12] Against the background of growing criticism of (the side effects of) hormonal contraception in Germany, there seems to be a new interest in the method. Rotthaus (2020) studied German users of NFP-apps, who turned to the method as a liberating alternative to the pill

the data for the user. Among research participants, only a very few employed this method in conjunction with their app.

Critical Dataflows of Intimate Data

In addition to failing to deliver on their promises of predictive power, apps are just as often, if not more so, criticised for their problematic data collection and sharing practices. For instance, Burke (2018) warned "Your Menstrual App Is Probably Selling Data about Your Body!" and drew on insights by a project of the Brazil-based feminist digital rights organisation *codingrights.org* on "How to turn your period into money (for others)" (Felizi & Varon, n.d.). Likewise, other NGOs advocating for digital rights, privacy, and data protection (e.g. the *Tactical Tech Collective*, the *Electronic Frontier Foundation* or *Privacy International*) have analysed the data-sending and security properties of period trackers in more detail and detected flaws in several of them, such as invasion of privacy and data protection issues. Most period trackers collect an enormous quantity of data and meta-data, share parts of this data without specifying with whom, or allow extensive third-party requests (Rizk & Othman, 2016; Quintin, 2017); some were even exposed as having automatically transferred data to *Facebook* or other third-party services (Privacy International, 2019). It is for reasons like these that period trackers have been in the limelight recently: as vampiric violators of privacy that profit from sensitive user data (Kresge et al., 2019). Subsequently, in response to this bad press, several app providers updated their privacy policies and some made changes to their data-sharing practices. However, machine learning-based systems can access an increasingly larger database of intimate data. It seems to be the case the creators of these apps would rather develop algorithms that process period-tracking data for commercial purposes (such as targeted advertising, data brokering, and analytics) than to improve those that predict fertility.

because it is hormone-free. For details on NFP, its reconfiguration of bodies, technologies, and gender relations in the 1980s as well as the involved mode of knowledge production see DeNora (1996).

Negotiating Data Insecurities—Pitfalls, Lessons Learned, and New Competences

Nonetheless, my interlocutors are among the many people who do use period-tracking apps. In view of the criticisms, one could easily sketch out a story of passive users delegating tasks to their apps and relying against better knowledge or judgement on false promises of security. Yet, this would neither do justice to the thoughts and practices of research participants nor help us in understanding why users tolerate app-related data insecurities, participate in sharing data, or find period trackers valuable (see also, Sharon & Zandbergen, 2017). So how do the interviewed app users encounter the two outlined forms of app-related data insecurities?

Data Insecurities 1: Understanding Menstruating Bodies with and Through Data

While interviewees changed their period tracker when they disliked changes to app features after an update or got annoyed with the way the app addressed them ("I am 29, I am not an *Ovy*-girl!"), they expressed hardly any concern with the ways in which the app actually arrives at its calculations. Just as in other cases of mundane technology, many simply trusted or assumed that their app would work properly. As mentioned above, interlocutors view period trackers not merely as calculators to which users delegate tasks but as "a new way" or as an additional instrument to deal with the insecurities of their menstruating bodies and feel more knowledgeable and certain about it. As an instrument of sorting, evaluating, and makings sense of individual bodily experiences, which guides participants in "having everything in black and white" and to help "sharpen" or "underpin" self-perception, the value of its use enfolds between the poles of self-empowerment and normalisation (see also Rotthaus, 2020) and the poles of security gained by self-awareness or by misguided certainty. Some interviewees reflect quite critically about the persuasive and normative power of data and its visualisation through the app: In particular, the constant display of the cycle (three months in advance) would easily eclipse inherent uncertainties of calculated probabilities. Yet in the accounts of research participants, insecurities are neither only caused by bodies nor simply resolved by data.

Rather, insecurities arise when participants align their embodied selves with their data(fied) bodies, that is, when they match their self-perception and own interpretations of tracking data with the predictions and the analysis of their apps. This process of "continuous synchronisation" [ptst7], to put it technically, is not a smooth process. Some took potential or occurring discrepancies lightly and the apps' output as a "rough orientation to think with", others found it "hugely unsettling". The expressed insecurities indicated and depended on a varying distance or proximity to the app/data-based mediation of one's body through default tracking categories. Nonetheless, and as already mentioned, all interviewees emphasised that their experience of the menstrual cycle has changed through app-based period tracking and generated learning processes, which resulted in an increased or new body competence.

Ironically, for several participants, this very competence included the possibility of emancipating themselves from the app to a certain extent. Moments of questioning the app's translation of intimate experiences into quantified data points were underpinned by self-awareness, everyday empiricism (e.g. when the app's predictions were implausible or repeatedly wrong), and newly acquired self-knowledge. For instance, one 20-year-old interviewee learned through her app and became interested in the variable nature of cervix mucus during the cycle. While she finds the symptom-thermal method "too much effort and inapplicable" for her, she started self-observing and logging cervix mucus shifts as well as searching for further information. As we met, she stated proudly that (now after a year) she would confidently assess the app's calculation of her ovulation. "It's like a game: does the app get me right?" she says. The fact that the app's prediction is not always correct according to the interviewee's self-observations has not yet caused her to change the app or to stop engaging with its calculations: "Do I get my cycle right, is [or remains] an important question, too." This and other examples in my material show that while period trackers cannot shake off all the insecurities involved in the taming or the making sense of menstruating bodies/selves, they include for some users the opportunity to move with the app beyond the app. Several participants decide on a case-to-case basis if they give more or less weight to the app's interpretation, while others simply ignore its estimations or recommendations.

One the one hand, interviewees find their period trackers convenient, helpful, and trust technical solutions, on the other hand do not passively place blind faith in mobile apps. For most of them, delegating tasks such as fertility calculations to an app does not mean to delegate personal responsibility, such as contraception, for example. At this point it should be noted that the accusation of a naïve or irresponsible behaviour (here the use of technology such as period trackers) is by no means new, especially in the culturally and politically charged area of contraception, and is once again directed solely at menstruating people (here users of period trackers). In sum, period trackers are another tool for research participants to get to know themselves and better live with menstruation, nothing more, and nothing less. What is new is that this tool makes users more knowledgeable through data but also "more knowable to an emerging set of data-driven interests" (Crawford et al., 2015: 495).

Data Insecurities 2: Sidestepping the Vague Privacy of Logged Data

While the non-users I met more often criticised or suspected privacy issues in sharing intimate data with period trackers, the app-using interviewees rarely brought up the issue themselves. Some noted that they made use of the option to password protect access to their app/data but did so without considering the cloud storage of these data, even though they all know, their "private dialogue" with the app is not limited to their mobile phone. Their attitudes and actions can be characterised by the "'privacy paradox' where intentions and behaviours around information disclosure often radically differ" (Shklovski et al., 2014: 2347). Quite a few waved off or shrugged their shoulders when asked about uncertainties emerging from sharing menstrual data and said this is neither a specific problem with period trackers, nor an occasion for them to worry. With more or less regret, they framed such privacy risks to be "part and parcel" of using mobile apps and devices. In regard to this tendency to normalise unanticipated data collection, use, and surveillance, Levy (2015) and Lupton (2015) have elaborately problematised the extraction of data on bodies, sexuality, and intimate relationships. It remains a central challenge in retaining and reclaiming privacy and autonomy in times of the databased commodification of users (see Véliz, 2020).

It was only when I followed up on the subject during our conversation that participants started to reflect on what they actually know or want to know about the whereabouts of their data, why it matters to them, or whether they should be more concerned. Lately, though, there seems to be slightly greater awareness. This also became apparent when I recently asked someone whether we could continue our talk in an interview and immediately got the response, "Ah, I bet you want to talk to me about data protection problems, don't you?" [ptsm24]. Like some others, this interlocutor was attentive to insecurities regarding privacy but was not too concerned as she "doesn't share everything" with her period tracker. To clarify her attitude, she remarked on occasions where it became obvious through the app's prompts and recommendations that the app "doesn't capture me completely correctly" and that it was "in some areas, in the dark". Like her, some decide with caution which information they share, omit, or enter incorrectly, "Why should I tell my period tracker about my hours of sleep, party life or libido?" [ptst21]. Yet, these forms of intentional non-disclosure were rare. More frequent were less intentional data gaps that resulted from irregular or partial tracking or from simply forgetting to enter personal data. Some pointed out these gaps when explaining that they do not view their random data points as particularly valuable to companies that try to make sense of their fragmented data doubles—distinguishing: "this is my data, but it's not me" [ptst13].

The most data-conscious interviewee I have met so far had distinctly looked for an "uncritical" period tracker. She chose one of the few that allows for the use of the app without a personal account (a precondition for cloud data storage, which others appreciated as backup of their data). Storing period data only locally on her phone meant for her that she was able to maintain a certain sense of control of her data. Albeit, though, once her phone had broken and made her feel quite insecure: "I was really crushed, I can tell you... I lost my menstrual history! Three years of properly documenting my cycle, [snaps her fingers] just gone" [ptiv2]. Starting over with a new phone and a "blank" app, she worried about the accuracy of predictions of an algorithm that would not yet know or, rather, had forgotten to know her. She almost regretted opting-out from cloud storage, but after a while realised that having a long record of cycles was less important than she thought—both for herself and for the app's calculations. Instead, the data loss led her to question the personalisation of the predictions she received from the app as well as the personal values of a "menstrual data history". "[Mostly] it's something the app suggests, [speaks in different voice]: 'give us data, [app name] is getting smarter...' I rarely use the app to look back. Thanks to my regular cycle, the averages displayed are not that surprising after all. I didn't keep my paper menstrual calendars either." With respect to addressing and approaching the second kind of app-related data insecurity, this interviewee was an exception. In contrast to the first form of data insecurity, research participants appeared to be less interested in gaining knowledge in this (ambivalent) aspect of their period trackers.

Conclusion

This chapter took period-tracking apps and menstrual self-observation as an entry point for exploring everyday data practices and the ways app users engage with insecurities of and through data. In juxtaposing the critical public discussion of period trackers with evaluations by my research participants, the chapter discussed two forms of insecurity with respect to period-tracking data and apps. These insecurities concern for one, the datafication of menstruating bodies, and for the other, the risks involved in sharing intimate data with an app and making it available to unknown others. In comparison, the former generates more agential possibilities and moments of confidence for users than the latter. The first form of data insecurity emerges where bodies are incalculable or where embodied selves and datafied bodies mismatch. It addresses the limits of period trackers' offer to counter the lack of control over the body with data and help users feel, if not in control of, at least less dependent on menstrual processes based on these data. The second form of data insecurity arises from the fact that what happens to data entered into the app is mostly beyond app users' control. The two kinds of data insecurities put emphasis on different app-related data practices—either, on data practices *with* mobile apps such as the ways in which users engage with apps and data to obtain their aims, or on the data practices *of* mobile apps, such as the ways apps allow app providers (or third parties) to gather and distribute data. In terms of both insecurities, the overall verdict on period trackers is that most of them are in a twofold way, not safe to use. This, however, does not stop users from using them. As far as the two insecurities are concerned, for my interviewees, the experienced benefits of period trackers outweigh their experienced or perceived harm. As I have shown, users deal with the ambivalences of data and app-related insecurities in their own way.

Such forms of engagement with data provide a good starting point for everyday perspectives in critical data studies. Examining how app users experience app-mediated menstrual cycles and negotiate the uncertainties of data-driven predictions and untrustworthy dataflows allows for the problematising of overly simplistic stories about processes of datafication. Surely, participants' gained body competence, their moving beyond the app's predictions, their evasion of sharing data, and even their deliberate ignoring of it cannot stand in for a counternarrative of subversive app users resisting the quantification of bodies, the persuasiveness of data (visualisation), or the flipside of using an app's cloud storage facility. What I told instead, based on my empirical material, was not a story of gullible users but of pragmatic uses. The accounts of interviewees' equally carefree, wayward, and reflexive use of period trackers tell of their being empowered and caught up through data as well as their enmeshment with and participation in the sociotechnical constellations of app-based data use. My findings show how people in their daily lives embrace data technologies such as apps while tackling the possibilities and imponderabilities of datafication but also point to the powerful arrangements and conditions in which their data encounters take place.

The practices, with which research participants put up with the deficiencies of period trackers in terms of data insecurities (in particular the second form), do not seem to leave much room for optimism. For this reason, I would like to end on a positive note. In 2019, the *Bloody Health Collective*, a feminist open-source project based in Berlin, launched the beta version of ***drip*** to provide a more secure, transparent, and non-commercial alternative to available period trackers. In January 2021, they released a redesigned and the first stable version of the app. Underlined by slogans like "your data, your choice" or "your body is not a black-box" the app only stores data locally on one's phone and is based on the symptom-thermal method. It invites users to look into the workings of its algorithm/calculations and reminds them that they do not need an app to understand their cycle. Regarding the two data insecurities, this, indeed, seems to be an app that is safe for users. I cannot predict how successful this app will be, or whether it will have any impact on the period tracker genre. Although such an intervention may only provide a safe period tracker for users actively seeking one, such new forms of feminist data activism also provide a possible alternative for imagining how life with data and its power could be otherwise. Everyday perspectives in critical data studies can contribute to such efforts of data activism by exploring people's experiences of and practices with and through data in order to understand how people live and could better live with data (to return to Kennedy's proposal quoted in the beginning of this chapter). Engaging with the mélange of the datafied everyday, life can not only empirically expand critical data studies, but also help reshape the circumstances in which data is used.

References

Berson, J. (2015). *Computable bodies: Instrumented life and the human somatic niche*. Bloomsbury.

Bertotti, A. M. (2013). Gendered divisions of fertility work: Socioeconomic predictors of female versus male sterilization. *Journal of Marriage and Family, 75*, 13–25.

Boston Women's Health Collective. (1970). *Women and their bodies: A course*. https://www.ourbodiesourselves.org/cms/assets/uploads/2014/04/Women-and-Their-Bodies-1970.pdf

boyd, d., & Crawford, K. (2012). Critical questions for big data. *Information, Communication & Society, 15*(5), 662–679.

Burke, S. (2018, May 11). Your menstrual app is probably selling data about your body. *Broadly (Vice)*. https://www.vice.com/en/article/8xe4yz/menstrual-app-period-tracker-data-cyber-security?

BVF. (2018, March 7). Alarmierende Zunahme von Schwangerschaftsabbrüchen: Pille danach und Verhütungs-Apps. *Press Release of the Professional Association of German Gynaecologists*. https://www.bvf.de/aktuelles/pressemitteilungen/meldung/news/alarmierende-zunahme-von-schwangerschaftsabbruechen-pille-danach-und-verhuetungs-apps/

Crawford, K., Lingle, J., & Karppi, T. (2015). Our metrics, ourselves: A hundred years of self-tracking from the weight scale to the wrist wearable device. *European Journal of Cultural Studies, 18*(4–5), 479–496.

Cukier, K., & Mayer-Schoenberger, V. (2013). The rise of big data: How it's changing the way we think about the world. *Foreign Affairs, 92*, 28–40.

DeNora, T. (1996). From physiology to feminism: Reconfiguring body, gender and expertise in natural fertility control. *International Sociology, 11*(3), 359–383.

Duane, M., Contreras, A., Jensen, E. T., & White, A. (2016). The performance of fertility awareness-based method apps marketed to avoid pregnancy. *The Journal of the American Board of Family Medicine, 29*(4), 508–511.

Epstein, D. A., Lee, N. B., Kang, J. H., Agapie, E., Schroeder, J., Pina, L. R., Fogarty, J., Kientz, J. A., Sean, A., Munson, S. A. (2017). Examining menstrual tracking to inform the design of personal informatics tools. *Proceedings of the SIGCHI conference on Human Factors in Computing Systems 2017 May 2*, 6876–6888.

Faulkner, W. (2001). The technology question in feminism: A view from feminist technology studies. *Women's Studies International Forum, 24*(1), 79–95.

Felizi, N., & Varon, J. (n.d.). Menstru apps: *How to turn your period into money* (for others). https://chupadados.codingrights.org/en/menstruapps-como-transformar-sua-menstruacao-em-dinheiro-para-os-outros/

Fiore-Gartland, B., & Neff, G. (2015). Communication, mediation, and the expectations of data: Data valences across health and wellness communities. *International Journal of Communication, 9*, 1466–1484.

Fluhrer, A. L. (2018, April). *Politics of fertility-tracking apps: A qualitative study with founders and developers.* Master thesis, Europäische Ethnologie HU Berlin.

Ford, A. (2019, November 12). *Period tracking apps: Something old, something new.* [Blog contribution]. http://blog.castac.org/2019/11/period-tracking-apps-something-old-something-new/

Foucault, M. (2012). *The birth of the clinic.* Routledge.

Freis, A., Freundl-Schütt, T., Wallwiener, L.-M., Baur, S., Strowitzki, T., Freundl, G., & Frank-Herrmann, P. (2018). Plausibility of menstrual cycle apps claiming to support conception. *Frontiers in Public Health, 6*, 98. https://www.frontiersin.org/article/10.3389/fpubh.2018.00098

Gerhard, U., & Hepp, A. (2018). Appropriating digital traces of self-quantification: Contextualising 'pragmatic' and 'enthusiast' self-trackers. *International Journal of Communication, 11*, 638–700.

Gusterson, H. (1997). Studying up revisited. *PoLAR, 20*(1), 114–119.

Hamper, J. (2020). 'Catching ovulation': Exploring women's use of fertility tracking apps as a reproductive technology. *Body & Society, 26*(3), 3–30.

Iliadis, A., & Russo, F. (2016). Critical data studies: An introduction. *Big Data & Society, 3*(2) 1–7.

Kennedy, H. (2018). Living with data: Aligning data studies and data activism through a focus on everyday experiences of datafication. *Krisis: Journal for Contemporary Philosophy, 1*, 18–30.

Kimport, K. (2018). More than a physical burden: Women's emotional and mental work in preventing pregnancy. *Journal of Sex Research, 55*(9), 1096–1105.

Kitchin, R. (2014). *The data revolution: Big data, open data, data infrastructures.* Sage.

Klein, J. (2020, March 31). *What it means for an app to be "for women": Is it any more than a Pink Coat of Paint.* https://www.inputmag.com/culture/what-it-means-for-app-to-be-for-women

Kresge, N., Khrennikov, I., & Ramli, D. (2019, January 24). Period-tracking apps are monetizing women's extremely personal data. *Bloomberg Businessweek.* https://www.bloomberg.com/news/articles/2019-01-24/how-period-tracking-apps-are-monetizing-women-s-extremely-personal-data

Leonard, K. (2018, September 13). Swedish authorities clear 'Natural Cycles' birth control app. *Washington Examiner.* https://www.washingtonexaminer.com/policy/healthcare/swedish-authorities-clear-natural-cycles-birth-control-app

Levy, K. E. C. (2015). Intimate surveillance. *Idaho Law Review, 51*, 679–693.

Light, B., Burgess, J., & Duguay, S. (2018). The walkthrough method: An approach to the study of apps. *New Media & Society, 20*(3), 881–900.

Lock, M., & Nguyen, V. K. (2010). *An anthropology of biomedicine*. Wiley-Blackwell.
Lupton, D. (2013). Quantifying the body: Monitoring and measuring health in the age of mHealth technologies. *Critical Public Health, 23*, 393–403.
Lupton, D. (2015). Quantified sex: A critical analysis of sexual and reproductive self-tracking using apps. *Culture, Health & Sexuality, 17*(4), 440–453.
Lupton, D. (2018). How data come to matter? Living and becoming with personal data. *Big Data & Society, 5*, 1–11.
Michael, M., & Lupton, D. (2016). Towards a manifesto for the 'public understanding of big data'. *Public Understanding of Science, 25*(1), 104–116.
Moglia, M. L., Nguyen, H. V., Chyjek, K., Chen, K. T., & Castaño, P. M. (2016). Evaluation of smartphone menstrual cycle tracking applications using an adapted APPLICATIONS scoring system. *Obstetrics & Gynecology, 127*(6), 1153–1160.
Mol, A., Moser, I., & Pols, J. (Eds.). (2010). *Care in practice: On tinkering in clinics, homes and farms*. Transcript Verlag.
Morris, J. W., & Murray, S. (Eds.). (2018). *Appified: Culture in the age of apps*. University of Michigan Press.
Pantzar, M., & Ruckenstein, M. (2017). Living the metrics: Self-tracking and situated objectivity. *Digital Health, 3*, 1–10.
Porter, T. M. (1995). *Trust in numbers: The pursuit of objectivity in science and public life*. Princeton University Press.
Privacy International. (2019, September 9). *No body's business but mine: How menstruation apps are sharing your data*. https://www.privacyinternational.org/long-read/3196/no-bodys-business-mine-how-menstruations-apps-are-sharing-your-data
Quintin, C. (2017, July). The pregnancy panopticon. *Electronic Frontier Foundation, White pape*. https://www.eff.org/files/2017/07/27/the_pregnancy_panopticon.pdf
Rizk, V., & Othman, D. (2016). Quantifying fertility and reproduction through mobile apps: A critical overview. *ARROW for Change, 22*(1), 13–21.
Rommes, E., Van Oost, E., & Oudshoorn, N. (1999). Gender in the design of the digital city of Amsterdam. *Information, Communication & Society, 2*(4), 476–495.
Rotthaus, H. (2020). Aushandlungen von Schwangerschaftsverhütung im Kontext digitaler Selbstbeobachtung. *Hamburger Journal für Kulturanthropologie, 11*, 1–9.
Ruckenstein, M. (2014). Visualized and interacted life: Personal analytics and engagements with data doubles. *Societies, 4*, 68–84.
Ruckenstein, M., & Schüll, N. D. (2017). The datafication of health. *Annual Review of Anthropology, 46*, 261–278.
Schlünder, M. (2005). Die Herren der Regel/n. Gynäkologen und der Menstruationskalender als Regulierungsinstrument weiblicher Natur. In

C. Borck, V. Hess, & H. Schmidgen (Eds.), *Mass und Eigensinn. Studien im Anschluss an Georges Canguilhem* (pp. 157–195). Wilhelm Fink.

Schmechel, C. (2016). “Der vermessene Mann?”: Vergeschlechtlichungsprozesse in und durch Selbstvermessungspraktiken. In S. Duttweiler, R. Gugutzer, G. Passoth, & I. Strübing (Eds.), *Leben nach Zahlen. Self-Tracking als Optimierungsprojekt?* (pp. 141–159). transcript.

Shklovski, I., Mainwaring, S. D., Skúladóttir, H. H., & Borgthorsson, H. (2014). Leakiness and creepiness in app space: Perceptions of privacy and mobile app use. CHI ‘14: Proceedings of the SIGCHI Conference on Human Factors in Computing Systems, April 2014, 2347–2356.

Sharon, T., & Zandbergen, D. (2017). From data fetishism to quantifying selves: Self-tracking practices and the other values of data. *New Media & Society, 19*, 1695–1709.

Stiftung Warentest. (2017, December). Zyklus Apps: im Test. Testergebnisse 12/2017. https://www.test.de/Zyklus-Apps-im-Test-5254377-0/

Van Dijck, J. (2014). Datafication, dataism and dataveillance: Big data between scientific paradigm and ideology. *Surveillance & Society, 12*(2), 197–208.

Véliz, C. (2020). *Privacy is power: Why and how you should take back control of your data*. Random House.

Weiner, K., Will, K., Henwood, F., & Williams, R. (2020). Everyday curation? Attending to data, records and record keeping in the practices of self-monitoring. *Big Data & Society, 7*, 1–15.

Wong, J. C. (2018, January 17). Birth control app reported to Swedish officials after 37 unwanted pregnancies. *The Guardian*. https://www.theguardian.com/technology/2018/jan/17/birth-control-app-natural-cycle-pregnancies

Open Access This chapter is licensed under the terms of the Creative Commons Attribution 4.0 International License (http://creativecommons.org/licenses/by/4.0/), which permits use, sharing, adaptation, distribution and reproduction in any medium or format, as long as you give appropriate credit to the original author(s) and the source, provide a link to the Creative Commons licence and indicate if changes were made.

The images or other third party material in this chapter are included in the chapter's Creative Commons licence, unless indicated otherwise in a credit line to the material. If material is not included in the chapter's Creative Commons licence and your intended use is not permitted by statutory regulation or exceeds the permitted use, you will need to obtain permission directly from the copyright holder.

Community Rankings and Affective Discipline: The Case of Fandometrics

Elena Maris and Nancy Baym

In 2015, the microblogging site and social network Tumblr launched Fandometrics, a project to track and rank fan engagement on the platform. The most public-facing aspect of Tumblr's Fandometrics is its weekly fandom rankings for everything from TV shows and movies to music and video games. Tumblr's (2020) "About Fandometrics" page describes the rankings as representing, "…each fandom's influence across Tumblr." In response to Fandometrics, one cultural observer predicted the rankings would result in fandoms that "duke it out for first place on the leaderboard" (Baker-Whitelaw, 2015). Tumblr is not alone in mobilising metrics to quantify and leverage fan communities. For example, fan-focused wiki site Wikia (2020) calculates a daily Wiki Activity Monitor (WAM) score, a similar ranking system that Wikia calls "an indicator of the

E. Maris (✉)
University of Illinois, Chicago, IL, USA
e-mail: emaris@uic.edu

N. Baym
Microsoft Research, Redmond, WA, USA
e-mail: baym@microsoft.com

© The Author(s) 2022

A. Hepp et al. (eds.), *New Perspectives in Critical Data Studies*,
Transforming Communications – Studies in Cross-Media Research,
https://doi.org/10.1007/978-3-030-96180-0_14

strength and momentum of a Fandom community." Fan fiction sites like AO3 publish fandom "Stats," and a number of fan-led fan data and fandom metrics projects also exist. In this chapter, we focus on Tumblr's Fandometrics, what it seeks to do, how it functions, and centrally, how the communities it measures are impacted by its rankings.

We conducted interviews with key fandom data/metrics workers and experts, and analysed Tumblr's Fandometrics site and other fandom metrics efforts, online user discourse, and trade and popular press. Building on work on audience measurement (Ang, 1991; Baym, 2013; Napoli, 2003) and the changing social role of metrics (Beer, 2016; Gillespie, 2016; Kennedy, 2016), we contextualise and locate Fandometrics' community rankings within larger traditions of audience and social media measurement. We demonstrate that Fandometrics encourages social jostling by online communities for relevance on the Tumblr platform, and within fandom and wider culture. By equating the strength of *communities* with their status as influencers or markets, these measurements and rankings usher fans towards subjectivities that put data and quantitative rankings at the centre of societal value and inter-community relationships. We argue that as metrics become more visible to users, some communities respond with a kind of *affective discipline*, at times exaggerating, restraining, cloaking, or reconfiguring positive and negative affect in their online engagement in line with algorithmic requirements for measurement. People tame themselves to tame the algorithms they know are at work, but which remain unknowable to them. These increasingly visible community metrics can affect users' everyday online practices and the subjectivities they engender.

We begin by locating Fandometrics relative to other forms of audience measurement. Following that, we identify and discuss the affective and social implications for the communities ranked by Tumblr's Fandometrics, including: (1) the need to be large and 'loud' to appear at all in the rankings and the affective discipline taken on by users due to Fandometrics' lack of sentiment measures; (2) that inevitably many communities will therefore feel (and effectively *be*) silenced within Fandometrics; and (3) that the rankings can represent industrial attempts at fostering competition between communities through understandings of social value based on quantification, leading to significant user anxiety about their standings. Finally, we discuss efforts by user communities to resist industrial measurement, including withdrawal from Fandometrics and/or the communities that value its rankings, and efforts to claim back their own data through self-measurement. These efforts further illustrate the social, political, cultural, and affective impacts of industrial measurement and ranking of online communities. Further, we argue that with platforms' increasing concentration of data power, critical data studies must attend to such community-driven alternative models of data and metrics. Overall, the Fandometrics phenomenon reflects larger societal anxieties about value, relevance, and power in increasingly metrified online spaces.

Fan Data and Fandom Metrics

What is today's Tumblr Fandometrics began simply as a "Year in Review" in 2013, an ambitious project thought up by Danielle Strle, the company's then Director of Community and Content. It was an exploratory attempt at representing the most reblogged tags on Tumblr. Amanda Brennan, a new hire tasked to put the content together explained that first ranking to us:

> And I got a big spreadsheet and it was just every tag used on Tumblr. And we sorted it by reblogs. And I read the spreadsheet by hand and made those lists just copying and pasting and lots of color coding. And it was my first month on the job and it was just the most wild project I've ever worked on.

Brennan (who asked to be identified and is currently Head of Editorial at Tumblr) told us that after the list was published, Strle wanted to produce a more regular ranking:

> Danielle was kind of like, okay, so how do we take this idea and make it something that's constantly there? Why should we wait a whole year to show off our fandoms? Because Tumblr is the home of fandom. It's where people go to really celebrate those interests.

As its name says, the weekly Fandometrics focused on fandom, in contrast to Tumblr's Year in Review tracking the most popular tags on Tumblr under a large variety of subject headings (e.g. Tumblr, DIY, Gif, etc.). Fandometrics also produces an annual Year in Review for Tumblr. Fandometrics weekly categories include Movies, TV Shows, Music, Ships, Anime & Manga, and other fandom-focused content. Each week, the results are ranked from 1–20, with marks indicating upward or downward movement on the charts from the previous week. Tumblr explicitly tells users that the Fandometric rankings are generated by a secret algorithm, explaining the engagement elements measured but not the weights given for each. The algorithmic nature of the rankings is emphasised and often referred to in the light voice of the Tumblr copy that accompanies Fandometrics posts, with one post declaring: "Hot off the algorithms, it's Fandometrics."

Locating Community Rankings in Social Media and Audience Measurement

Fandometrics' algorithmic measurement is part of a longer history of efforts at buying and selling audiences for commercial purposes (Ang, 1991; Napoli, 2003). It can be distinguished from those efforts in terms of what it measures and its visibility to users. Tumblr describes Fandometrics as a measure of various *fandoms*' "influence." The focus on measuring a community's influence is distinct from measuring users or viewers in atomistic demographic categories, measuring networks in order to assess influential audience members, measuring affect in online chatter, and even from trending topics.

The Fandometrics rankings may in many ways most resemble social media "Trending" lists that publicly display a ranking of the most discussed topics on a platform in near real-time. And indeed, much of what we have learned about such lists (see Gillespie, 2016) is extremely applicable to understanding the social implications of the Fandometrics rankings. However, there are key differences between typical social media trending lists and the data collection, measurement, and discursive work involved in executing Tumblr's Fandometrics. Gillespie defined trending algorithms as inclusive of "the myriad ways in which platforms offer quick, calculated glimpses of what 'we' are looking at and talking about" (2016, p. 56). Fandometrics differs in that it could more accurately be said to measure the "we's" doing the talking. Trends are metrics of social activity (ibid). Fandometrics might be considered metrics of social *communities*. The distinction between what is trending and what Fandometrics measures can also be illustrated through example. While the Fandometrics Movies ranking may list the animated film *Zootopia* in the top 10, users know the high rank does not necessarily indicate the film is having broad influence or doing huge viewership numbers. Indeed, Zootopia often makes the weekly Tumblr rankings years after it was released—films on

Twitter, for example, would be most likely to trend on their release date. Rather, the film's placement on Fandometrics demonstrates the high activity of *Zootopia* superfans on Tumblr, and thus *their* influence on-platform. Tumblr's encouragement of these communities to propel themselves up the rankings cements this intent. Fandometrics is meant to measure and represent the "influence" or "strength" of particular communities that cohere around topics, rather than the topics themselves.

Fandometrics also resembles trending—and is distinct from demographic, influencer, and affect strategies for measuring audiences—in its visibility to users. This user-facing side of trending lists and similar social media metrics can obscure the tracking and trading of audiences/users that is core to algorithmic social media (Baym, 2013). Gillespie explains, "We might think of trends as a user-facing tip of an immense back-end iceberg, the enormous amount of user analytics run by platforms for their own benefit and for the benefit of advertisers and partners, the results of which users rarely see" (2016, p. 64). Fandometrics takes a step out of the murkiness of social media data collection efforts to quite candidly make it known to users that their communities are what are being measured, insinuating value to users almost solely in the act of being quantified (as opposed to typical algorithmic sells that engagement will lead to more relevant content).

Platforms navigate tensions in serving multiple constituencies (Gillespie, 2010). Indeed, Tumblr's Fandometrics, Wikia's WAM, and other attempts to measure fan activity are often touted as benefitting multiple stakeholders. Fandometrics is framed as being first and foremost for the fans. Bea Vantapool, a Senior Editorial Strategist at Tumblr (who asked to be identified), told us about the rankings, "They are for Tumblr users...We want them to feel represented, and we want them to know that we love the same things they do." Hearn argues about rankings and ratings systems, "...it is crucial to note that what is extracted from the expression of...feeling is valuable only to those who develop, control and license the mechanisms of extraction, measurement and representation, not for the people doing the expressing" (2010, p. 423). And Fandometrics does serve stakeholders other than fans. The data collected and represented tell Tumblr about its own users and potentially, their content preferences. Indeed, Vantapool told us Fandometrics is: "For us as well...so we know what people like so we can gear our social posts toward that type of thing."

Fandometrics' placement in the Tumblr organisation may indicate who it is *most* for. Fandometrics is part of the Partnerships division of Tumblr, a marketing side of operations. In a news interview about the launch of Fandometrics, Tumblr's head of media Sima Sistani explained about the metrics' market value, "[S]mart social marketers are moving away from measuring success in terms of real-time conversations, instead focusing on building momentum through influential fan communities that serve as powerful brand advocates" (Jarvey, 2015). Thus, it becomes clear the fan communities themselves are what hold value for the platform and outside commercial actors. It is not clear all of the ways Tumblr might partner with outside media and brands through Fandometrics data and metrics, but it is certainly framed as an important data-driven opportunity that delivers particular data about highly invested and digitally active, self-organized communities. Powers notes that, "…trends course at warp speed through our social media platforms and evermore sophisticated analytics aim to interpret their signals" (2018, p. 16). Indeed, Sistani framed Tumblr's fan data as a key analytic meant to provide important cultural insights: "…our Fandometrics provides a colorful and meaningful glimpse into the zeitgeist" (Jarvey, 2015).

Fandometrics thus offers an interesting blend of the claims and implications of traditional audience measurement, big data and metrics, and social media monitoring and tracking. Relevant then to understanding the Fandometrics phenomenon is our prior knowledge about quantification: social media data and metrics, like all efforts at classification (Bowker & Star, 2000) are not natural, but are constructed (Beer, 2016; boyd & Crawford, 2012; Gitelman, 2013), they are not objective, but contain assumptions and biases (Beer, 2016; boyd & Crawford, 2012), and are skewed (Baym, 2013). Further, "because these are affective measures, they lead individuals to self-monitor, to pre-empt the systems, to play the game, to act before being measured" (Beer, 2016, p. 210). These behavioural impacts may indeed be amplified with Fandometrics rankings that say outright its measures are meant to demonstrate the value of users and their on-platform activities. Gillespie notes that when metrics are "delivered back to audiences," "There is evidence that metrics not only describe popularity, they also amplify it, a Matthew Effect with real economic consequences for the winners and losers" (2016, p. 60). Similarly, when discussing institutional drives towards increased classification and measurement, Gane argued Foucault's work on biopolitics, "remind us that neoliberalism is not simply about deregulation, privatization or governing through freedom, but also about intervention and regulation with the aim of injecting market principles of competition into all forms of social and cultural life" (2012, p. 629). Fandometrics then, provides a window into how the metrification of communities can impact those communities' everyday social and cultural lives. We turn now to a more detailed analysis of how Tumblr explains Fandometrics' secret algorithm to users, and how users interpret that algorithm and manage their own affective displays in response.

Large and Loud...Without Sentiment

While never providing complete information about how various forms of engagement on-platform are weighted towards the eventual public Fandometrics rankings, Tumblr's descriptions of their measurements have changed over time. Around 2018, Tumblr's description of the rankings still reads: "Tumblr's Fandometrics is the result of our efforts to compile a database of Tumblr's *favorite* entertainers and entertainments, and track the shifts in our users' collective *affection*" (emphases ours). In 2020, the sentence read: "Fandometrics is the result of our efforts to compile a database of Tumblr's *most talked-about* entertainers and entertainments, and track the shifts in our users' collective *conversations*" (emphases ours). Though only a few words had changed, the 2020 description was more accurate: "favorite" had been replaced with "most talked about" and "affection" was replaced with "conversations." Often, Tumblr describes Fandometrics as measuring different fan communities' *influence* across the platform. However, such influence is inevitably a quantitative metric and Tumblr's measures do not account for sentiment. More recently, Tumblr has stated this clearly. Its current description of the Fandometrics algorithm reads: "To make a long story short: We weight and normalize the number of actions to create a more accurate picture of each fandom's influence across Tumblr, without sentiment" (Tumblr, 2020).

Online audience and user research increasingly attempt to measure or account for sentiment in their data collection and measurement. Sentiment analysis is a quantitative measure of emotion that uses Natural Language Processing (NLP) to "measure" the degrees of intensity of a positive/negative emotional binary that is imposed on social media posters' language use. The method is thus limited in a number of ways (see Hearn, 2010; Andrejevic, 2011; Arvidsson, 2012; and Kennedy, 2012, 2016 for useful accounts of sentiment analysis and critiques of its use). Despite the limitations of sentiment analysis (and its inevitable implications for social life), *not* accounting for sentiment, emotion, or affect in user measurement has its own implications. In the case of Fandometrics, user understandings of the blunt quantitative nature of the rankings have led to disagreement about the meanings of those metrics, the value of various on-platform activities and communities, and behavioural changes meant to surface more 'correct' counts in the eventual rankings.

Indeed, Tumblr users have noted that *quantity* of engagement rather than actual enthusiasm or fannishness of particular fan objects/subjects often accounts for their high rankings on Fandometrics. Many fans believe that frequent mentions, comments, reblogs, and so on of controversial or heavily disliked content or entertainers, or even toxic or particularly competitive fan objects that encourage intra- and inter-fandom fighting, are likely propelled to the top rankings simply due to all of the negative 'engagement.' Fans especially discuss the dynamics of this in relation to traditional fan culture activities like hate-posting, antifandom, and other online engagement related to disliked fandoms and fan objects. This particularly comes into play with "ship wars." "Ships" (from the word relationships) are preferred romantic pairings between two characters or celebrities; "shippers" are those fans dedicated to a particular ship. The Ships rankings are some of the most popular and hotly contested on Fandometrics, with Tumblr (2019) stating in its 2019 annual rankings, "Shipping is Tumblr's favorite sport and this is the Big Game." A "ship war" is defined by Fanlore (2020) thus:

> A ship war is a heated disagreement between two or more groups of shippers… Ship wars span a long time (often years) and involve many people in their fandom. Symptoms of a ship war include: rants, …long-winded essays trying to prove canonicity or superiority of the preferred ship… or pointing the flaws in similar essays by rival shippers, a refusal to quiet down till well after the canon is closed, anti-ship/per posts appearing in that ship's Tumblr tag.

Some Tumblr users lament the salience the Fandometrics algorithms lend to such behaviours that they would consider negative.

Others find it humorous when such negative engagement seems to benefit their fandom or ship in the rankings. There were a number of examples of this in Tumblr conversations around the *Star Wars* 'Reylo' ship wars (Reylo is a particularly controversial pairing of the characters Rey and Kylo Ren). One user's *Star Wars* fan account posted a question they had been asked using Tumblr's Ask function:

> Question—as far as how tumblr Fandometrics for ships list that is going around is concerned, is it just based by how much a specific ship is used/tagged? Because, if so, aren't antis[1] talking about reylo just helping it go up the list? That would be kind of hilarious TBH[2]

The fan account user posted this answer: "I'm no authority, but I'm pretty sure that the antis' incessant conversations about Reylo contribute towards its popularity on Fandometrics. This is, of course, absolutely hilarious." Indeed, users often framed those engaging in such 'anti' posting as unsavvy and uninformed. One Reylo shipper wrote a post saying "My aesthetic": followed by images of ants representing anti-Reylo posters, continuing, "tagging their hate as 'reylo' and unknowingly making the ship go higher in the fandometrics." The poster clearly found it amusing that those who disliked Reylo were likely actually responsible for Reylo ranking highly on the Fandometrics lists.

The lack of sentiment in Fandometrics' algorithmic logic seems to invite a certain type of affective discipline in fans who wish to place well in the rankings, and importantly, wish for those they dislike to rank lower. Fans often call on their communities to refrain from mentioning rival fandoms and groups so as to avoid this unintentional boost to their adversaries. However, user behaviour changes meant to avoid "negative" measurement outcomes can mean disruption of longstanding core fan activities, namely discussing and interacting with various fan objects and communities. While fandom has long been engaged in competitive practices, algorithmic rankings like Fandometrics constrain traditional modes of discourse, community, and competition and, perhaps unwittingly, may train fan communities in new cultural practices. Further, the murkiness around the affective impulses behind the rankings means various, and often competing, narratives emerge about who has (or has not) made the rankings and why. These narratives and hypotheses about Tumblr's algorithmic practices (Bucher, 2015; Maris, 2018) can lead to distrust of the metrics and platform, but just as easily to distrust and resentment of other users and user communities.

[1] "Antis" refers to Star Wars fans that are anti, or against, this romantic pairing of characters.
[2] TBH = To Be Honest.

Who Is Silenced?

It is important to ask (but impossible to fully know) who is silenced, or made to feel silenced, by the measurement logics made visible in Tumblr's Fandometrics. Certainly, numerically small or niche fan objects and communities have little chance of appearing in the rankings. The same is likely true of communities whose norms, and thus on-platform activities, do not count due to Tumblr policies or count less to Fandometrics algorithms. It is also potentially the case for those communities who do not invest energy into performing "properly" for the algorithm. Invisibility (or its threat) is key to the structure of Fandometrics itself; if your community does not add up enough to place in the top 20 spaces of a category (or top 100 for the annual Year in Review), it does not exist in Fandometrics. While there is certainly user anxiety about the threat of invisibility on Fandometrics, we also found Tumblr workers who felt constrained by the rankings' inability to represent smaller fan communities. Quantitative measures focused on the largest numbers inevitably leave out many, and despite Tumblr's claim that the rankings are for the fans, a way to have their voices heard—the Fandometrics architecture means only the loudest will be.

Latina and Docherty (2014) argue organising logics of metrification like Twitter hashtags inevitably exclude. Specifically, they note that platform user bases are often small in comparison to wider populations and thus not representative in any meaningful way; numerous potential users cannot access platforms due to lack of access to internet service, technical devices, and/or digital literacy; and many lack the platform literacy necessary to sufficiently engage in on-site discourse and community. Gillespie explains that trending algorithms, "…start with a measure of popularity, for instance how many users are favouriting a particular image or using a particular hashtag. But this entails deciding first who counts" (2016, p. 55). As with most algorithmically sorted social media, policy-prescribed human and machine content moderation (Gerrard, 2020; Gillespie, 2018; Roberts, 2019) will inevitably ensure an unknown amount of user content never surfaces on Tumblr. Those who cohere around content deemed offensive by Tumblr policies are likely to be invisible in the published metrics, while those who skirt the borderlines of such policies or even respectability on-site or in larger society also run the risk of having their communities' engagement rendered invisible.

In our interviews with those working at Tumblr (conducted before Tumblr's 2018 policy change banning adult content), one worker told us that trending topics on the platform are monitored throughout the day to "make sure that that's all kosher for public consumption," explaining that content labelled as pornography "...wouldn't even end up in our ... (Fandometrics) list. If we see something porn-related, it goes into a not-safe-for-work tag." Thus, fan objects, fan engagement, fan communities, and/or fan-created content considered porn or otherwise sexually "indecent" by Tumblr have no chance at being made visible in the rankings. As with other forms of user-generated content on social media, what we do not see, and what we do not know we are not seeing, represent highly political corporate decision-making (Gillespie, 2010). And indeed, when Tumblr banned adult content in 2018, it publicly became very clear that much of the content labelled porn or indecent was indeed not porn at all, or that such labelling and subsequent moderation especially harmed already marginalised communities (Romano, 2018).

The limits of visibility imposed by the structure of Fandometrics is not lost on those who work on it. Brennan spoke of their attempts to algorithmically give niche fandoms a chance at making the rankings:

> We kind of thought about that when we were building the algorithm for it and how do we normalize a little bit? And we took that into account. So, the niche fandoms do tend to make it in if they have enough—if they're spikey enough, if you will. Like if conversation goes from 0 to 100, we try to account for that spikiness. *The Get Down* is a good example. They were in Fandometrics once and it was just like "Okay, how do we do this? How do we get there again?" ...And you'll see weird stuff because we do account for that kind of spikey—that spike in volume, things will trend and then they'll just go away because it doesn't have that sustainability.

Tumblr workers often spoke of the diverse fan communities on the platform with affection. When asked about niche interests that might not make it into the Fandometrics rankings, Vantapool noted that she wished books could become a ranked category but that they would fail to make the cut:

> So, people really love books on Tumblr, and we've thought about making a books list, but there's just not enough data. People aren't talking about it enough, so we regularly would not be able to get 20 different books in that category to make a list, which is sad. I feel really bad, because that's one of the most frequent asks we get, is like "People love books. Why don't we have a books list?" And I don't want to tell them like, "You guys aren't doing

> a good enough job," because *they are*. They're talking about it at the rate that they're talking about it, but it's not on the scale of movies or television, so the numbers just aren't there.

"In Depth" has been a less quantitatively determined feature of Fandometrics. In Depths are special features where Tumblr focuses in on one fandom or fan topic, discussing it in detail and displaying various associated metrics. To qualify for an In Depth, a topic must still be deemed popular enough to generate interest. And repeatedly, the constraints of metrics and of resources were cited by workers as limitations in providing visibility to more communities. Brennan explained: "[W]ith In Depth we can really explore other sorts of presentations of data because we have more time. But…we're a small strappy team and getting design support can sometimes be hard." Vantapool told us, "I love Fandometrics…but I do wish there was a way to include …stuff that doesn't have as big of metrics…I think a lot of people would really appreciate that on a very personal level, and *I* feel that very personally."

Some optimistic accounts of the potentials of Fandometrics posit that such rankings will allow fan communities to have more influence in the production of the media they enjoy (Baker-Whitelaw, 2015). Ostensibly, fan communities could propel themselves up the rankings in order to get the attention of media production for save-our-show type campaigns and other fan-requests. While the internet and social media have long been used astutely by fan communities to do just this (Maris, 2018, 2020), the use of Fandometrics in this regard will likely be limited to certain fan communities and content—those that have the numerical strength to become visible in the rankings. Bucher argued that social media's algorithmic logics present a "threat of invisibility," the "…possibility of constantly disappearing, of not being considered important enough. In order to appear, to become visible, one needs to follow a certain platform logic…" (2018, p. 84). Indeed, Fandometrics is meant to empower fan communities, but as a tool for empowerment it can also represent a threat to those who may not wield it.

"Drown Them Out!" Industry-Encouraged Competition and Quantification Anxiety

Efforts at quantifying social life are often central to neoliberal projects. Beer explains that, "[M]etrics are used to manufacture uncertainty and to drive entrepreneurialism and self-training" (2016, p. 210). And indeed, Tumblr encourages fan communities to engage on-site in order to matter to Fandometrics and the larger platform community. In a light but taunting tone, Fandometrics sometimes frames drops in rankings as failures of user communities. While it is impossible to know how closely all fans attend to these prompts, there is evidence that many become quite invested in their communities' Fandometrics placement. Much of this investment manifests as friendly competition, but much also reveals user concerns about their standings in the rankings and associated anxieties about the size and value of their communities. Indeed, some also evaluate other communities based on quantitative data. Beer argues we should strive to understand "...how measurement is felt, how it is embodied, and how it can be seen to be experienced emotionally" (ibid: 196). How user communities engage with Tumblr's Fandometrics, and with one another, points to some of these affective implications of community measurement.

The weekly Fandometrics rankings visually highlight upward and downward movement. If something has moved up the rankings from the previous week, a small plus or minus sign next to a number indicates how many places it has risen or fallen. Often, along with the weekly rankings, Tumblr includes some bullets with commentary about each category. The text is often humorous and notes new arrivals or dramatic movements in the rankings. It sometimes seems to poke fun at the media/objects being ranked as in this 2018 post on the Celebs category: "Our Condolences to Adam Driver (No. 16), as evidently no one is talking about him." This light-hearted teasing can lead to some fans feeling the pressure themselves. One Adam Driver fan reblogged the Tumblr post, commenting: "Uh, wtf Fandometrics? Like, EVERYONE on my feed can't shut up about him!! Ok, Driver fans, not cool. Let's do something about this!" That user went on to suggest ways fans could propel Adam Driver up the rankings. Fandometrics itself often directly shifts its focus from the content in the rankings to the content supporters themselves, urging users to engage more. For example, Fandometrics posted the following with a 2018 weekly ranking in the Music category, "Beyoncé falls five spots to No. 15.

Beyhive, the queen needs your help!" Indeed, Fandometrics often places direct responsibility on users to do the work of engagement if they truly love their fan object enough. In a 2018 Ships ranking, the text read about fans' preferred ships (here called OTP, an acronym for One True Pairing), "Remember: If your OTP didn't make the list, its okay. It just means you are directly responsible and should've made more posts about them." These nudges towards particular types of engagement frame the rankings as malleable; making clear Fandometrics is not meant simply as an entertaining representation of naturally occurring on-site fan behaviour, but instead are competitive metrics as users algorithmically *perform* them.

Competition is central to many users' experiences of Fandometrics, whether they enthusiastically engage in line with Fandometrics' urgings, or begin to value their own and other communities by their quantitative data. Van Dijck describes social media's culture of connectivity as:

> [...] a culture where the organization of social exchange is staked on neoliberal economic principles. Connectivity derives from a continuous pressure—both from peers and from technologies—to expand through competition and gain power through strategic alliances. Platform tactics such as the popularity principle and ranking mechanisms...are firmly rooted in an ideology that values hierarchy, competition, and a winner-takes-all mind-set. (2013, p. 21)

The competition can have very clear affective impacts on community members. Users often ridicule other communities for their standings in the rankings or express disappointment in their own. That disappointment may spill over from concerns about value on-platform to the value of their communities more generally, which becomes increasingly equated with numerical strength. For instance, one user posted their disappointment with their favourite anime's standing by equating it to the anime's fan community itself fading away, "Guys, hetalia isn't even in the last place on the fandometrics top-twenty anime of every week. The fandom is seriously dying..." Such sentiments are in line with how Beer describes the ways systems of measurement operate affectively: "They target, cajole, and provoke. They are aimed at stimulating anticipation and uncertainty-often coupling these with senses of insecurity and precarity" (2016, p. 210).

The fan described above, worried about their favourite anime, described how other fan groups engaged on Tumblr, and suggested if *Hetalia* fans behaved similarly, they might grow the fan community's numbers.

Kennedy notes, "In the digital reputation economy...we see ourselves as brands, as saleable, exchangeable commodities" (2016, p. 59). That fan communities invested in the Tumblr platform and culture might take on such self-branding communally may seem quite natural. After all, fan communities tend to cohere around commercial (entertainment) products. And indeed, competition and various forms of antagonism have long been central to fan cultures (Johnson, 2007). However, online fan cultures also traditionally operate within a sharing culture or gift economy (Hellekson, 2009; Scott, 2009; Turk, 2014). Further, fan communities have often been concerned with interests considered niche or specialty; the value for many fans often *is* the perceived smallness of their community, their distance from "the mainstream" (Hills, 2002). Tumblr's Fandometrics ushers fans towards other measures of value, encouraging them to equate their community's relevance with its size, with community size defined as its algorithmically prescribed and measurable engagement on-platform.

Leaving Metrics, Reclaiming Data

Fandometrics serves as a useful case study for how online communities/audiences react and interact in the face of their own everyday experiences of public measurement and ranking. Indeed, the public and ranked nature of Fandometrics may be an example of industrial movement towards blurring or eliding the boundaries between backend and user-facing metrics such that the packaging and privileging/marginalising of audiences is increasingly explicit and normalised. Our results certainly show some of this normalisation. We witnessed many Tumblr users with affective investments in their own and other communities' (in)visibility in Fandometrics' rankings. With users increasingly aware of and attuned to algorithmic (Bucher, 2015) and industrial (Maris, 2018) imaginaries, and with their—and their communities'—algorithmically assigned values increasingly displayed back to them, affective impacts are likely inescapable. However, in the face of increasingly concentrated platform data power, it is important for critical data studies to attend to resistance and other models of data and metrics presented by those very communities being tracked and measured.

Over the years that Fandometrics has existed, fans on Tumblr continue to create communities, consume content, and perform productive practices as usual. However, there are signs that some have already become frustrated with Tumblr's Fandometrics and other industrial efforts at the quantification of their communities, and especially the affective discipline seemingly required to endure their own public ranking. Some opt-out of such tracking or the online cultures that value it. Others work to reclaim their own data through more fair and transparent measurement for their own communities. These user efforts fall in line with what Van Dijck notes are characteristics of users who are also "value creators":

> Network communities that collectively define popularity may be used for their evaluative labor or as deliverers of metadata, but they cannot be held captive to the attention industry. When users are no longer interested or when they feel manipulated, they may simply leave. (2013, p. 63)

And as Tumblr communities' value is made clear in Fandometrics' focus on them, many wield that power to resist commodification and/or reclaim their communities without the weight of industry-defined value assignments.

Indeed, some fans indicate very clearly that they do not want to play Tumblr's metrics "game." For example, one user posted to other members of their ship community, "Who gives a fuck about fandometrics when we basically just got canon confirmation that they're both thirsting after one another like crazy." The user celebrated a textual "win" for the ship community: seeing their favourite relationship blossom on-screen, highlighting its importance over any online rankings. Indeed, some fans similarly discuss returning to the object of their fandom for pleasure versus looking to their communities' place in Tumblr's rankings. Some users also air concerns over Fandometrics' potential amplification of fan culture practices that are seen as anti-social (like intense competition), or problematic. For example, the common fan practice of shipping real people (like actors, musicians, YouTubers, and other celebrities) versus fictional characters has increasingly come under fire in fandom as a disrespectful practice that can cause discomfort for those people whose personal/sexual lives become the focus of huge communities of online strangers. Some Tumblr users oppose Fandometrics' inclusion of real people ships in the rankings, with one user posting:

> Why are 'ships' that involve real people included in fandometrics?… Those are real people being shipped, They're not cartoon characters you can shove together just cuz you think they're cute…It's a little unsettling that their love lives are being treated like that.

Thus, fan communities interrogate which fan cultures are represented by Tumblr and to what ends. These responses are in line with Gillespie's claim that users grasp and contend with algorithmic representations of their cultures: "Users will be concerned about the politics of algorithms, not in the abstract, but when they see themselves and their knowledge, culture, and community reflected back to them in particular ways, and those representations themselves become points of contention" (2016, p. 70).

Those points of contention can also become community-led projects aimed at self-representation. Fans are increasingly conducting their own data and metrics projects. One fan we spoke with runs a fandom data site as a hobby. She and other fandom data enthusiasts come together online to answer data questions that have been bothering them to "prove" things about fandom that are in question, or simply play with the numbers for fun. Her efforts, along with other fan data projects, point to displays of data and algorithmic literacy that work to self-represent through data without underlying market logics. She is very aware of the limits of quantitative measurement and crude rankings, but pointed to her efforts to include accompanying data when she displays data as rankings:

> Along with my ranking I will give the actual numbers. And I'll point things out and often include a bar graph or a pie graph or whatever is the appropriate way to... visualize things so that you can also see, "Wow, after the top three, like the top three actually make up half the data all by themselves. And then there's this huge dip, and why is that?" It leads to more interesting questions as well as a better picture of things... It's just like there's a lot more that I want to know there than just the ranking.

Fan data enthusiasts don't necessarily dislike Fandometrics, but do see limits in transparency around how and why data and metrics appear as they do on the site and in other commercial measures of fandom. For example, she told us about Fandometrics:

> Don't get me wrong. I respect what all of those folks are doing, and I'm not like "Wow, you have a shitty service that doesn't tell us anything real," or something. It's not like that at all. It's just like well, I don't totally know what their goals are. I can't totally see how they're generating things and I would love to know more about this, and it's not there. So, I'm going to keep on doing my own looking at things as well because it doesn't answer all of my questions.

Part of this project is representation, and on one very popular "Fandom Stats" site, the site creator makes clear they work towards full transparency regarding how metrics are calculated and the potentials and limits of data for fandom. The creator notes on the site:

> I don't think my fandom stats tell deep truths about fandom. They can provide some insights into some aspects of fanworks…But trying to figure out what exactly that data means, and *why* fans are producing/consuming the things they are, is beyond the scope of the numbers… I don't ascribe any moral judgments to my fandom stats—that is, I don't intend to imply any opinions about whether fans are doing good or bad things. …The answer to almost every interesting question about fandom (or any complex system) is "It's complicated/nuanced, and the answer depends on the details of how you ask the question." I try to explain my starting assumptions and to map out some of the complexity of the data, where I can. There's always more to the story, though…Data is good food for thought and discussion fodder, but can't tell us what to do. (Destination: Toast!, 2020)

Such understandings of transparency and ethical data use represent alternative uses of quantification being embraced by some fan communities. These data projects are not always responses to specific commercial metrics projects (like Fandometrics), but do represent some communities' understandings of, and experiences with, the affective and larger sociopolitical impacts of being publicly measured and ranked. Indeed, the algorithmic literacy of these and other fan community responses to Tumblr's user-facing rankings demonstrate how some communities are already struggling to rebalance social relations in the face of their outright valuation and commodification in their "home" platform spaces. And as some of these communities have shown, other models are possible.

References

Andrejevic, M. (2011). The work that affective economics does. *Cultural Studies, 4–5*, 604–620.

Ang, I. (1991). *Desperately seeking the audience*. Routledge.

Arvidsson, A. (2012). General sentiment: How value and affect converge in the information economy. *The Sociological Review, 59*(2), 39–59.

Baker-Whitelaw, G. (2015, January 21). Tumblr launches tool to measure the most popular fandoms. *The Daily Dot*. https://www.dailydot.com/parsec/tumblr-fandometrics-trends/

Baym, N. K. (2013). Data not seen: The uses and shortcomings of social media metrics. *First Monday, 18*(10) https://firstmonday.org/article/view/4873/3752

Beer, D. (2016). *Metric power*. Palgrave Macmillan.

boyd, D., & Crawford, K. (2012). *Six provocations for Big Data. Paper presented at A Decade in Internet Time: Symposium on the Dynamics of the Internet and Society*, Oxford, September. https://papers.ssrn.com/sol3/papers.cfm?abstract_id=1926431

Bowker, G.C., & Star, S.L. (2000). *Sorting things out: Classification and its consequences*. MIT Press.

Bucher, T. (2015). The algorithmic imaginary: Exploring the ordinary affects of Facebook algorithms. *Information, Communication & Society, 20*(1), 30–44.

Destination: Toast! (2020). *Some notes on my fandom stats*. https://destinationtoast.tumblr.com/post/166518286054/some-notes-on-my-fandom-stats

Fanlore. (2020). *Ship War*. https://fanlore.org/wiki/Ship_War

Gane, N. (2012). The governmentalities of neoliberalism: Panopticism, post-panopticism and beyond. *The Sociological Review, 60*, 611–634.

Gerrard, Y. (2020). Social media content moderation: Six opportunities for feminist intervention. *Feminist Media Studies, 20*(5), 748–751.

Gillespie, T. (2010). The politics of 'platforms'. *New Media & Society, 12*(3), 347–364.

Gillespie, T. (2016). #trendingistrending: When algorithms become culture. In R. Seyfert & J. Roberge (Eds.), *Algorithmic cultures: Essays on meaning, performance and new technologies* (pp. 52–75). Routledge.

Gillespie, T. (2018). *Custodians of the internet: Platforms, content moderation, and the hidden decisions that shape social media*. Yale University Press.

Gitelman, L. (2013). *Raw data is an oxymoron*. MIT Press.

Hearn, A. (2010). Structuring feeling: Web 2.0, online ranking and rating, and the digital 'reputation' economy. *Ephemera: Theory & Politics in Organization, 10*(3/4), 421–438.

Hellekson, K. (2009). A fannish field of value: Online fan gift culture. *Cinema Journal, 48*(4), 113–118.

Hills, M. (2002). *Fan cultures*. Routledge.

Jarvey, N. (2015, January 20). Tumblr launches pop culture destination Fandometrics (exclusive). *Hollywood Reporter*. https://www.hollywoodreporter.com/news/tumblr-launches-pop-culture-destination-764841

Johnson, D. (2007). Fan-tagonism: Factions, Institutions, and constitutive hegemonies of fandom. In J. Gray, C. Sandvoss, & C. Lee Harrington (Eds.), *Fandom: Identities and communities in a mediated world* (pp. 285–300). New York University Press.

Kennedy, H. (2012). Perspectives on sentiment analysis. *Journal of Broadcasting & Electronic Media, 56*(4), 435–450.

Kennedy, H. (2016). *Post, Mine, Repeat: Social Media Data Mining becomes Ordinary*. Palgrave Macmillan.

Latina, D., & Docherty, S. (2014). Trending participation, trending exclusion? *Feminist Media Studies, 14*(6), 1103–1105.

Maris, E. (2018). *Desperately seeking the producer: Audiences, identity, and the margins of the internet*. ProQuest Dissertations Publishing.

Maris, E. (2020). Fan cultures as analytic nexus of media audiences and industries. In M. Filimowicz & V. Tzankova (Eds.), *Reimagining communication: Experience* (pp. 209–225). Routledge.

Napoli, P. (2003). *Audience economics: Media institutions and the audience marketplace*. Columbia University Press.

Powers, D. (2018). Thinking in trends: The rise of trend forecasting in the United States. *Journal of Historical Research in Marketing, 10*(1), 2–20.

Roberts, S. (2019). *Behind the screen: Content moderation in the shadows of social media*. Yale University Press.

Romano, A. (2018, December 17). Tumblr is banning adult content. It's about so much more than porn. *Vox*. https://www.vox.com/2018/12/4/18124120/tumblr-porn-adult-content-ban-user-backlash

Scott, S. (2009). Repackaging fan culture: The regifting economy of ancillary content models. *Transformative Works and Cultures, 3*. https://doi.org/10.3983/twc.2009.0150

Tumblr. (2019). *Tumblr year in review: 2019's top 100 ships*. https://fandom.tumblr.com/post/189431747929/2019-ships

Tumblr. (2020). *About fandometrics*. https://fandom.tumblr.com/about

Turk, T. (2014). Fan work: Labor, worth, and participation in fandom's gift economy. *Transformative Works and Cultures, 15*. https://doi.org/10.3983/twc.2014.0518

Van Dijck, J. (2013). *The culture of connectivity: A critical history of social media*. Oxford University Press.

Wikia. (2020). WAM score: Wikiw Activity Monitor (WAM) Score is an indicator of the strength and momentum of a FANDOM community. *Fandom*. https://community.fandom.com/wiki/WAM

Open Access This chapter is licensed under the terms of the Creative Commons Attribution 4.0 International License (http://creativecommons.org/licenses/by/4.0/), which permits use, sharing, adaptation, distribution and reproduction in any medium or format, as long as you give appropriate credit to the original author(s) and the source, provide a link to the Creative Commons licence and indicate if changes were made.

The images or other third party material in this chapter are included in the chapter's Creative Commons licence, unless indicated otherwise in a credit line to the material. If material is not included in the chapter's Creative Commons licence and your intended use is not permitted by statutory regulation or exceeds the permitted use, you will need to obtain permission directly from the copyright holder.

Open Access This chapter is licensed under the terms of the Creative Commons Attribution 4.0 International License (http://creativecommons.org/licenses/by/4.0/), which permits use, sharing, adaptation, distribution and reproduction in any medium or format, as long as you give appropriate credit to the original author(s) and the source, provide a link to the Creative Commons licence and indicate if changes were made.

The images or other third party material in this chapter are included in the chapter's Creative Commons licence, unless indicated otherwise in a credit line to the material. If material is not included in the chapter's Creative Commons licence and your intended use is not permitted by statutory regulation or exceeds the permitted use, you will need to obtain permission directly from the copyright holder.

Affinity Spaces as an Analytical Lens for Attending to Temporality in Critical Data Studies: The Case of COVID-19-Related, Educational Twitter Communication

Irina Zakharova, Juliane Jarke, and Andreas Breiter

Introduction

The ambiguity of data power and related in/securities acquired a new meaning in 2020 as the COVID-19 pandemic gripped our globalised societies. The educational domain was particularly affected by COVID-19 by early lockdowns, a transition to online schooling and eventually hybrid modes of teaching. Educational scholars observed inequalities and transformations taking place in the educational domain (e.g. Black et al., 2020; Selwyn & Jandrić, 2020). A public debate ignited around online teaching,

I. Zakharova (✉) • J. Jarke • A. Breiter
ZeMKI, Centre for Media, Communication and Information Research, University of Bremen, Bremen, Germany

ifib, Institute for Information Management Bremen, University of Bremen, Bremen, Germany
e-mail: izakharova@ifib.de; jjarke@ifib.de; abreiter@ifib.de

© The Author(s) 2022

A. Hepp et al. (eds.), *New Perspectives in Critical Data Studies*, Transforming Communications – Studies in Cross-Media Research, https://doi.org/10.1007/978-3-030-96180-0_15

digital learning infrastructures, digital literacy (for both teachers and students), and digital content. As educators in Germany and other countries increasingly use Twitter for professional communication (e.g. Carpenter et al., 2020; Greenhalgh et al., 2020; Staudt Willet, 2019), Twitter became one of the spaces where the educational crisis was publicly discussed during the pandemic. To conceptualise the implications of datafication for education, educational scholars often turn to critical data studies (Breiter & Hepp, 2018; Jarke & Breiter, 2019).

Many scholars have turned to social media such as Twitter or Facebook to study public discourses during times of crisis (e.g. Marres & Moats, 2015). A manifold of digital tools has been developed to support such analysis. In particular, when studying crisis, social media researchers focus on Twitter for "real-time data" on public communication and conduct so-called hashtag studies for disaster and crisis analysis (Bruns & Burgess, 2016, p. 23). In such studies, hashtags associated with a particular disaster or moments of crisis are the point of reference for further analysis. However, as critical data studies scholars have widely argued, (social media) data are performative in relation to the definitions of the phenomena—in this case the crises—they are supposed to merely represent (Crawford & Finn, 2015). For example, at times a disaster (or crisis) is so pervasive in public discourse that users cease to use hashtags ascribed to it, because almost all communication relates to this disaster (Tufekci, 2014).

We take these key insights from critical data studies and apply them to the analysis of Twitter communication. We propose to address the recursive and temporal character of platforms as objects of study (Baygi et al., 2021; Ruppert, 2013; Williamson, 2016) by attending to hashtags not as stabilised networks representing public discourse, but rather, as an unfolding process, a continuous flow of action. We argue that a hashtag is more than the sum of the human actors contributing to a particular topic. Rather hashtags

> should be understood not simply as 'gadgets' that do things but as complex and unstable assemblages that draw together a diversity of people, things and concepts in the pursuit of particular purposes, aims, and objectives. (Harvey et al., 2013, p. 294)

This "drawing together" through hashtags not only provides a representation of discourse about a crisis, but rather, it performs the crisis in particular ways in that they "generate new attentional flows" (Baygi et al., 2021). In this chapter, we propose to study the education-related discourse on the Corona pandemic not through a Corona-related education hashtag but rather through the study of the unfolding discourse *within* (and *along*) an education-related hashtag that serves as an "affinity space" (Gee, 2005) for German educators. Our study is, therefore, different from "hashtag studies" that follow the most popular hashtags emerging in a crisis to explore the dynamics of content (re-)distribution (e.g. Gruzd & Mai, 2020) in that we demonstrate how the analysis of a hashtag as an "affinity space" (Gee, 2005) is well suited to the examination of the unfolding of a crisis across Twitter. Understanding a hashtag as an affinity space—as a learning environment—allows us to attend to the recursivity of social media data by highlighting how the hashtag becomes reconfigured in the wake of and during times of crisis. Affinity spaces are a useful analytical lens through which to study the temporality and unfolding of a "controversy" (Marres & Moats, 2015) and for attending to the attentional flows of its participants.

Our study is based on the hashtag #twitterlehrerzimmer (or #twlz), Germany's biggest affinity space for educators covering topics such as teachers' everyday lives, pedagogy, and educational technologies. #twlz[1] can be translated as "Twitter staff room", signifying the room in which teachers meet in between classes and find time to update each other on important school-related matters or simply chat among themselves. Among the #twlz hashtag users are not only teachers but also the general public, collectives, and commercial organisations from the educational domain, as well as scholars, parents, and some pupils. Some of the #twlz users occupy multiple roles. We collected tweets related to the #twlz hashtag from November 2019 until the start of the summer holidays in mid-July 2020. A mixed-method explorative approach, including data science methods and qualitative content analysis, was applied to the dataset.

We examine our dataset in regard to two arguments developed by education scholars as responses to the COVID-19 crisis: (1) educational technology (ed tech) providers and political actors increasingly use social media to mediate their COVID-19 crisis management; (2) at the same time, educational technologies are being increasingly positioned as solutions to the educational challenges posed by the pandemic (e.g. Johns, 2020; Selwyn, 2020; Teräs et al., 2020; Williamson et al., 2020). Starting with these arguments, we ask (1) how have ed tech providers and political

[1] To improve the readability for non-German-speaking readers, we will use the shorter version of the hashtag throughout the text.

actors reconfigured communication via #twlz on Twitter and (2) how suitable is the concept of affinity space for studying controversies in times of crisis through Twitter hashtags.

In the following, we first provide an account of work related to the study of Twitter communication in times of crisis. Subsequently, we introduce the context in which our study took place and provide a timeline of education-related events during the COVID-19 crisis in Germany as well as results of studies considering the impact of ed tech providers and political actors on education. In the next section, we present our case study and research design. Subsequently, we identify shifts in the topics and actors mentioned in tweets and retweets in the #twlz affinity space as a reaction to the national and regional educational crisis management over time. Finally, we reflect on how the concept of affinity spaces contributes to new perspectives in critical data studies on three levels: conceptually, methodologically, and to critical data studies in education.

Analysing Twitter in Times of Crisis

Educators' Twitter use has been examined by educational researchers from a variety of perspectives: the professional development of teachers (Britt & Paulus, 2016; Carpenter & Krutka, 2014; Visser et al., 2014) and school leaders (Sauers & Richardson, 2015), education activism (Thapliyal, 2018), and Twitter's role in teaching practices (Tang & Hew, 2017). Methodologically, earlier studies of educators' Twitter practices were based on self-reports in surveys (see Tang & Hew, 2017), while later studies focus on the data available through Twitter platform affordances such as account information (Kimmons et al., 2018) or specific education-related hashtags. Most educational hashtags are specific to a language and/or region (Greenhalgh et al., 2020). Through hashtags, educators share experiences and resources based on their geographical (Carpenter et al., 2020) or subject-related (Larsen & Parrish, 2019) interests. Educational stakeholders (e.g. teachers, public administration, and ed tech providers) use hashtags to communicate their diverging interests about topics such as digital education and to, varying degrees, collaborate.

Studying various stakeholder groups on Twitter, boyd (2010) proposed to approach spaces and collectives emerging through social media as "networked publics", a rather stable set of actors embedded in a similarly stable space, the social media infrastructure. The notion of (networked) publics can cover a great variety of human actors, connected either tightly or loosely based on their common identities, endeavours, and practices (ibid.). Despite being sensitive to the platform affordances, publics as an analytical concept delineates people (social media users) from the infrastructural and material properties of the space where they communicate, primarily focusing on content production and consumption. The tweets, retweets, likes, and replies can be understood as redistribution practices, allowing for the circulation of content and increasing the visibility of popular topics (e.g. Theocharis et al., 2015, p. 205).

However, as studies into educational Twitter communication have shown (e.g. Carpenter et al., 2020), hashtags such as #twlz not only serve to redistribute content but to also facilitate the interactions between users and enhance mutual learning. It is for this reason that some scholars have argued that social media and educational hashtags specifically should be understood as communities (see e.g. Britt & Paulus, 2016). For example, teachers can be described as a "community of practice" (Lave & Wenger, 1991). The #twlz hashtag serves as a space in which community members learn about and discuss education-related matters. However, in hashtags studies, relying on the concept of "belonging" to a hashtag community is challenging; hashtag studies need to include people who use hashtags passively, stop using hashtags as they become "obvious" to others, or change hashtags used to describe phenomena over time (Baygi et al., 2021; Tufekci, 2014).

To circumvent that challenge, we propose that contributors to hashtags such as #twlz do not form a community or a "networked public", but rather, they create an "affinity space" (Gee, 2005). The membership in an affinity space is not defined by (institutional) boundaries as is the case for teachers meeting in a school's staff room, but through the performance of platform-specific ways of interacting and is not restricted to tightly connected actors; rather, the space allows for different kinds of participation. Platform-specific dynamics and affordances in the affinity space can be approached first as varying resources, skills, and forms of participation available to different actors (e.g. differences between Twitter accounts used privately or by organisations or diverging access to further information beyond social media sites). Second, affinity spaces can be accessed through "portals" (Gee, 2005, p. 226), which do not simply open the spaces for new participants, but rather co-produce the resulting spaces, for example, learning from each other. Finally, affinity spaces illustrate the recursive character of social media (data), as the notions of internal and external "grammar"—the organisation of the elements and practices in an affinity space—suggest mutual transformation (ibid., p. 226).

In the case of the #twlz as an affinity space, we observe how users change and create new hashtags or use @-mentions to invite specific accounts, thus shifting over time both the common endeavours and the sets of actors. It is in this way that attending to hashtags as purposefully created entry points to a particular space and a flow of action, we stay sensitive to the challenges of Twitter and particularly hashtag research. We extend on the work of Marres and Gerlitz (2016) in that our starting point is an affinity space through which we aim to understand the dynamics of the problematisation (Callon et al., 1983) of the COVID-19 pandemic for German education. For example, we demonstrate what can be gained by attending to the temporality and the changing dynamics of hashtag and topic authorship rather than the shifting frequency of hashtags. We make an analytical move away from studying the differences between hashtags as several stabilised networks of topics and actors. By investigating processes of reconfiguration and establishing how the #twlz affinity space emerges and unfolds in a time of crisis we contribute to the new perspectives to critical data studies.

The COVID-19 Pandemic and German Education in Context

Education in Germany is run by the federal states (Laender). This means that there is no coherent national strategy, but rather different approaches across the 16 Laender. States' Ministries of Education were responsible for educational crisis management and governance as the COVID-19 pandemic began to take hold. The 16 Education Ministers build a Standing Conference of the Ministers of Education and Cultural Affairs of the Laender of Germany which enables exchange and results in multilateral agreements. An exception from the Laender-led educational politics represents the so-called DigitalPakt Schule. This national funding programme passed in April 2019 in accordance with the Federal Department of Education and the 16 Laender and was designed to provide 5 billion Euros to cover school districts' digital infrastructure expenses. During the COVID-19 pandemic, additional funds have been made available within the "DigitalPakt Schule" scheme to support content and infrastructure development for remote teaching and learning. In the early days of the pandemic, many school districts were still preparing their funding applications and by the time of the first lockdown they were left without the necessary ICT equipment. Twitter communication surrounding digital education both before and during the pandemic has been tightly entangled with political topics such as the "DigitalPakt Schule" scheme. Moreover, while some political actors such as the Ministries of Education of the Laender have served as ed tech providers, local school districts have been responsible for maintenance and technical support.

To be able to contextualise our analysis of the #twlz affinity space, we developed an education-related COVID-19 timeline of events in Germany (Table 1). It covers the events related to the recognition of the COVID-19 crisis and the national and regional measures taken in Germany with particular focus on educational crisis management. Overall, the educational domain has not received much attention from political actors. However, the political debates and the actions that followed around school reopenings in spring 2020 generated strong opposition from educators and other stakeholders. In North-Rhine Westphalia (short NRW) the lack of straightforward communication between the responsible ministry of education and schools resulted in appeals (and hashtags such as #schulboykotnrw—school boycott NRW) in and beyond the #twlz affinity space to boycott school openings, followed by an official clarification of the situation by the state's premier of the federal state. In the context of remote and hybrid schooling, affinity spaces such as #twlz, already used for the exchange of information and resources among educators before the pandemic, acquired additional significance for a broad swathe of actors in the educational domain. As we will demonstrate in what follows, certain actors addressed the #twlz hashtag much more frequently during the pandemic than before as new topics emerged.

Educational researchers have analysed the implications of the COVID-19 pandemic for education within and across different countries. One focus was on *ed tech providers*' communication and self-presentation (Selwyn, 2020; Teräs et al., 2020; Williamson et al., 2020). Many companies offering ed tech provided their services during the pandemic "for free" while continuing to collect data and pushing towards a longer-term transition to digital learning. Another group that also receives scholarly attention in reflections on the pandemic are *political actors*. Questions rise about the ways in which political actors (and governments) are engaged in publicly mediating their COVID-19 crisis management differently with many increasingly turning to digital media to communicate with various stakeholders (Johns, 2020). Through our analysis, we explore whether

Table 1 COVID-19 timeline in Germany

1	27 January 2020	Evidence of the COVID-19 virus in Germany first comes to light and public discussions take place as the epidemic spreads in other countries. At that time there were yet no clear signs of potential policy responses and related physical distancing measures.
2	6 March 2020	A Council session of the European ministers of health within the Employment, Social Policy, Health and Consumer Affairs Council (Health) on COVID-19 leading to public awareness of the situation and potential ramifications.
3	11 March 2020	Declaration of the pandemic by the WHO followed by Germany's Federal Chancellor and the Prime Ministers of the Laender' declarations of local measures in response to the possible COVID-19 outbreaks.
4	15 March 2020	Germany's federal government issues school closures starting immediately or the next day depending on the states and continue nationwide until April 19th (the so-called prolonged Easter break).
5	22 March 2020	Federal government issues nationwide social distancing measures and closure of some branches of the economy ("lockdown").
6	12 April 2020	Easter break. (The time span for Easter school holidays in Germany differs across the federal states.)
7	13 April 2020	Ad-hoc statement by the German National Academy of Sciences Leopoldina on the implications of the COVID-19 pandemic for the educational domain.
8	15 April 2020	The Standing Conference of the Ministers of Education and Cultural Affairs of the Laender comes to an agreement to gradually reopen schools on 4 May in all federal states at the earliest.
9	16 April 2020	North-Rhine Westphalia (NRW) issues an early, gradual reopening of schools starting 23 April for students preparing for final examinations. The reopening strategy provoked public opposition that was coined school boycott (#schulboykott) on social media.
10	6 May 2020	The Prime Ministers of the Laender and the federal Chancellor agree upon individual regional strategies for the reopening of schools in the federal states.
11	18 June 2020	The Standing Conference of the Ministers of Education and Cultural Affairs of the Laender comes to an agreement to pursue in-person learning for the next academic year 2020/2021.
12	Starting 22 June 2020	School summer holidays in Germany begin at different times across the Laender. The table includes the earliest day marking the beginning of summer holidays.

An overview of selected core (educational) measures issued on the international, national, and regional levels and school holidays in the summer term 2020.

and how the #twlz affinity space reflects these reports in the German context. To do so we trace how the COVID-19 pandemic is problematised in one education-related affinity space and how that, in turn, reconfigured that space.

Case Study and Research Design

The history of the hashtag #twlz goes back (at least) seven years. Our analysis covers the first months of the COVID-19 pandemic and spans the period beginning 10 November 2019 until the start of the school summer holidays on 31 July 2020. We used the free Twitter streaming API to collect tweets and retweets including one of the selected hashtags #twitterlehrerzimmer, #twlz, #edchatde (historically the earliest German education-related hashtag, later superseded by #twlz). Replies and other (possibly relevant) tweets which did not use any of the listed hashtags were not included in the dataset and subsequent analysis. In total, we collected 131,394 individual posts (39,011 tweets and 92,383 retweets). Around 25,003 accounts participated in the #twlz affinity space. Bots, deleted, or blocked accounts were excluded from the analysis. By participation, we understand both producing tweets and retweets, but also being introduced to the affinity space by others through the Twitter @-mention functionality. Applying the framework of affinity space, we understand @-mentions as "portals" that open and reconfigure the #twlz affinity space. Examining the mentioned accounts enables us to identify how shifts in the affinity space emerge as a reaction to the pandemic and its crisis management over time. We manually coded all user accounts belonging to those who participated in the affinity space more than three times (N=5840) to assign them to an inductively generated actor group. Following previous pandemic-related educational research, we, too, mainly focus on two actor groups in this chapter: ed tech providers (N=218) and political actors (N=323). Moreover, both groups experienced a rise in participation in the #twlz affinity space during the COVID-19 pandemic. The accounts that participated through tweets or @-mentions less than three times that we did not analyse made up the majority ("long tail") of accounts in the affinity space.

In total, the users generated both with their tweets and retweets 13,749 additional hashtags to the ones we had identified for collection (#twlz, #twitterlehrerzimmer). To determine hashtag-based topics, we manually compiled hashtags used 25 times or more (N=1134) in the dataset into 14 distinctive, inductively generated topics of varying size and complexity. In approaching these topics, we aimed to trace the shifts in the dynamics of problematisation. Certainly, our personal and professional backgrounds as

German-based researchers at different stages of our careers, living in differing family contexts, and our research interest in datafied education were performative to the resulting list of topics (for the detailed discussion on the performative agency of the data scientists and interpreter, see e.g. D'Ignazio & Klein, 2020). Studying Twitter datasets is related to a number of challenges: the opaque Twitter API algorithm (Bruns & Burgess, 2016, pp. 21–22); the challenges of differentiation between studies of users activities on Twitter, platform affordances, and sociality in general (see also Marres & Gerlitz, 2016); focus on active tweeters; and a wide range of ethical challenges. For example, we encountered some accounts ($N = 406$) to which we could often clearly assign a particular actor group despite the user's statement in their bio about tweeting "privately" with this account. Simply by using a hashtag on Twitter, users cannot foresee research that uses the platform where their Tweets may be included as part of a larger dataset. We primarily focused, therefore, on aggregated data. Usually when a study is not centred around a content analysis of Tweets and instead focuses on hashtag-based topics, as we did, the next step should be to interview users and make their voices heard in the research process. This circumvents the challenge of carrying out research and talking about people rather than with them.

#twlz as an Affinity Space

#twlz has no specific local boundaries, although it mostly covers German issues and, therefore, German users while there is also a relatively small number of users from other German-speaking countries. Both before and during the pandemic, the hashtag #twlz was promoted by some pioneer educators (see, in regard to pioneer communities, Hepp, 2016) on Twitter and their personal websites, blogs, vlogs, or podcasts and even by a number of commercial educational companies. Many of the active #twlz contributors are engaged in educators' professional development as facilitators at the local and regional level, particularly in the domain of educational media and technology. These groups accounted for up to 39 per cent of the accounts we analysed. Respectively, the most frequently used hashtag was #digitalebildung (digital education); it was used 7082 times across the whole dataset. Before the pandemic and over the first few months of the COVID-19 crisis digital education was a core hashtag (and topic) within the #twlz affinity space and directed us to the starting point of our analysis. The digital education hashtag and those similar to it (e.g. #zeitgemäßebildung—contemporary education or #digitalesklassenzimmer—digital classroom) both attend to the educational technologies required for providing "contemporary" teaching and imply a political appeal about perceptions of what constitutes quality school education. Besides educators actively using the #twlz affinity space, the hashtag was able to bring in political actors and the subject of educational technologies (and their providers).

For the purpose of our analysis, we divided the dataset into two periods of varying length: pre-crisis—which marked the first part of our data collection from 10 November 2019 until 5 March 2020—and during the crisis period until the end of our data collection 31 July 2020. The most active group throughout the whole period were *educators* (which is not surprising given the affinity space we researched, see Fig. 1). The two actor groups experiencing a rather continuous participation boost were *political actors* and *ed tech providers*, even though both groups were among the least present in the #twlz affinity space. The *other actors* who were not included in the manual coding process (because they participated only one or two times in the #twlz hashtag discourse) make up the second biggest

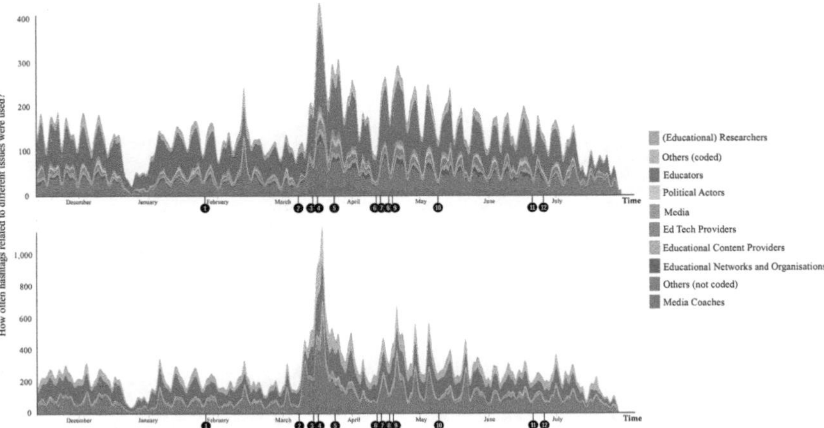

Fig. 1 Different actor groups' tweeting (above) and retweeting (below) activities over time. The vertical axis shows the number of tweets and retweets generated daily by each actor group. The horizontal axis shows the COVID-19 and #twlz timeline's overlaid. Numbers indicate the rows of Table 1

group contributing to the #twlz affinity space. However, these actors were much more actively retweeting #twlz-related content and, therefore, can be considered as observers rather than drivers of the discussion. The charts below illustrate how our manual coding covered the majority of actors creating original tweets with the #twlz hashtag and thus constitute the #twlz affinity space to the widest extent.

Shifting from an analysis of frequency to one of the dynamics of problematisation, we proceeded with our analysis by examining the topics the identified actor groups discussed before and during the COVID-19 pandemic. Figure 2 illustrates 14 topics that emerged through manual coding and the remaining/unsorted hashtags grouped together. Any tweet or retweet may have included more than one hashtag; therefore, the number of hashtags used in the #twlz affinity space differs from the number of tweets and retweets produced in the same affinity space over the course of data collection. The topic concerning educational technologies received much more attention since the beginning of the pandemic both in tweets and in retweets (Fig. 2). By contrast to topics addressed in original tweets, retweets since the beginning of the COVID-19 pandemic were dominated by COVID-19-related issues, followed by topics such as *digital education* or *general topics* in education. During the pandemic, the #twlz affinity

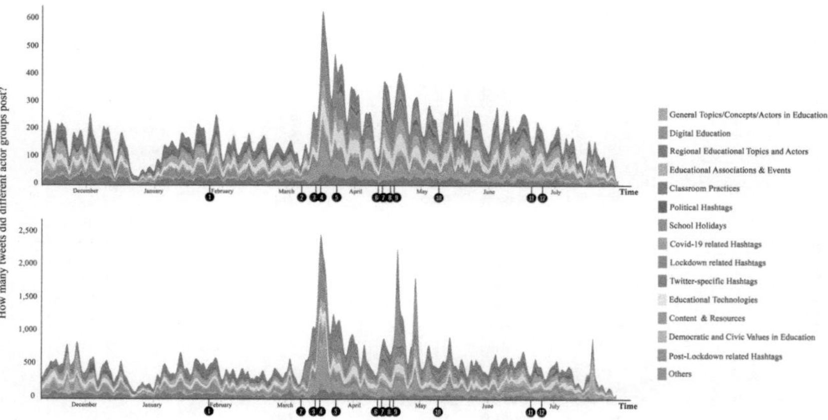

Fig. 2 Development of the #twlz topics over time in tweets (above) and retweets (below). The vertical axis shows the number of hashtags used in tweets and retweets. The horizontal axis shows the COVID-19 and #twlz timeline's overlaid. Numbers indicate the rows of Table 1

space opened up to new actors, possibly from "outside" the educational domain. As the topics (re)tweeted by these accounts suggest, they were concerned with the state of German education during the pandemic and shared the information they came across in the #twlz affinity space with their personal Twitter networks.

In our analysis of the affinity space before and during COVID-19, we now turn to two actor groups, which experienced a boost in their participation: *political actors* and *ed tech providers.* We are interested in understanding the extent to which the dynamics of problematisation during COVID-19, as observed by other scholars during the pandemic, are visible in the #twlz affinity space.

Educational Technologies and Their Providers in #twlz

The first issue we examined through our dataset concerned ed tech providers and their role in the #twlz affinity space. In our analysis, ed tech providers constituted one of the smallest actor groups participating in the affinity space both before and during the pandemic. Among the reasons for the small number of ed tech providers in the affinity space may be the particularities of the German market, state ed tech providers (Laender), strong legal regulations (e.g. data protection, privacy), big stakeholders among educational media providers, also supported through strong regulation (e.g. accreditation of textbooks). Even though we could observe a growing number of tweets by ed tech providers, in general, their participation in #twlz communication remained lower than that of other actor groups (see Fig. 1). In contrast to the ed tech providers' accounts, hashtags and topics relating to *educational technologies* experienced a gradual growth not only in the general peaks of #twlz communication activity, but until the beginning of the summer holidays (Fig. 2). These hashtags can be seen as entry points to the #twlz affinity space, where all interested actors could exchange and build their knowledge about specific software or hardware. According to the topics used in the twlz affinity space referring to educational technologies, we assume that at least before the pandemic, the #twlz users were mostly interested in the expertise and knowledge of educators and facilitators, that is, those who knew how to support teaching and learning processes with educational technologies with the hindsight of the #twlz core topic, digital education. However, during the pandemic, both the number of educational technologies and the methods of their application altered through the practices of remote and hybrid teaching and learning. Through ed tech-related hashtags, actors from "outside" the educational domain could enter the affinity space to contribute with their experiences of, for example, video conferencing or remote team communication and organisation.

To examine these dynamics further, we focused on the *educational technologies* issue and identified, after an additional round of qualitative coding, four sub-topics: hashtags related to ed tech providers, learning apps, technologies appropriated for education, and hardware. The sub-topic *ed tech providers* included 59 hashtags ranging from solutions offered by Microsoft (e.g. teamsedu) to solutions offered by companies such as *its-learning!*—an international learning management system used by a number of Laender—or solutions developed by the states themselves such as *LogineoNRW* (a learning management system used in one of the Laender). The sub-topic *learning apps* included hashtags for apps that have been genuinely developed for learning (such as the reading app Antolin). The sub-topic *technology appropriated for education* included all those hashtags that refer to technologies that had not been developed for educational settings originally but came to be appropriated as tools for (social) learning (e.g. Padlet, Twitch, YouTube). Diverse actors, including the media, ed tech providers, and educational associations used hashtags related to the technologies appropriated for learning more often during the pandemic. Remarkably, since 6 March, ed tech providers alongside educational content providers increasingly tweeted, not only about their own products, but also about other technologies appropriated for education. With the aim of maintaining a high demand for the educational products brought about by remote teaching and learning, technology providers and educational content publishers may have an interest in partnerships with bigger technology corporations (such as YouTube and other streaming platforms).

With this investigation, we were interested in whether the #twlz affinity space was increasingly focused on by the ed tech providers during the COVID-19 pandemic. Our analysis shows that pioneering educators were not increasingly "targeted" by ed tech providers via Twitter in their "own" affinity space #twlz. These data contrast, then, with the claim that ed tech providers increased their online activity during the pandemic. This increased activity may be observed in other communication spaces on Twitter (e.g. other associated hashtags and affinity spaces) and beyond (e.g. direct communication with educational decision makers). A further content analysis of tweets is required in order to investigate the extent to which educators and other #twlz users attended to the pedagogical and didactic issues of digital technologies in their tweets with ed tech-related hashtags. Observing how ed tech-related topics shifted during the pandemic and how other actors became involved in the knowledge exchange with educators, we notice how digital technologies became a particularly distinct part of the initially central topic of "digital education". Approaching #twlz as an affinity space renders visible the ways in which hashtags related to digital (educational) technologies facilitate learning about these technologies among different actor groups. The analytical lens of the affinity space allows attendance to other hashtags and not just those most frequently used and tracing how these become more important and reconfigure the affinity space over time.

The Quest for Dialogue with Political Actors

The second argument we explored in our dataset is about political actors' crisis mediation in their acts of public communication. Results from current scholarly work (e.g. Johns, 2020) are similar to what we observe in the German educational domain. Public administrations acted as facilitators in public events dedicated to digital education and schooling, not least within the "DigitalPakt Schule" funding programme. Recent examples include the "#WirVsVirus" hackathon initiated by the German government or other education-related events such as an online barcamp "#DIGITALITAET20", organised under the auspices of the Federal Government Commissioner for Digital Affairs. Others included the hackathon, "wirfuerschule" organised by the volunteers from the educational community and initiated under the patronage of the Federal Ministry of Education and Research and other public education organisations. Both hackathons were conceived as a crowd-sourced solution to the challenges posed to the education sector by the COVID-19 pandemic.

Many of the national and regional political actors and organisations have had Twitter accounts since before the pandemic; however, they were not a part nor had they even been on the margins of the #twlz affinity space at the time. The epidemiological and political insecurities brought about with the spread of the virus demanded speedy reactions to and communication centred around the changing circumstances and the details of COVID-19 crisis management. Even though our dataset includes different time spans pre- and post-pandemic—around four and five months respectively—we assume that the growth in the amount of political actors' participation illustrates how #twlz became a space for mediating educational politics during the crisis. Politicians and federal or Laender ministries were addressed by other actor groups in the #twlz affinity space more often during the COVID-19 pandemic (Table 2). The disparity between the number of political actors being mentioned by other #twlz users and themselves tweeting actively before the pandemic also indicates a continuous quest for dialogue with educational policymakers.

The scatter plot in Fig. 3 maps political actors according to the frequency of their participation in the #twlz affinity space actively (tweeting) or passively (@-mentions). Overall, the Laender Ministries of Education and the Federal Ministry of Education (account names in red) were mentioned more often than they themselves used the #twlz hashtag. Some of them however, engaged with the affinity space (e.g. the Federal Ministry of Education). Politicians such as the Federal Minister of Education or the state premier of NRW (account names in black with the "*"-symbol) were mentioned relatively often. However, they did not directly engage with the affinity space. Most political actors (light grey dots), however, only played a marginal role in the communication that took place in the affinity space. The differences between public accounts and the organisational accounts of political actors and the individual Twitter accounts of educators and other stakeholders are notable if we consider the distribution of resources in the affinity space. Political actors have (access to) different resources than educators and other #twlz users, both in regard to the maintenance work required to sustain a Twitter account and the skills of professional social media content creators. Despite lacking access to these

Table 2 Political actors' participation in #twlz before and after COVID-19 pandemic

	Total number of political actors (with tweets or mentions ≥3) in affinity space	*Participated only through @-mentions, never tweeted using the hashtag #twlz (tweets = 0; mentions ≥3)*	*Participated only through tweets, were never mentioned by others (mentions = 0; tweets ≥3)*
Before 5 March	173	106	26
After 6 March	246	135	37

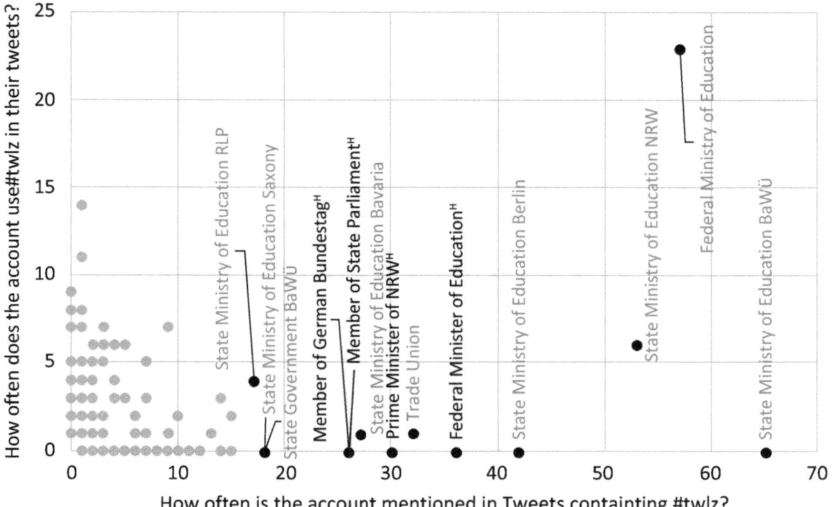

Fig. 3 Political actors' participation in the #twlz affinity space during the COVID-19 pandemic. Account names in red are federal or state governmental organisations; account names in black with the "*" symbol are politicians

resources, educators and other groups reached out to politicians much more often and created additional entry points to the affinity space through Twitter affordances such as @-mentions.

With the development of the COVID-19 pandemic and the necessity to communicate prevention strategies, that quest became ever more important for educators and other actors including politicians and political organisations themselves. The educators and other actor groups were addressing politicians and political organisations predominantly with political topics, presumably reacting to the rapid changes in political strategies. On the other hand, political actors turned most often to the topics covering general topics in education and COVID-19 related hashtags during the pandemic (Fig. 4). Political actors were using political hashtags as well, however at different times than the other actor groups addressed their political appeals to policymakers. This leads us to observe how political actors attempt to "change the subject" (and shift the dynamics of problematisation) of their COVID-19 related communication and pursue their own strategies in the educational crisis mediation on Twitter. Our

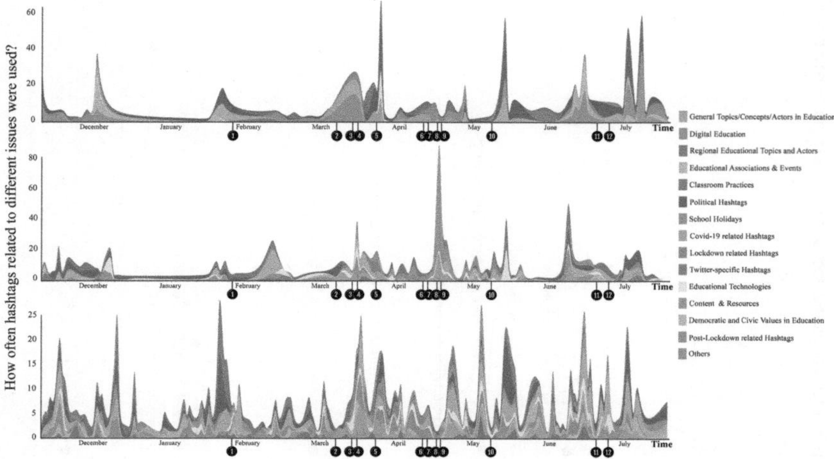

Fig. 4 Political actors in relation to topics. Topics co-occurring with @mentions of federal (above) and state (middle) political actors over time. Political actors' contributions to all topics in their original tweets over time (below). Numbers on the horizontal axis indicate the rows of Table 1. Y-axes are aligned with the numbers of hashtags used, which differ among the graphs

findings show that political actors problematise the pandemic-related educational crisis differently to other actor groups within the #twlz affinity space, despite the #twlz users' attempts to introduce the political actors to their discussions. In sum, an analysis of #twlz as an affinity space directs our attention not to the stabilised network of actors, but rather to the reconfigurations brought about by the COVID-19 pandemic, such as the introduction of a greater number of political actors into the affinity space, and to the dynamics of problematisation of pandemic-related topics.

Towards a Process View in Critical Data Studies

Critical data studies have developed various theoretical and methodological tools that help to make sense and disentangle the complex relationships of human agency, platform affordances, and corporate interests intertwined in digital data and software. Following changes in affinity spaces over time provides an attractive theoretical and methodological perspective for critical data studies, enabling a processual analysis of platforms.

This approach allows for the observation of affinity spaces and how they are being configured while eschewing a "bird's eye" view on social media communication as a sequence of stable states enacted by a well-defined group of actors. Dynamic, temporal changes in affinity spaces illustrate not only the stabilisations of platform communication but also how and why these stabilisations come to be over time. Overall, our chapter contributes to new perspectives in critical data studies in the three ways: (1) *conceptually*, we reflect on the dynamic configuration of affinity spaces as an analytical lens for critical data studies; (2) *methodologically*, we reflect on the ways in which the temporality of a platform can be "captured"; and (3) *thematically*, we reflect on how our analysis contributes to critical data studies in education.

Approaching affinity spaces as processes of their reconfiguration over time contributes conceptually to critical data studies, as it circumvents some common critique of other approaches—for example, actor-network theory (see, for extensive critical analysis, Couldry, 2020)—such as its "flatness" and little attention to the intentions of human agency. Affinity spaces preserve the focus on the intentionality (as well as identities, knowledge, and values) of human actors and the practices of their ongoing reciprocal learning, driving the dynamics of problematisation forward and enacting change over time. Moreover, similar to the theoretical approach of controversy mapping (e.g. Marres & Moats, 2015), affinity spaces-based analysis allows for the circumvention of the focus on multiple, but seemingly stable networks and instead guides our attention to the practices and temporalities of problematisation and its subsequent stabilisations, staying sensitive to the platform-specific dynamics and affordances. The identification of particular non-human actors (e.g. @-mentions or hashtags) as entry points to affinity spaces makes a conceptualisation of their agency more accessible. Applying the framework of affinity space to a hashtag study such as ours renders visible the ways of configuring the affinity space that drive changes in a recursive relationship between the events, people, issues, and topics that come to be associated in the affinity space.

Methodologically, the attention to the affinity space #twlz is very different to hashtag studies (of crisis communication) that usually tend to pick those hashtags describing the crisis or most visible content. For example, in our dataset a long list of hashtags was related not to education but to COVID-19 (e.g. #COVID2019, #coronavirus), to the lockdown (e.g. #stayhomechallenge, #shutdowngermany, #onlineteaching), and to the "new normal" (e.g. #hybridkonzepte, #lernentrotzcorona, #hybridunterricht). We argue here that rather than attending to these hashtags that are symptomatic during a time of crisis, we should turn to an affinity space (such as #twlz) and study its dynamics of problematisation. This means that we need to find ways to depict and study the process in which a problematisation of the crisis takes place. Hence, rather than depicting stabilised networks of such an affinity space, we are interested in the unfolding dynamics that constitute it. Computational methods and dynamic visualisations of empirical data over longer periods of time enable us to illustrate the breakdowns and frictions, signifying changes in the dynamics of problematisation. Currently, a number of new methodological approaches within critical data studies and beyond are being developed that enact a temporal understanding of data. For example, Bates et al. (2016) propose data journeys as a methodological tool to follow the data movements and frictions within organisations. Baygi et al. (2021) encourage us to "re-orientat[e] our theoretical gaze from spatial relationality to the temporal qualities, conditionalities, and directionalities of flows of action". Such approaches draw on a variety of data retraction and processing methods as well as on new ways of data visualisation. Affinity spaces as a methodological tool contribute to that rapidly emerging academic discourse.

Our findings contribute to critical data studies in education. We show that the dynamics of problematisation in times of crisis can be examined by attending to the changing reconfigurations of affinity spaces. At the beginning of our analysis, we identified the topic of digital education as central to the #twlz affinity space. At that point, both political actors and ed tech providers were participating in the affinity space, however, following different interests, for example, coupled with the political endeavour to provide national funding as a part of the "DigitalPakt Schule" funding scheme. This funding scheme is obviously highly relevant for both actor groups: for political reception and for businesses. However, as the virus spread all over the world, not only the new modes of teaching and learning, but also new modes of coping with the pandemic were required. During the pandemic, the main topics picked up on Twitter covered COVID-19, lockdown, and post-lockdown related hashtags, reconfiguring educational communication to the challenges of the present crisis. Through the lens of the affinity spaces, however, we identified further topics and dynamics of problematisation. We could show the growing role of political actors, who were directly invited (through @-mentions) to enter a direct dialogue with other #twlz contributors. In the case of educational technologies, no dialogue with technology providers was expected. Instead, a variety of technology-related hashtags served as points of access to the spaces where educators, parents, students, and other actors exchanged their new knowledge and resources.

With our empirical examination of #twlz hashtag-related communication, we were primarily interested in how educational actors conceptualise and problematise the COVID-19 pandemic and reconfigure the affinity space in which the (re-)negotiation of the educational crisis happens and is linked to technological solutions. Overall, our empirical study of the #twlz as an affinity space contributes to the new perspectives of critical data studies as we attend to the #twlz hashtag not as a sum of keywords with which actors describe a topic. It is, rather, a continuous practice of associating actors, topics, and things through which the implications of the COVID-19 pandemic for the German educational domain are problematised. Future cross-platform research needs to address the #twlz affinity space in the broader context of public media discourses about the role of education in times of crisis, an in-depth content analysis of tweets (and media coverage) is required to understand these dynamics and the role of data therein.

Acknowledgements This work was conducted as part of the DATAFIED research project funded by the German Federal Ministry of Education and Research (project number 01JD1803A). The authors are grateful to the DATAFIED team members Annekatrin Bock, Vito Dabisch, Sigrid Hartong, Sieglinde Jornitz, Angelina Lange, Felicitas Macgilchrist, Ben Mayer, Tjark Raabe, Jasmin Tröger for their crucial feedback and lively discussions. We also thank our student assistants Hendrik Meyer, Maximilian Spliethöver, and Yan Brick.

References

Bates, J., Lin, Y.-W., & Goodale, P. (2016). Data journeys: Capturing the socio-material constitution of data objects and flows. *Big Data & Society, 3*(2). https://doi.org/10.1177/2053951716654502

Baygi, R. M., Introna, L. D., & Hultin, L. (2021). Everything flows: Studying continuous socio-technological transformation in a fluid and dynamic digital world. *MIS Quarterly, 45*(1). https://aisel.aisnet.org/misq/vol45/iss1/16/

Black, S., Spreen, C. A., & Vally, S. (2020). *Education, COVID-19 and care: Social inequality and social relations of value in South Africa and the United States.* http://hdl.handle.net/10210/440221

Boyd, D. (2010). Chapter 2: Social network sites and networked publics: Affordances, dynamics and implications. In N. Self & Z. Papacharissi (Eds.), *A networked self: Identity, community and culture on social network sites* (pp. 39–58). Routledge.

Breiter, A., & Hepp, A. (2018). The complexity of datafication: Putting digital traces in context. In A. Hepp, A. Breiter, & U. Hasebrink (Eds.), *Communicative figurations: Transforming communications in times of deep mediatization* (pp. 387–405). Springer International Publishing. https://doi.org/10.1007/978-3-319-65584-0_16

Britt, V. G., & Paulus, T. (2016). "Beyond the Four Walls of My Building": A case study of #Edchat as a community of practice. *American Journal of Distance Education, 30*(1), 48–59. https://doi.org/10.1080/08923647.2016.1119609

Bruns, A., & Burgess, J. (2016). Methodological innovation in precarious spaces: The case of Twitter. In H. Snee, C. Hine, Y. Morey, S. Roberts, & H. Watson (Eds.), *Digital methods for social science: An interdisciplinary guide to research innovation* (pp. 17–33). Palgrave Macmillan. https://doi.org/10.1057/9781137453662_2

Callon, M., Courtial, J.-P., Turner, W. A., & Bauin, S. (1983). From translations to problematic networks: An introduction to co-word analysis. *Social Science Information, 22*(2), 191–235. https://doi.org/10.1177/053901883022002003

Carpenter, J. P., & Krutka, D. G. (2014). How and why educators use Twitter: A survey of the field. *Journal of Research on Technology in Education, 46*(4), 414–434. https://doi.org/10.1080/15391523.2014.925701

Carpenter, J. P., Tani, T., Morrison, S., & Keane, J. (2020). Exploring the landscape of educator professional activity on Twitter: An analysis of 16 education-related Twitter hashtags. *Professional Development in Education, 1–22.* https://doi.org/10.1080/19415257.2020.1752287

Couldry, N. (2020). Recovering critique in an age of datafication. *New Media & Society, 22*(7), 1135–1151. https://doi.org/10.1177/1461444820912536

Crawford, K., & Finn, M. (2015). The limits of crisis data: Analytical and ethical challenges of using social and mobile data to understand disasters. *GeoJournal, 80*(4), 491–502. https://doi.org/10.1007/s10708-014-9597-z

D'Ignazio, C., & Klein, F. L. (2020). Seven intersectional feminist principles for equitable and actionable COVID-19 data. *Big Data & Society, 7*(2). https://doi.org/10.1177/2053951720942544

Gee, J. P. (2005). Semiotic social spaces and affinity spaces: From the age of mythology to today's schools. In D. Barton & K. Tusting (Eds.), *Beyond communities of practice: Language, power, and social context* (pp. 214–232). Cambridge University Press.

Greenhalgh, S. P., Rosenberg, J. M., Staudt Willet, K. B., Koehler, M. J., & Akcaoglu, M. (2020). Identifying multiple learning spaces within a single teacher-focused Twitter hashtag. *Computers & Education, 148*. https://doi.org/10.1016/j.compedu.2020.103809

Gruzd, A., & Mai, P. (2020). Going viral: How a single tweet spawned a COVID-19 conspiracy theory on Twitter. *Big Data & Society, 7*(2). https://doi.org/10.1177/2053951720938405

Harvey, P., Reeves, M., & Ruppert, E. (2013). Anticipating failure. *Journal of Cultural Economy, 6*(3), 294–312. https://doi.org/10.1080/17530350.2012.739973

Hepp, A. (2016). Pioneer communities: Collective actors in deep mediatisation. *Media, Culture & Society, 38*(6). https://doi.org/10.1177/0163443716664484

Jarke, J., & Breiter, A. (2019). Editorial: The datafication of education. *Learning, Media and Technology, 44*(1), 1–6. https://doi.org/10.1080/17439884.2019.1573833

Johns, F. (2020). Counting, countering and claiming the pandemic: Digital practices, players, policies. In L. Taylor, G. Sharma, A. Martin, & S. Jameson (Eds.), *Data justice and COVID-19: Global perspectives* (pp. 90–99). Meatspace Press.

Kimmons, R., Carpenter, J. P., Veletsianos, G., & Krutka, D. G. (2018). Mining social media divides: An analysis of K-12 U.S. School uses of Twitter. *Learning, Media and Technology, 43*(3), 307–325. https://doi.org/10.1080/17439884.2018.1504791

Larsen, J., & Parrish, C. W. (2019). Community building in the MTBoS: Mathematics educators establishing value in resources exchanged in an online practitioner community. *Educational Media International, 56*(4), 313–327. Scopus. https://doi.org/10.1080/09523987.2019.1681105

Lave, J., & Wenger, E. (1991). *Situated learning: Legitimate peripheral participation*. Cambridge University Press.

Marres, N., & Gerlitz, C. (2016). Interface methods: Renegotiating relations between digital social research. *STS and Sociology, 64*(1), 21–46. https://doi.org/10.1111/1467-954X.12314

Marres, N., & Moats, D. (2015). Mapping controversies with social media: The case for symmetry. *Social Media + Society, 1*(2). https://doi.org/10.1177/2056305115604176

Ruppert, E. (2013). Rethinking empirical social sciences. *Dialogues in Human Geography, 3*(3), 268–273. https://doi.org/10.1177/2043820613514321

Sauers, N. J., & Richardson, J. W. (2015). Leading by following: An analysis of how K-12 School leaders use Twitter. *NASSP Bulletin, 99*(2), 127–146. https://doi.org/10.1177/0192636515583869

Selwyn, N. (2020). Digital education in the aftermath of COVID-19: Critical concerns & hopes. Selwyn, N., Macgilchrist, F., Williamson, B. (eds.). *Digital Education After COVID-19. TECHLASH, 1*, 6–10.

Selwyn, N., & Jandrić, P. (2020). Postdigital living in the age of COVID-19: Unsettling what we see as possible. *Postdigital Science and Education, 2*, 989–1005. https://doi.org/10.1007/s42438-020-00166-9

Staudt Willet, K. B. (2019). Revisiting how and why educators use Twitter: Tweet types and purposes in #Edchat. *Journal of Research on Technology in Education, 51*(3), 273–289. https://doi.org/10.1080/15391523.2019.1611507

Tang, Y., & Hew, K. F. (2017). Using Twitter for education: Beneficial or simply a waste of time? *Computers and Education, 106*, 97–118. https://doi.org/10.1016/j.compedu.2016.12.004

Teräs, M., Suoranta, J., Teräs, H., & Curcher, M. (2020). Post-COVID-19 education and education technology 'Solutionism': A seller's market. *Postdigital Science and Education, 2*, 863–878. https://doi.org/10.1007/s42438-020-00164-x

Thapliyal, N. (2018). #Eduresistance: A critical analysis of the role of digital media in collective struggles for public education in the USA. *Globalisation, Societies and Education, 16*(1), 49–65. https://doi.org/10.1080/14767724.2017.1356701

Theocharis, Y., Lowe, W., van Deth, J. W., & García-Albacete, G. (2015). Using Twitter to mobilize protest action: Online mobilization patterns and action repertoires in the Occupy Wall Street, Indignados, and Aganaktismenoi movements. *Information, Communication & Society, 18*(2), 202–220. https://doi.org/10.1080/1369118X.2014.948035

Tufekci, Z. (2014). Big questions for social media big data: Representativeness, validity and other methodological pitfalls. In *ICWSM '14: Proceedings of the 8th International AAAI Conference on Weblogs and Social Media*, 2014. https://arxiv.org/abs/1403.7400

Visser, R. D., Evering, L. C., & Barrett, D. E. (2014). #TwitterforTeachers: The implications of Twitter as a self-directed professional development tool for K–12 teachers. *Journal of Research on Technology in Education, 46*(4), 396–413. https://doi.org/10.1080/15391523.2014.925694

Williamson, B. (2016). Digital methodologies of education governance: Pearson plc and the remediation of methods. *European Educational Research Journal, 15*(1), 34–53. https://doi.org/10.1177/1474904115612485

Williamson, B., Eynon, R., & Potter, J. (2020). Pandemic politics, pedagogies and practices: Digital technologies and distance education during the coronavirus emergency. *Learning, Media and Technology, 45*(2), 107–114. https://doi.org/10.1080/17439884.2020.1761641

Open Access This chapter is licensed under the terms of the Creative Commons Attribution 4.0 International License (http://creativecommons.org/licenses/by/4.0/), which permits use, sharing, adaptation, distribution and reproduction in any medium or format, as long as you give appropriate credit to the original author(s) and the source, provide a link to the Creative Commons licence and indicate if changes were made.

The images or other third party material in this chapter are included in the chapter's Creative Commons licence, unless indicated otherwise in a credit line to the material. If material is not included in the chapter's Creative Commons licence and your intended use is not permitted by statutory regulation or exceeds the permitted use, you will need to obtain permission directly from the copyright holder.

"Party like it's December 31, 1983": Supporting Data Literacy at CryptoParties

Sigrid Kannengießer

Introduction

Today's datafied societies are characterised by processes of datafication that render "into data many aspects of the world that have never been quantified before" (Cukier and Mayer-Schoenberger, 2013: 29). Critical data studies have been deconstructing datafication and point to the problems and challenges posed by datafied societies, such as risks to privacy (e.g. Kitchin & Lauriault, 2014; Iliadis & Russo, 2016). Concepts like dataveillance (van Dijck, 2014) and surveillance capitalism (Zuboff, 2019) are two of many concepts that can be used to define datafication's central challenges.

Research on civic tech (e.g. Gordon and Lopez, 2019; Saldivar et al. 2018; May and Ross 2018) and data activism (Gutierrez, 2018; Milan and

---

www.crpytoparty.in

S. Kannengießer (✉)
Center for Media, Communication and Information Research,
University of Bremen, Bremen, Germany
e-mail: sigrid.kannengiesser@uni-bremen.de

© The Author(s) 2022

A. Hepp et al. (eds.), *New Perspectives in Critical Data Studies*,
Transforming Communications – Studies in Cross-Media Research,
https://doi.org/10.1007/978-3-030-96180-0_16

van der Velden, 2016; Milan and Gutierrez, 2015) demonstrates the ways in which different actors reflect on and face the challenges of datafication and how they aim at empowering citizens to take informed decisions about their data. Data literacy (e.g. Mandinach and Gummar, 2013) becomes a crucial competence citizens require to face the challenges of datafied societies.

Competences in relation to media have been discussed in different fields of academia under the umbrella term of "media literacy" for a long time (e.g. Aufderheide, 1993; Kubey, 1997). Data literacy is a more recent term used to capture the abilities and necessities required to address datafication and dealing with the pitfalls of sharing and protecting one's data. Discussing different concepts of literacy, I characterise data literacy in this chapter through four different criteria: (1) citizens possess *knowledge* of datafication, the ambivalences and challenges they are forced to confront, (2) people have *access* to their personal data, and (3) they have the *skills* which are required to engage with data's specific *materiality*.

I developed my understanding of data literacy through an empirical study of CryptoParties.[1] CryptoParties are events where people meet to pass on their knowledge about or learn about critical data practices which allow secure online communication, e.g. encrypting online communication, internet browsers, or hard drives. While some people offer support in realising these processes, others attend with their devices to learn how to encrypt their data. CryptoParties are organised by different people in different locations; they are a global phenomenon.

In a qualitative study, I analysed the events of CryptoParties according to the following questions: What does a CryptoParty look like? Who is organising and attending these events? What do people do at CryptoParties? What are the aims of the organisers, people offering help and those seeking support? As case studies, two CryptoParties in Germany were analysed: one event was organised at a well-known hackerspace in Berlin, Germany, and the second at the University of Bremen, Germany, organised by students in cooperation with the German non-governmental organisation DigitalCourage. During these events, observations were carried out as were interviews with organisers, people offering help, and people seeking support. The results of this study are interpreted through the lens of data literacy, discussing CryptoParties as an example of how civil

[1] The capital P within the term CryptoParty is used by the organisers of these events and the administrators of the online platform to stress the combination of the two words.

society initiatives support citizens in developing data literacy.[2] A critical perspective has been applied that acknowledges the constraints and ambivalences between the practices that take place at these events and the ambitions of the actors.

In its discussion of CryptoParties, this chapter contributes to the fields of critical data studies and data literacy, in exploring the ways in which civil society initiatives outside institutionalised settings reflect critically on datafication and privacy risks and how they support the development of data literacy—which is considered an essential competence for citizens in a datafied society.

The chapter is structured as follows: first, I briefly sketch an interdisciplinary research field dealing with literacy in relation to media and data. I will then describe CryptoParties and my methodology. On the basis of this theoretical debate and my methodological reflections, I will present the findings of my study. Finally, I will show how this study contributes to critical data studies in general and studies on data literacy in particular.

From Media Literacy to Data Literacy

Questions of literacy in relation to media have been discussed for a long time across different disciplines (e.g. Aufderheide, 1993; Potter, 1998; Kubey, 1997). The definition of *media literacy* mainly focused on people's *skills*, defining media literacy as "the ability to access, analyse, evaluate and create messages across a variety of contexts" (Livingstone, 2004: 3). A media literate person was perceived as someone who "can decode, evaluate, analyse and produce both print and electronic media" (Aufderheide, 1993: 1).

While media literacy has been of significant importance in the era of mediatisation (Hjarvard, 2008; Hepp, 2013; Lundby, 2014), which was characterised by media's increasing ubiquity and saturation into our everyday lives (Krotz, 2007), processes of digitisation made research to reconfigure the understanding of media literacy (Tyner, 1998; Gurak, 2001; Kellner, 2002). This process has led Livingstone (2004: 8) to argue that "as people engage with a diversity of ICTs, we must consider the possibility of literacies in the plural, defined through their relations with different media rather than defined independently of them".

[2] The results of the study presented here have been published before, discussing CryptoParties as examples of re-active data activism (Kannengießer, 2019). For this chapter, the results of the study are discussed from the theoretical angle of data literacy.

While computers and then the internet gained an importance in all societies around the world, terms like "computer literacy" (e.g. Horton, 1983; critically Goodson and Mangan, 1996), "internet literacy" (e.g. Livingstone 2008), "cyber literacy" (Gurak, 2001), or "digital literacy" (Gilster, 1997; Lankshear and Knobel, 2008; Bawden, 2008) have been conceptualised. In an age of datafication, characterised by processes that render "into data many aspects of the world that have never been quantified before" (Cukier and Mayer-Schoenberger, 2013: 29; see above), the concept of "data literacy" altered the discourse on media. Mandinach and Gummar (2013, 30) define data literacy "as the ability to understand and use data effectively to inform decisions". While this definition again focuses on people's *skills*, Livingstone argues that alongside a skills-based approach which comprises access, analysis, evaluation, and content creation, we also need to acknowledge the "textuality and technology that mediates communication" (Livingstone, 2004: 8) when conceptualizing media literacy. A similar assumption also has to be acknowledged in the concept of data literacy as we need to consider data and digital media's materiality and the ways in which users interact with this materiality. In bringing together these concepts, I characterise the concept of data literacy through four criteria: (1) citizens possess *knowledge* of datafication, the ambivalences and challenges they are forced to confront, (2) people have *access* to their personal data, and (3) they have the *skills* which are required to engage with data's specific *materiality*. Similar to Livingstone's comments on media literacy (see above), it is also important to stress that there is not *one* singular data literacy, but data *literacies* (Fotopoulou, 2020, 1).

Discussing the event format of CryptoParties, this chapter contributes to the research field of data literacy, showing how civil society initiatives (try to) empower citizens to take informed decisions about their data in online communication. It also adds to the broader research field of critical data studies in pointing out the challenges posed by a datafied society from the perspective of civil society actors, demonstrating how they address perceived challenges and their attempts to shape datafication by focussing on the citizens' competences and promoting critical data practices and the development of data literacy.

Case Studies and Methods

Before discussing CryptoParties from a data literacy perspective, the study presented in this chapter will be described in more detail. The case studies used are described as well as the methods used to analyse these cases (see also Kannengießer, 2019). As mentioned in the introduction, CryptoParties are events in which people meet to pass on their knowledge or to learn about critical data practics that allow secure online data practices such as encrypting online communication, the use of the internet, or hard drives. While some people offer help in realising these practices, others attend with their devices to learn how to encrypt their data. CryptoParties are organised by different people in different locations. Asher Wolf stakes a claim as being the "founder" of the CryptoParty phenomenon after organising an event in Melbourne in 2012 (Wolf, 2012). CryptoParties are a global phenomenon—being organised on all continents, in different cultures and national contexts (for the wide range of locations, see www.cryptoparty.in/location). Many CryptoParty organisers register on the online platform www.cryptoparty.in to advertise their events. The administrators of the platform support the organisation of CryptoParties with the aim of building a CryptoParty movement.

For the qualitative study presented in this chapter, two CryptoParties have been analysed as case studies to gain a "rich picture" (Thomas, 2016: 23). Although these cases do not afford a comparison of CryptoParties according to national or cultural differences, they allow for an in-depth analysis that differ in their settings and the backgrounds of the organisers: one of the CryptoParties took place in the hackerspace c-base in the centre of Berlin, Germany.[3] The second one took place at the University of Bremen, Germany, and was organised by students in collaboration with the local non-governmental organisation DigitalCourage[4] based in Bielefeld, a city in the Northwest of Germany. DigitalCourage lobbies for secure online communication and organises data literacy projects. In what follows, I will provide some background information about the events, which are necessary for a full understanding of my results.

The hacker organisation c-base, that hosted the CryptoParty I visited in Berlin, was founded in 1995 as a non-profit organisation focusing on education in hardware, software, and network technology (c-base, n.d.). Members of c-base invited the organisers of the CryptoParty to set up these events in their hackerspace. Before the Corona pandemic, the CryptoParty took place one evening a month at c-base's location and was hosted by two people (male and female in their early 30s and 40s) who were not members of c-base but still affiliated to the local hacker scene. Members of c-base tell a story about their hackerspace: the hackerspace is

[3] https://c-base.org/

[4] https://digitalcourage.de/en

designed using elements of a space-shuttle to represent a narrative in which this space-shuttle crashed and went back in time. Coming from the future, the hackers pretend to work on technological solutions in the present that will make the current society "fit" for the future—which they pretend to already have knowledge of since they have come back from the future—(c-base, n.d.). Being designed as a space-shuttle, the interior of the location is silver and black, the light is bluish, puppets representing "aliens" are exhibited in glass cabinets, miniature space-shuttles are hanging from the wall, and computer monitors are fixed under the ceiling. In the basement, there are workshops where members of the organisation can develop their "technological solutions for the future" (c-base, n.d.), meaning that everybody works on whatever technological project they are interested in.

The CryptoParty took place on the ground floor of the building, which also includes a bar. For the CryptoParty event, there were two bigger tables and some smaller ones arranged in the room, and a screen for the presentation which was the introduction into the CryptoParty and that was given by one of the organizers. Some people entering the room were welcomed by the organisers, while others just found a chair and waited until the event started. The party then began with the short presentation already mentioned in which one of the organisers pointing to the problems of data generation and surveillance, and the CryptoParty's efforts to act on datafication (see below).

After this presentation, the organisers of the event formed groups—asking "experts" on different critical data practices (e.g. email encryption, Linux, or on discussing and explaining how the internet works) to sit on separate tables, and people wanting to learn one of these practices to sit down at the tables that interested them the most. During the event, some people moved from table to table to switch between the different practices. There was no real end to each session but people left whenever their problems were solved. The CryptoParty finished at 01:00, when the organisers started cleaning the room.

The CryptoParty in Bremen was the first one organised by this group and did not advertise the event on www.cryptoparty.in. Nevertheless, people organising the event referred to the online platform and the general movement during the event. This CryptoParty was organised by students in collaboration with the non-governmental organisation DigitalCourage and took place in a student-run café called "Souterrain" at the University of Bremen one evening in January 2019. This was the first event they had organised although some of them had already participated at other CryptoParties in the role of advisers.

The event started at 6 p.m. and finished at 21:00. Similar to the event in Berlin, the organisers gave an introduction to explain the concept of CryptoParties and the critical data practices that would be taught during that evening. After that introduction, the seventeen participants formed groups (consisting of "teachers" and "students") to deal with these different encryption practices. The interior of the "Souterrain" consists of old sofas and some tables and there is a bar where drinks are served which are paid for by donation. On the walls, there are many posters advertising left-wing political events. After a while, some participants switched among the groups and at 9.30 p.m. the CryptoParty closed with one of the organisers thanking everybody for attending. Many people stayed to chat.

To analyse these events, I used a *focused* ethnographic approach (Knoblauch, 2001), which allows the researcher to examine a particular part of culture—for this study this is the CryptoParties. As participatory observations are central to ethnographic studies (Ayaß, 2016: 337), I conducted participatory observations at the two CryptoParties mentioned above in November 2018 and January 2019, taking part as a participant seeking help but at the same time making transparent that I was participating for the purpose of academic research and that I would conduct several interviews.

During these two events, I conducted eight qualitative semi-structured interviews (Hopf, 2004) with organisers, people offering help, and those wanting to learn different encryption practices. The latter were laypersons that I define as those not being professionals in the fields of digital media technologies and datafication. The interview partners differed not only in their roles at these evenets but also in age and gender as well as educational background. But, as the observations and interviews revealed, most of the organisers and participants were male. While at the event in Berlin, people from different age groups participated, it was mainly people in their twenties and thirties participating in the event in Bremen—which was most likely because the event took place at the student café at the university. As the organisers, as well as people helping and seeking help at the CryptoParties I visited were very sensitive about privacy and anonymity, some interviews could only be recorded in a separate room and were transcribed afterwards, while others I needed to be protocolled during the interview as I was not granted permission to record them.

As the organisers of the CryptoParty in Berlin have registered on the online platform www.cryptoparty.in, and the organisers in Bremen referred to this platform several times and perceive themselves as part of this movement, I also conducted a virtual ethnography (Hine, 2000) of the online platform, focusing particularly on the promotional content for the events at c-base.

All research data (protocols for the observation, interview transcripts and protocols, and the protocols of the virtual ethnography) were analysed using a Grounded Theory approach (Corbin and Strauss, 2008) by developing a list of categories across all data to compare the two different initiatives. Through this open coding process, different categories were developed out of the data, grasping the actors who are involved in the CryptoParties, the roles they take, and their aims, as well as the practices that are conducted at these events. Below, I present the results of the study according to these different categories, acknowledging that the categories are interconnected, as the roles and aims of the actors shape the practices that they conduct. Presenting the results according to these categories, CryptoParties are discussed through the theoretical lens of data literacy and finally contextualised to apply to the broader field of critical data studies.

Supporting Data Literacy at CryptoParties

Discussing the CryptoParty events through the theoretical lens of data literacy—a concept that stresses the importance of people having *knowledge* about datafication, the ability of people to *access* their data, to have *skills* to deal with data and to engage with the *materiality* of data (see above)—data literacy appears to be fundamentally central to CryptoParties.

One of the main goals of CryptoParties is to share *knowledge*: people participate to help others encrypt their data; others are often keen to learn how to manage these processes themselves.[5] The organisers of CryptoParties distinguish between different roles, "teachers" and "students", which the participants of these events adopt (c-base, 2018b). "Teachers" are also referred to as "CryptoParty angels" as one of the organisers of the event in Berlin explains—adapting the term from the ChaosComputerClub.[6]

[5] See also Kannengießer (2019) for a discussion on the relevance of education within CryptoParties but from the theoretical angle of data activism.

[6] The ChaosComputerClub is Europe's largest organisation of hackers organising a congress on any hacker-relevant topic after Christmas each year (ChaosComputerClub 2019). See Kubitschko (2015) for an analysis of the ChaosComputerClub.

Participants take on different roles: while some people offer support in "encrypted communication, preventing being tracked while browsing the web, and general security advice regarding computers and smartphones" (c-base, 2018a), others bring their devices to learn about different encryption practices. Many volunteers help in dealing with concrete encryption practices, others explain the background of these actions.

The organisers and helpers at both events examined in the ethnographic study stressed while being interviewed that it is the exchange of knowledge that is their principal aim through the provision of a space to do so. They refer to knowledge as information about datafication and online risks to privacy, as well as knowledge about different encryption practices that they want participants to develop in pursuing different encryption practices themselves. I define these encryption practices as critical data practices in which users of digital media technologies critically reflect on datafication and try to protect their data in online communication processes by encrpyting.

To share knowledge about datafication and privacy risks, the organisers give short presentations at the beginning of each CryptoParty: the organiser in Berlin started by problematising a Facebook advertisement in which the company states that the user's data is safe by presenting pictures of users asking questions about data security. Showing a picture of one of these advertisements, the organiser compared the advertisement with a picture of a milk bottle on which a cow is presented standing on a green lawn with some flowers in front of it. Her argument was that the Facebook advertisement is as much a lie as the milk advertisement pretending that the cow giving the bottled milk had a good life. She continued problematising different aspects of datafication—criticising companies that collect data of their users, selling those data and not being transparent about those processes. After her ten-minute presentation she asked who would like to serve as "teachers" during the event and which encryption practices could be dealt with. Some of the people volunteered to encrypt emails, use Linux, encrypt hard drives, and explain the workings of the internet. The introduction at the CryptoParty in Bremen was much shorter: two of the organisers were explaining the concept of the format and the different encryption practices which could be learned at the event. At both parties, people formed groups after the introduction—each group working on one of the encryption practices—either encrypting emails or hard drives, safe browsing, Linux (in Berlin) and safe mobile phone use (in Bremen). During the events, people sit at tables in small groups working on these different issues.

In addition to learning concrete encryption practices, broad information on datafication was shared within groups: one woman offered to "explain the internet", as she states, at the Berlin event. She brought approximately twenty cards on which different icons were presented symbolising, for example, computers, routers, companies, and firewalls. People participating in this group were asked to put these cards on the table in order of the extent to which they felt they were connected to them, thereby explaining the processes of online communication. While combining these cards, people described the reasons for the order that they chose and the woman facilitating this group asked questions to provoke explanations, while also answering questions from participants. She made very sure to underline the fact that she was very new in this field, having only participated in one CryptoParty. Her first experience at a similar event motivated her to volunteer for future CryptoParties to share the knowledge she gained. In stressing that she is new to this, the woman invites the participants to share their knowledge and thereby enriches her knowledge as a "teacher". The discussion shows that there are experts as well as non-experts in this field and that people learn from one another by sharing their knowledge.

This moderator's actions correspond with the claim of the organisers, who state, "A successful CryptoParty is a CryptoParty where each person learned and taught at least one new thing" (c-base, 2018b). This claim is fulfilled when the group discusses the way the infrastructure of the internet is designed. Yet, in regard to actual encrypting practices, this claim must be viewed critically, as the observations at both the events and during the interviews showed that firm hierarchies persist between "teachers" and "students" (see below).

Still, the organisers invite "newcomers, beginners and the curious" (c-base, 2018b), in particular, and stress that "[a]bsolutely no prior knowledge is required and all questions are beautiful!" (c-base, 2018b). This is also something that the organisers of the event in Bremen underline. They try to destroy any assumptions that CryptoParties are only for technophiles or experts on datafication. This is what the organisers of the CryptoParties repeat during the event—inviting anyone to pose questions. They acknowledge that there are people who hesitate to engage with their digital media technologies, who are not data literate or possess only little knowledge about datafication and different encryption practices:

> The main objective is to tear down the mental walls which prohibit people from even thinking about these topics or picking them up as they occur throughout their lives. […] Sadly, many people don't consider themselves able to process them and don't even start. That's what we want to change. Take away the fear of cryptic and technical things (two properties inherent to cryptographic tools) so they can continue educating themselves and others" (c-base, 2018b)

CryptoParties aim to support non-experts in becoming data literate. They try to do so though the sharing of ***knowledge*** about the problems of datafication and critical data practices which allow encryption. They aiming at equipping the "students" participating in the CryptoParties with the appropriate knowledge required to ***access*** and protect their data on their digital media devices (smartphones, tablets, and laptops—depending on what the people bring to the CryptoParties). It is in this way that participants engage with the ***materiality*** of their digital media technologies and their data. These four characteristics: knowledge, access, skills, and materiality are the central components of data literacy as defined above, showing that non-experts are supported to develop data literacy at events such as CryptoParties.

Trying to support non-experts develop their data literacy, CryptoParties claim to be open and inclusive. CryptoParties are organised by people from different backgrounds in different locations. They are organised by hacker organisations, adult education centres, people working in libraries (as described in Belveze, 2017), students, and others. People organising CryptoParties claim to host open and inclusive events: "CryptoParty is a free and open format" (c-base, 2018a). Organisers claim that everybody is welcome to the events: "We, the organisers and participants of CryptoParties, pledge our dedication to making our events open and welcoming to everyone who shares our guiding principles: Being excellent to each other, and doing things. […] We would like people to be able to teach and learn from each other regardless of background or level of expertise" (c-base, 2018b).

Moreover, CryptoParties aim at allowing inclusion, meaning that disabled people should also be able to participate in the events (c-base, 2018b). To realise an open and inclusive setting, CryptoParty organisers have developed a "code of conduct" (c-base, 2018b) which provides the guidelines for openness and inclusiveness. The "code of conduct" could be found on the tables at the Berlin event. To guarantee an open and inclusive event, it is argued that "[p]eople who act in discriminatory or otherwise excluding ways [...] and who are not able or willing to change their behaviour, may be excluded to preserve a welcoming atmosphere for everybody else" (c-base, 2018b).

The observations revealed, however, that levels of inclusion are variable depending on the setting of the CryptoParty and the background of its organisers. It is the openness and inclusiveness, which is at the same time one of the major aims of the CryptoParties while at the same time being once their major challenges. As one of the organisers of the Berlin event stresses, it is diversity, which is one of the major problems for CryptoParties. She explained that it is very difficult to bring in more of a diverse group, particularly those who do not have much technical experience. She found that the core audience for the events were predominantly university educated men.

During my participatory observation at the event in Berlin, I perceived that nearly all of the participants had specific questions regarding encryption, which suggests they were attending with some degree of previous technical knowledge. This does not only mean that people participating are already sensitive about questions of online privacy and surveillance but also that they are technophiles in that they were already quite engaged with their technologies before the event.

Moreover, regarding hierarchy, only men (with one exception) took on the role of "teacher" at both events; only one woman took on the role of "teacher" at both events, and only few women participated as "students". The one exception in Berlin was the woman who was not teaching concrete encryption practices but instead organised a discussion using cards to explain the infrastructure of the internet (see above). Interestingly, although in the position of the "teacher" or "expert", she several times stressed that she was "new to this", that she was not an expert, something none of the male "teachers" stated.

The gender gap was perpetuated by the setting of the CryptoParty in Berlin at the c-base hackerspace (see Kannengießer, 2020 for a discussion on gender at CryptoParties). While the party itself took place on the ground floor, participants could take part in a "tour for aliens" visiting the basement of the hackerspace. This tour revealed an insight into the hackerspace. Hackers taking part at this event repeatedly constructed the binary between insiders—the members of the hacker organisation and outsiders—people visiting the hackerspace and the CryptoParty. It is this binary which perpetuates the hierarchy of hackers and non-hackers, "teachers" and "students" and "male" and "female" during the CryptoParty as nearly all of the members of the hacker organisation (with one exception) who were there that evening were male.

The organisers are aware of these produced hierarchies: "While we acknowledge the implicit and practical hierarchies and power-relations within the CryptoParty community, extra effort may be put into resolving them" (c-base, 2018b). Nevertheless, these hierarchies have not been deconstructed. It is the location chosen, the hackerspace, which attracts people from a special, technophile scene and implies inhibitions for others. The setting of the CryptoParty regulates the target group—those CryptoParties taking place in libraries or adult education centres attract other people than those taking place in hackerspaces. In this way the potential to learn data literacy is regulated by the events.

At the Bremen event, it was mostly students who participated, mainly because of the location—the university. Only two "teachers" were not students but part of the local DigitalCourage organisation.

The target groups are not only a direct consequence of the venue choice but are also a consequence of the ways in which organisers promote the CryptoParties. One of the organisers of the Berlin event explains that they advertise using posters or flyers although the best public relations for her is still word-of-mouth, which means for the Berlin event that most participants were associated with the hackerspace. Looking at the public relations the organisers of the CryptoParty in Berlin conduct, one of the constraints to the aims and practices of the CryptoParties can be found. As one of the organisers of the Berlin event explains, they also use Twitter for advertising the event, admitting in the interview that "Twitter is also evil", but that they still use it to reach out to more people.

Nevertheless, there are some people involved in CryptoParties who switch roles from "teachers" to "students" thereby disrupting the hierarchies. Several people organising CryptoParties or acting as "teachers" had been previous participants and their positive experiences within the first CryptoParties they attended encouraged them to get more involved as organisers and helpers. Organisers and "teachers" offer "train the trainers" seminars between CryptoParty events. During these training sessions people are taught how to explain basic knowledge on the internet and rudimentary approaches to software and hardware encryption.

CryptoParties are spaces where support is provided to participants so that they may develop their data literacy. The aim is not to encrypt *for* other people but to teach people to encrypt their data themselves: "Even when there are weird problems with a computer, take your time to dictate and explain even complicated procedures and commands. The student will learn more, and if consequent problems arise from these actions maybe even weeks after the CryptoParty, the person who owns the computer might remember what was done, and what might be a source of a problem" (c-base, 2018b). To make people engage with their media technologies, the organisers of both events stress that participants' keyboards are "lava" and therefore nobody is allowed to touch somebody else's device: "Other people's keyboards are lava. Don't touch anyone's keyboard, but your own" (c-base, 2018b).

Because of these goals and the observed practices during the events, CryptoParties can be described as events in which organisers and helpers aim at empowering people to become data literate; non-experts should become empowered in their use of digital media technologies and their knowledge of datafication. In this context, empowerment can be defined as "knowing more about technology and making more informed choices around technology as a result" (Rosner & Ames, 2014: 326). This definition of empowerment plays to the definition of data literacy cited above, understanding data literacy "as the ability to understand and use data effectively to inform decisions" (Mandinach & Gummar, 2013: 30).

The aim of empowerment is part of the organisers' self-understanding as one of the Berlin organisers puts it: "I give people the possibility to get back a piece of their privacy during this one evening". She thinks that she changes the lives of a small number of people at every CryptoParty that she organises: this is her motivation for organising these events. Throughout the interview and while observing her during the event, it became clear that this feeling of being able to change something in people's lives, of knowing more than others and passing on that knowledge is what drives her. This self-efficacy is one of the key motivations for people organising these events or serving as "teachers", as other interviews reveal.

The feeling of empowerment through sharing and gaining knowledge is also stressed on the CryptoParty network's online platform: "People come together and learn from one another how to protect their privacy online in times of pervasive commercial tracking and state surveillance. […] After a few hours everybody has learned something and leaves with new ideas and a sense of empowerment" (c-base, 2018a).

This empowerment aims at what is defined above as data literacy—the possibility of gaining *knowledge* about datafication, people's ability to *access* their data, to gain the *skills* (here different encryption skills), to deal with data, and to engage with data's *materiality* as they engage with the materiality of the media technologies they bring to these events. Non-experts gain (critical) knowledge about datafication, for example, about privacy risks, at the events and learn how to encrypt their digital media technologies and online communication. This knowledge should enable them to take informed decisions (encrypting or even not encrypting) and become data literate people who "can decode, evaluate, analyse and produce [...] media" (Aufderheide, 1993: 1)—in this case digital media.

Still, the process of becoming data literate is gradual, it depends not only on the knowledge people possess on datafication and privacy risks but also on the knowledge about encryption practices and the abilities to engage with digital media technologies and encrypt those. Moreover, across the encryption practices that take place at the CryptoParties, it becomes apparent that there is not *one* data literacy, but *many* data literacies (Fotopoulou, 2020, 1), and that different people focus on different competencies.

Conclusion

Today's datafied societies are characterised by the phenomena of dataveillance (van Dijck, 2014) and surveillance capitalism (Zuboff, 2019), exploiting users of online media in respect to their data while implying complex privacy risks. Data literacy characterised in this chapter became, therefore, a crucial competence for citizens wanting to make informed decisions about (sharing or protecting) their data.

In this chapter, CryptoParties have been discussed as examples of how civil society initiatives reflect on the challenges of datafication and (try to) support non-experts in developing their data literacy, that is, gaining *knowledge* about privacy risks and aspects of surveillance through presentations and informal discussions, having *access* to their data and engaging with the *materiality* of data and digital media technologies while learning the *skills* required for different critical data practices, mainly encryption practices, so that they might be able to protect their data.

Although the actors aim at "data justice" (Dencik et al., 2019) through supporting people in developing their data literacy, the results of the study also showed that the events are not necessarily free of power relations and hierarchies (e.g. in regard to gender). Nevertheless, CryptoParties are efforts to support people who want to improve their data literacy in informal settings. The slogan "Party like it's December 31, 1983" (www.crpytoparty.in) stresses this informal setting and the fun aspect of the learning process (although not all CryptoParties necessarily happen in a party atmosphere). The date in the slogan refers to Orwell's (1992 [1949]) "*Nineteen Eighty-Four*" implying the desire for a society free of surveillance. It is in this way that data literacy performs the premise of a novel utopia.

This chapter aims to make a contribution to the field of data literacy in showing how data literacy is developed and supported in informal learning environments. It became apparent that being knowledgeable of and gaining *knowledge* on datafication and its challenges is as crucial as possessing and learning the *skills* needed to face these challenges. The skills are different critical data practices, mainly encryption practices, in this context.

Analysing civil society initiatives such as CryptoParties provides revealing insights into how different actors critically reflect on the challenges of datafication and how they try to shape it. Through a reconstruction of the participants' perspective, we not only learn about the challenges of datafication, we can also (critically) reflect on solutions that are developed in pursuit of a more "data just" society. Critical data studies has a responsibility to addressing both: the risks and challenges posed by datafication *and* the ambitions and practices developed to deal with them.

References

Aufderheide, P. (1993). *Media literacy: A report of the national leadership conference on media literacy*. Aspen Institute.

Ayaß, R. (2016). Medienethnographie. In S. Averbeck-Lietz & M. Meyen (Eds.), *Handbuch nicht standardisierter Methoden in der Kommunikationswissenschaft* (pp. 335–346). Springer.

Bawden, D. (2008). Origins and concepts of digital literacy. In C. Lankshear & M. Knobel (Eds.), *Digital literacies. Concepts, policies and practices* (pp. 17–32). Peter Lang.

Belveze, D. (2017). *Programming cryptoparties in Libraries: How Librarians can contribute to Students and Citizens empowerment against tracking and mass-surveillance*. https://hal.archives-ouvertes.fr/hal-01504076/document

c-base. (2018a). *Cryptoparty @ c-base*. https://www.cryptoparty.in/berlin/c-base

c-base. (2018b). *c-base (Berlin)—Code of conduct*. https://www.cryptoparty.in/berlin/c-base/code-of-conduct-en

c-base. (n.d.). Official handout. https://www.c-base.org/presse/pressemappe.pdf

ChaosComputerClub. (2019). *36th Chaos Communication Congress to take place in Leipzig*. https://www.ccc.de/en/updates/2019/36c3-in-leipzig

Corbin, J., & Strauss, A. (2008). *Basics of qualitative research. Techniques and procedures for developing grounded theory* (3rd ed.). Sage.

CryptoParty. (2018a). *What is a cryptoparty?* www.cryptoparty.in

CryptoParty. (2018b). *How to organise a cryptoparty*. https://www.cryptoparty.in/organise/howto

Cukier, K., & Mayer-Schoenberger, V. (2013). The rise of big data: How it's changing the way we think about the world. *Foreign Affairs, 92*(3), 28–40.

Dencik, L., Hintz, A., Redden, J., & Treré, E. (2019). Exploring data justice: Conceptions, applications and directions. *Information, Communication and Society, 22*(7), 873–881.

Fotopoulou, A. (2020). Conceptualising critical data literacies for civil society organisations: Agency, care, and social responsibility. *Information, Communication & Society*, 1–20, https://doi.org/10.1080/1369118X.2020.1716041

Gilster, P. (1997). *Digital Literacy*. Wiley.

Goodson, I. F., & Mangan, J. M. (1996). Computer literacy as ideology. *British Journal of Sociology of Education., 17*(1), 65–79.

Gordon, E., & Lopez, R. A. (2019). The practice of civic tech: Tensions in the adoption and use of new technologies in community based organizations. *Media and Communication, 7*(3), 57–68.

Gurak, L. J. (2001). *Cyberliteracy: Navigating the Internet with awareness*. Yale University Press.

Gutierrez, M. (2018). *Data Activism and Social Change*. Palgrave.

Hepp, A. (2013). *Cultures of mediatization*. Polity.

Hine, C. (2000). *Virtual ethnography*. Sage.

Hjarvard, S. (2008). The mediatization of society. A theory of the media as agents of social and cultural change. *Nordicom Review, 29*(2), 105–134.

Hopf, C. (2004). Qualitative interviews. In U. Flick, E. von Kardoff, & I. Steinke (Eds.), *A companion to qualitative research* (pp. 334–339). Sage.

Horton, F. W. (1983). Information literacy vs. Computer literacy. *Bulletin of the American Society for Information Science, 9*(4), 14–16.

Iliadis, A., & Russo, F. (2016). Critical data studies: An introduction. *Big Data & Society, 3*(2), 1–7. https://doi.org/10.1177/2053951716674238

Kannengießer, S. (2019). Reflecting and acting on datafication—CryptoParties as an example of re-active data activism. *Convergence: The International Journal of Research into New Media Technologies*. https://doi.org/10.1177/1354856519893357

Kannengießer, S. (2020). Ungleichheit und Ermächtigung. Zum Verhältnis von Medientechnologie und Geschlecht. *Medien und Kommunikationswissenschaft, 68*(3), 239–254.

Kellner, D. (2002). New media and new literacies: Reconstructing education for the new millennium. In L. Lievrouw & S. Livingstone (Eds.), *The handbook of new media* (pp. 90–104). Sage.

Kitchin, R., & Lauriault, T. (2014). Towards critical data studies: Charting and unpacking data assemblages and their work. *The Programmable City Working Paper*, 2. Forthcoming, https://papers.ssrn.com/sol3/papers.cfm?abstract_id=2474112

Knoblauch, H. (2001). Fokussierte Ethnographie: Soziologie, Ethnologie und die neue Welle der Ethnographie. *Sozialer Sinn, 2*(1), 123–141.

Krotz, F. (2007). The meta-process of mediatization as a conceptual frame. *Global Media and Communication, 3*(3), 256–260.

Kubey, R. (Ed.). (1997). *Media literacy in the information age: Current perspectives.* Transaction Publishers.

Kubitschko, S. (2015). Hackers' media practices demonstrating and articulating expertise as interlocking arrangements. *Convergence—The International Journal of Research into New Media Technologies, 21*(3), 388–402.

Lankshear, C., & Knobel, M. (Eds.). (2008). *Digital literacies. Concepts, policies and practices.* Peter Lang.

Livingstone, S. (2004). Media literacy and the challenge of new information and communication technologies. *The Communication Review, 7*(1), 3–14.

Livingstone, S. (2008). Internet literacy: Young people's negotiation of new online opportunities. Digital youth, innovation, and the unexpected. In T. McPherson (Ed.), *Digital youth, innovation, and the unexpected* (pp. 101–122). MIT Press.

Lundby, K. (2014). *Mediatization of communication. Handbooks of communication science.* De Gruyter Mouton.

Mandinach, E. B., & Gummar, E. S. (2013). A systemic view of implementing data literacy in educator preparation. *Educational Researcher, 42*(1), 30–37.

May, A., & Ross, T. (2018). The design of civic technology: Factors that influence public participation and impact. *Ergonomics, 61*(2), 214–225.

Milan, S., & Gutierrez, M. (2015). Citizens′ media meets big data: The emergence of data activism. *Mediaciones, 14*, 120–133.

Milan, S., & Van der Velden, L. (2016). The alternative epistemologies of data activism. *Digital Culture & Society, 2*(2), 57–74.

Orwell, G. (1992[1949]). *Nineteen eighty-four.* : Random House.

Potter, W. J. (1998). *Media literacy.* Sage.

Rosner, D. K., & Ames, M. G. (2014). Designing for repair? Infrastructures and materialities of breakdown. *Paper presented at 17th ACM Conference on Computer Supported Cooperative Work and Social Computing*, Baltimore, MD, February 15–19, 319–331.

Saldivar, J., Parra, C., Alcaraz, M., Arteta, R., & Cernuzi, L. (2018). Civic technology for social innovation. *Computer Supported Coop Work, 28*, 169–207. https://doi.org/10.1007/s10606-018-9311-7

Thomas, G. (2016). *How to do your case study* (2nd ed.). Sage.
Tyner, K. (1998). *Literacy in a digital world*. Lawrence Erlbaum.
van Dijck, J. (2014). Datafication, dataism, and dataveillance. *Surveillance & Society, 12*(2), *197–208*.
Wolf, A. (2012). Twitter profile of Asher Wolf. https://twitter.com/Asher_Wolf
Zuboff, S. (2019). *The age of surveillance capitalism. The fight for a human future at the new frontier of power*. Public Affairs.

Open Access This chapter is licensed under the terms of the Creative Commons Attribution 4.0 International License (http://creativecommons.org/licenses/by/4.0/), which permits use, sharing, adaptation, distribution and reproduction in any medium or format, as long as you give appropriate credit to the original author(s) and the source, provide a link to the Creative Commons licence and indicate if changes were made.

The images or other third party material in this chapter are included in the chapter's Creative Commons licence, unless indicated otherwise in a credit line to the material. If material is not included in the chapter's Creative Commons licence and your intended use is not permitted by statutory regulation or exceeds the permitted use, you will need to obtain permission directly from the copyright holder.

Researching Public Trust in Datafication: Reflections on the Deliberative Citizen Jury as Method

Helen Kennedy, Robin Steedman, and Rhianne Jones

Introduction: Citizen Juries as Research Methods

Understanding public perceptions of data-driven systems is an essential component of ensuring that data and related practices work "for people and society", to quote the strapline of the UK Ada Lovelace Institute. In other words, engaging with publics about issues relating to data plays an important role in working towards data justice. And yet, data-related

H. Kennedy (✉)
University of Sheffield, Sheffield, UK
e-mail: h.kennedy@sheffield.ac.uk

R. Steedman
Department of Management, Society and Communication, Copenhagen Business School, Frederiksberg, Denmark
e-mail: rst.msc@cbs.dk

R. Jones
BBC Research & Development, London, UK
e-mail: rhia.jones@bbc.co.uk

© The Author(s) 2022

A. Hepp et al. (eds.), *New Perspectives in Critical Data Studies*, Transforming Communications – Studies in Cross-Media Research,
https://doi.org/10.1007/978-3-030-96180-0_17

matters are complex and not easy to understand. The citizen jury offers a solution to the challenge of asking members of the public their views about complex data practices. Citizen juries are policy making aids, where diverse citizens are brought together to debate a complex issue of social importance and make a policy recommendation. They are increasingly used in research contexts (e.g. by Roberts et al., 2020) because, it is argued, they value citizens' experiences and give citizens the opportunity to contribute their informed opinions about issues that materially impact their lives. In citizen juries, citizens are seen to have valuable experiential knowledge that they contribute to "a dynamic process of critical scrutiny of expert authority" (Moore, 2016).

A further strength of the citizen jury is that it gives access to collective views which are formed and given expression through the citizen jury process, something which is not possible through methods which produce individual accounts, like interviews. In contrast, the citizen jury format allows for the expression of community values, some writers claim (e.g. Geleta et al., 2018). This is achieved through the dialogic, deliberative process which is at the heart of citizen juries, through which, advocates argue, "participants can come to appreciate the concerns of others" (Evans & Kotchetkova, 2009, p. 628). Thus citizen juries move beyond the expression of multiple opinions; instead, they synthesise opinions through a deliberative process.

Central to this deliberative process are expert witnesses, who are brought in to present evidence and so facilitate an informed discussion. In the literature on citizen juries, expert witness selection is acknowledged as important. Roberts et al. (2020), who ran citizen juries about wind farms, argue that "the basis of witness recruitment for evidence-giving [...] should be the level and relevance of expertise, and inclusion of a diversity of relevant perspectives" (Roberts et al., 2020, p. 9). They note that who experts are, their institutional affiliation and how clearly they can communicate and answer questions about complex subjects within a short and accessible presentation all matter. Evans and Kotchetkova (2009) argue that having the wrong experts can skew deliberations. Roberts et al. (2020) concur, noting that experts, the expertise they present and the manner in which they present it can sometimes have "too much influence" on how issues are framed and therefore how they are considered by participants.

Citizen juries are of growing interest to researchers and other stakeholders interested in understanding public perceptions of data-driven systems and what might make them trustworthy. In the UK, citizen juries have been used to research public opinion on matters such as ethical AI or fair data-sharing (e.g. the Information Commissioner's Office (ICO, 2019), the Royal Society for the Encouragement of Arts, Manufacture and Commerce (RSA, 2018) and the Ada Lovelace Institute (2020), working with Understanding Patient Data and the Wellcome Trust). Like the literature discussed above, reports on these citizen juries also go some way towards acknowledging the role that experts play in shaping discussion and deliberation. For example, reporting on research into explanations of AI decisions, the ICO (2019) notes that emphasising the accuracy of AI decision systems and not acknowledging their limitations may have led jurors to trust AI decisions to be accurate and not give adequate consideration to the potential utility of explanations. The RSA conclude their report on their Forum on Ethical AI by noting that citizen jury discussions tend to be "framed from the top down, not reflecting the most pertinent questions to participants" (2018, p. 48).

These reflections notwithstanding, there is broad enthusiasm about the potential of citizen juries for capturing public perceptions of the ambivalences of data power, as witnessed in their growing use and claims about what they enable. In this chapter, we argue that this enthusiasm needs to be somewhat tempered. We propose that citizen juries can be usefully conceived through the lens of two sub-fields of sociology: the sociology of knowledge and expertise and the social life of methods, or SLOM. In the former, knowledge and expertise are seen as far from neutral, despite assumptions to the contrary. As Harding bluntly put it, they emerge from science which is shaped by "the institutionalised, normalised politics of male supremacy, class exploitation, racism and imperialism" (Harding, 1992, p. 568). In SLOM, methods are understood to be "shaped *by* the social world in which they are located" and to "help to *shape* that social world" (Law et al., 2011, p. 2). Methods constitute the things they claim to represent: "they have effects; they make differences; they enact realities; and they can help to bring into being what they also discover" (Law & Urry, 2004, pp. 392–3). The citizen jury as research method is no exception.

We build on these schools of thought to argue that citizen juries, like all methods, shape their own outcomes, not least because the expertise which informs deliberation is itself socially shaped. This is not to write off citizen juries, but rather to recognise their limitations alongside their strengths. In this chapter, we tell the story of a citizen jury that we held in the summer of 2019, to explore the usefulness of this approach for eliciting public views on data-driven systems and data management models. We argue that the synthesis of participants' opinions which results from the deliberative approach is a strength that is unique to the citizen jury as a method for researching public perceptions of data power. At the same time, we propose that the expertise which informed deliberation was shaped by the experts called in to provide it and by broader social structures, and it shaped the way that deliberation proceeded and the conclusions that citizen jurors drew. We also argue that the citizen jury facilitator played a role in shaping its process.

Our chapter opens up two new perspectives on critical data studies. First, we propose the citizen jury as a mechanism to foster informed public participation in discussions about data power. Citizen juries can also contribute to data justice, because they enable civic engagement in data-related decision-making. Second, we call for more critical attention to methods and to the role that critical data studies researchers themselves play in framing and shaping their research. We conclude that there is a need for more reflection and greater transparency about researcher positionality in critical data studies and the ways in which it shapes how we understand data power, data justice and related matters.

The chapter proceeds with a brief discussion of literature about public perceptions of datafication in which we situate our research, which highlights the gap that our research aimed to fill. This is followed by a discussion of our citizen jury process, the conclusions that participants drew and reflection on the citizen jury as method.

Public Perceptions of Datafication

Interest in how the public perceives datafication has grown in recent years, amongst academic researchers, policy-makers and practitioners keen to understand citizens' views of the new role of data in society (see Kennedy et al., 2020a for an extensive review of research in this area). Understanding public perceptions is seen as increasingly pressing, in order to address datafication's trust problem (Royal Statistical Society, 2014) and to advance data justice. A major theme in recent research into public perceptions, therefore, is whether people trust data practices, by which we mean the systematic collection, analysis and sharing of data and the outcomes of these processes. Often this is examined by surveying whom people trust with their data (e.g. Dodds, 2018; ICO/Harris Interactive, 2019;

Robinson & Dolk, 2015). Research into why people trust or distrust different institutions (such as Ipsos Mori, 2018) finds that feeling a lack of control over personal data sometimes leads to distrust. Where people do not trust organisations, this is often because of concern that organisations will sell or share data without consent in the case of the private sector or that they are not secure in the case of the public sector. Some research concludes that the public need to be informed about data practices in order to trust them and that "appropriate safeguards, accountability and transparency" are a way of building trust in data practices (Hopkins Van Mil, 2015, p. 1).

In contrast to the findings of surveys, qualitative research challenges simplistic understandings of trust and distrust as clearly distinct from each other. Such research draws attention to the multiple, interrelated, context-dependent layers of trust and distrust that people feel in their interactions with data practices. For example, in an article about focus group research that we carried out with BBC audiences, we, the authors of this chapter, highlight the complex range of factors that come together to engender or undermine trust in data practices (Steedman et al., 2020). These relate to whether people trust the institution that is gathering data in general, whether they trust it specifically to manage their data securely, degrees of trust in the broader data ecosystem and even whether they trust themselves to manage their own data carefully and thoughtfully. As a result, trust, scepticism and distrust sometimes co-exist. We argue that distrust is often appropriate, if organisational data practices are not deemed trustworthy, as in the case of scandals about data breaches (see also Pink et al., 2018 for another qualitative exploration of trust and data).

Qualitative research also calls into question the assumed relationship between trust and understanding, which is implied in the belief that clear information will result in greater trust found in some survey research (e.g. a report by Doteveryone (2018) proposes that without understanding, "it is likely that distrust of technologies may grow"). Pink et al. (2018) show that trust has affective dimensions which will not necessarily be addressed by clear and legible information. Exploring this relationship between trust, understanding and feelings about data practices is necessary to understand public perceptions and, in turn, move towards greater data justice, so we did just that in our citizen jury.

We focused on two areas: (1) trust in data-driven practices and (2) trust in data management models, the latter because they form an important part of the infrastructural arrangements within which data practices take place. There is increasing experimentation with alternative approaches to data management as a result of the individual rights to access and the portability of personal data that are enshrined in GDPR, and yet, there has been little attention paid to what the public thinks about these models and whether they are deemed just and trustworthy. We addressed this gap in a survey of the UK public undertaken in May 2019 (Hartman et al., 2020), which found that approaches that give people control over data about them, that include oversight from regulatory bodies or that enable people to opt out of data gathering were preferred. We carried out our citizen jury to explore the thinking behind these preferences (the why behind the what), whether, after informed deliberation, these characteristics remain important, and the role of feelings and understandings in the formation of preferences. We discuss the citizen jury in more detail in the next section.

Our Citizen Jury Process

Citizen juries often last for several days and bring in diverse experts to present evidence from different perspectives. In citizen juries, experts are understood to have specialist knowledge of the domain and issues under consideration, although we acknowledge that citizens are experts on their own lives, bringing valuable experiential knowledge to the deliberation (see Moore, 2016). Including experts in citizen juries is costly, requiring significant human and financial resources. We adapted the citizen jury model to fit with our limited resources and experimental aims, whilst also ensuring that we incorporated the key element of informed deliberation. Our citizen jury lasted for one day and, more importantly for our argument here, the experts who spoke to the participants were two of the authors of this chapter, Helen, who presented on the benefits and risks of data-driven systems, and Rhianne, who presented on the advantages and disadvantages of different data management models.

Helen and Rhianne are experts in the topics that they spoke about. Helen has been researching and teaching about the social implications of digital and data-driven systems in society for over 20 years and has played a key role in establishing the field of critical data studies in which this edited collection is situated. Rhianne is research lead in Human Data Interaction at BBC R&D and has been immersed in debates, developments and practices relating to data-driven technology for over 10 years. Roberts et al. argue that the task of being a citizen jury expert is difficult, and experts sometimes need to be trained in how to present their material effectively. This was not the case for us: as organisers of the citizen jury and familiar with the format, we knew what was required. Roberts et al. go on to distinguish between what they call "neutral experts", who "explain the wider context and cover the range of issues that are relevant to the topic, rather as a teacher might" (2020, p. 17), and "advocate experts", who present detailed information from their own stance on the topic under consideration. As our citizen jury lasted for one day and included only two experts, we needed neutral, not advocate, experts, so Helen and Rhianne drew on their knowledge of relevant debates to present a balance of perspectives. Given our extensive teaching experience, we had the skills, experience and breadth of knowledge required for this task.

However, as we note above, expertise is never neutral. It is shaped both by social structures and by the individual experts providing it—in our case, two white, (now if not always) middle-class professional women. As part of an ongoing debate about the relationship of expertise and political context, Jasanoff (2003) argues that expertise is neither neutral nor innocent. So who Helen and Rhianne are, our institutional affiliations, communication styles, how clearly we communicate and answer questions—to paraphrase the characteristics of experts that Roberts et al. (2020) claim matter—played a role in shaping the deliberation that took place and the conclusions that jurors reached. But because methods shape research findings, as SLOM literature proposes (Law et al., 2011), this would have been the case with all experts, regardless of their degree of involvement in the research. In one sense, it was not a problem that two of us were the presenting experts, because the deliberative process would have been shaped by any other experts we might have selected.

The role of the facilitator in citizen juries is also important, yet it is rarely acknowledged. Smith (2009) draws attention to the impact that different facilitation styles can have on the deliberative process, yet he notes that the values, principles and philosophy that underpin facilitator practice are seldom considered in the literature. In our case, Robin, the second author of this chapter, was the facilitator. Facilitators, like the methods they deploy, are "shaped *by* the social world" and "help to *shape* that social world" (Law et al., 2011, p. 2). Like methods and experts, they have effects. Robin, a white, middle-class, early career researcher interested in diversity in the media industries, was present for the whole of the citizen jury, whereas Helen and Rhianne only attended the expert sessions in which they presented. Thus Robin played a role in "bringing into being" the data that emerged from the citizen jury (Law & Urry, 2004). We say more about our roles as experts and facilitator below.

Twelve people participated in our citizen jury. They were from a range of socio-economic backgrounds and ethnicities, were of diverse ages, a mix of genders, and some of them had disabilities or health conditions. We selected these particular demographics because they have been shown to be important in shaping views on datafication in previous research (Kennedy et al., 2020b). Jurors worked in a range of industries including the service industry, healthcare, financial services, and travel, and some were retirees and students. Thus we included a diverse mix of people in our citizen jury. We used a market research company to recruit them, as this recruitment method has been shown to be effective for recruiting diverse participants for citizen juries (Street et al., 2014). Their task was to come up with criteria for trusted interactions with data-driven systems and for building trustworthy models for managing data. The day was divided into three sessions: in the morning participants discussed criteria for trusted interactions with data-driven systems and in the afternoon they discussed criteria for trusted ways of managing data. These two sessions included discussion amongst jurors, a presentation by and question and answer slot with an expert, and drafting and re-drafting of criteria. At the end of the day, in the third session, we asked participants to bring their two sets of criteria together to answer the question: what are the most important criteria for the design of ethical, just and trusted data-driven systems? Table 1 provides further detail on how we structured the citizen jury. The jury was recorded and transcribed for analysis, and participants were each given a £70 voucher to thank them for their contributions.

To address our interest in the role of feelings in trust in data practices, participants were asked to use what we called "feelings notes" to track how they felt at key moments and to trace whether their feelings changed over the course of the day. This involved writing their feelings on Post-it Notes at structured moments during the citizen jury. Each participant was assigned a number so the feelings of individuals could be traced throughout the day. We felt that it was important to attend to emotions because the citizen jury approach has been criticised for sidelining feelings and emphasising expertise and rational discussion, much like a legal jury (e.g. by Escobar, 2011). In our previous research (Kennedy et al., 2020b), we have found that thoughts and feelings about data practices are connected

Table 1 The structure of the citizen jury

What	Duration
Introduction and consent	15 mins.
Session 1: Answering the question: What are your criteria for trusted *interactions* with data-driven systems?	
Discuss & evaluate existing data-driven systems	60 mins.
First draft of criteria for trusted interactions with data-driven systems	15 mins.
Coffee break	15 mins.
Expert witness: Benefits and risks of data-driven systems	20 mins.
Question the expert	15 mins.
Second draft of criteria for trusted interactions with data-driven systems	20 mins.
Rank final criteria for trusted interactions with data-driven systems	10 mins.
Lunch break	40 mins.
Session 2: Answering the question: What are your criteria for a trusted way of *managing* data?	
Expert witness: Advantages and disadvantages of five different data management models	20 mins
Question the expert	15 mins.
Discuss five models for managing data	60 mins.
Coffee break	15 mins.
First draft of criteria for a trusted way of managing data	20 mins.
Rank final criteria for a trusted way of managing data	10 mins.
Session 3: Answering the question: What are the most important criteria for the design of ethical, just and trusted data-driven systems?	
Comparing criteria from the morning and afternoon sessions and producing final set of criteria	15 mins.
END	

and understanding how people feel is important in comprehending their views about more just data futures. Moreover, Barnes (2008) argues that bringing emotion into deliberation makes the process more inclusive of diverse groups. Our participants, who we call jurors hereafter, started the day with a wide range of feelings—anxiety about or support for data practices, doubts about their own understanding and contradictory combinations of these emotions. We map their feelings throughout the day in the discussion of proceedings that follows. These feelings notes were collated in a table and analysed in conjunction with the transcripts to add an additional layer of contextual analysis regarding participants' feelings throughout the day.

Session 1: What Are Your Criteria for Trusted Interactions with Data-Driven Systems?

We began the citizen jury with explanations of key terms that would surface during the day, such as "data-driven", "artificial intelligence" and "automated decision-making". We gave participants a handout explaining these terms that they could consult as needed throughout the day. We then proceeded to ground our discussion of data-driven technologies in concrete contexts, discussing four types of data-driven systems, giving examples in practice and explaining how they work. These were personalisation, voice assistants, data scoring and facial recognition technology. Jurors made lists of the benefits and risks of each type of data-driven system and then responded to these questions that Robin posed to them: "to what extent do those benefits/risks lead you to trust or not trust the data-driven system?" and "what would make it trustworthy for you?"

In regard to the trustworthiness of data-driven systems, most jurors agreed that it depends on context. Some jurors accepted personalisation in some contexts, whereas others did not trust it in any context. Jurors tended to be suspicious of voice assistant devices like Alexa, used in the home. They made Alexander, for example, feel "uneasy" because "[o]bviously with Alexa, to activate it you have to say 'Alexa', so it's obviously always listening for that". Some jurors were concerned about how voice assistant technology could evolve. For example, Allyssa said, "I think I trust it right now, but in the future I'm not sure, depending on how much they develop". This was also a concern in relation to data scoring. There was some trust amongst jurors because it was perceived to be less biased than human decision-making. However, Lizzy noted that data scoring systems seem trustworthy "at the moment" but "in the future it scares me what it could become". Jurors noted a relationship between trust in a data-driven system and regulation. Voice assistants would be more trustworthy if there was a strong regulatory framework, some jurors noted.

After this discussion, jurors were asked to draw up criteria for trusted interactions with data-driven systems, which they could modify later in the day after they had heard from an expert witness. Every criterion that a jury member suggested was added to a list. The open format aimed to encourage the free sharing of ideas. The interaction between the jurors and the facilitator was critical to this process. Jurors stated criteria and the facilitator clarified what they meant and formulated the phrasing to be used in the notes that were taken. In this process, Robin—as facilitator—tried not to add interpretation or meaning, but nonetheless, she played a role in shaping how criteria were recorded. We present the criteria that jurors came up with at this and subsequent points throughout the day in Table 2.

At the end of this initial discussion, jurors felt more concerned about and less trusting of data-driven services than they had felt beforehand. Jurors who began the day expressing positive feelings nuanced these feelings, with comments such as "feel comfortable in this moment in time but a little nervous for the future". Jurors who began the day with strong negative feelings noted that these negative feelings had strengthened, and no juror reported feeling more positive after the first discussion. Deliberating data-driven services led jurors to feel more negatively towards them.

After a break, we held our first expert session. Acknowledging, like Harding and others, that neutral expertise is not possible, we aimed for balance in these sessions. Helen outlined five benefits and five risks that have been identified by other experts on data-driven systems in the first session, and later in the day, Rhianne summarised the advantages and disadvantages that have been noted in relation to different data management models. (In the interests of transparency, we have shared the slides and notes that we used.)[1]

The benefits of data-driven systems that other experts have identified and that Helen discussed were enhanced human capability, enhanced understanding, enhanced communication, removal of error and human bias and wide-ranging economic benefits. The risks were concerns relating to ownership and control; less privacy and more surveillance; error and inaccuracies; bias, inequality, discrimination; and technological dependency (i.e. the belief that data-driven systems are accurate because they appear objective and scientific and subsequent deferral to them). The presentation was followed by a question and answer session, most of which was devoted to discussing the final risk Helen presented. Helen's decision to talk about this risk shaped jurors' discussion and eventually their criteria, as we describe below.

After this session, jurors revisited their criteria for trusted interactions with data-driven systems. They added six new criteria, shown in the middle row of the first column of Table 2, four of which were directly related to the final risk that Helen identified, such as "data should inform decision-making but not make decisions", and "data systems should

[1] https://livingwithdata.org/previous-research/trust-in-data/.

Table 2 Jurors criteria throughout the day

Session 1. What are your criteria for trusted interactions with data-driven systems?	*Session 2. What are your criteria for a trusted way of managing data?*	*Session 3. What are the most important criteria for the design of ethical, just and trusted data-driven systems?*
Five most important criteria 1. Individuals retain an element of control 2. How the data will be collected and used is transparent 3. Opting out is possible, clear and easy 4. A platform to individually view and delete [this should exist] 5. Potential security risks are clear to users	Five most important criteria 1. Transparency in who the data is going to and what the data is used for throughout the whole process 2. Safeguards in place for the security of my data 3. Easy to update or delete the data 4. Speedy sanctions for misuse of data 5. Statutory regulations for data management across data management model	Five most important criteria 1. Transparency and accountability, how the data will be collected and used is transparent and companies are held accountable 2. Safety and security, safeguards in place for security of my data 3. Individuals retain an element of control, easy to update, delete and opt out 4. Regulations and sanctions, countries need to collaborate and work together to regulate 5. Despite further developments there are always human fail-safes
Added after expert presentation • Data should inform decision-making but not make decisions • Data-driven system should be tested out on diverse groups and prove suitable before they go public • Pros and cons of systems and their outcomes are made clear • Companies that profit from our data should give back to society • The system should always have human oversight subjective to the system size/number of users • There need to be human fail-safes	After expert presentation • Transparency—in who the data is going to and what the data is used for throughout the whole process • Safeguards in place for security of my data • Easy to update or delete the data • Limit how long data can be stored for depending on the context • Consumers are given updates about changes to the services • Have ability to be told about the outcomes of providing their data • Speedy sanctions for misuse of data • Statutory regulations for data management across data management model types • Simplest and most fit for purpose model is used every time	

After discussion of DDS

- It is transparent
- Motives of companies are clear
- Companies are held accountable and there is redress
- I retain an element of control
- Opting out is possible, clear and easy
- A platform to individually view and delete of your data
- They can continuously update users how they work as the system changes and develops
- Potential security risks are clear to users at the start and throughout
- There is active policing and legal sanctions for bad actors
- Regulation bodies for international cooperation/international court
- Countries need to work together to regulate

always have human oversight". This demonstrates how the expert presentation and related discussion shaped the deliberative process. In the first draft of criteria, transparency, personal control, explanations, regulation and sanctions were identified as important. After the expert talk, jurors started thinking about the process of data-driven decision-making. They concluded that decisions should not be based on data alone and that human oversight of data-driven systems is needed.

Jurors then ranked their full list of 17 criteria, tweaking and rephrasing them in the process. Once the list was complete, each juror was asked to rank them from most to least important. After the jury, we compared individual rankings and produced a top five list, shown in the top row of the first column in Table 2. Interestingly, none of the criteria added after the expert talk were among the overall top five criteria at the end of the morning session. This suggests that the expert talk had some influence on jurors' views, but that their own initial views remained important to them as the jury progressed.

At the end of the morning session, jurors produced further feelings notes. Most jurors expressed ambivalent feelings about data-driven systems. One felt that "they are useful if used ethically and securely" and another that they "can do some good but only in the right hands and with the right controls in place". Jurors felt that having control over data-driven systems was important—either personal control or having "the right controls in place". These feelings appear to reflect the nuance that was presented in the expert talk: data-driven systems offer some benefits, but they also pose some risks. The feelings expressed might also be described as deliberative—they reflect the thoughtful weighing of options that had taken place.

Session 2: What Are Your Criteria for a Trusted Way of Managing Data?

In the second session of the day, jurors discussed criteria for trusted ways of managing data. They completed feelings notes about data management models before the expert talk on this topic, most of which acknowledged a lack of knowledge. This is something we had expected, and it informed our decision to hold the expert presentation of data management models before asking jurors to discuss them.

In this session, Rhianne gave an expert talk on five approaches to data management. In debates about data management, approaches which are subject to discussion and experimentation include personal data stores and data trusts, along with more community-based or commons-based approaches such as data collectives or cooperatives (Lehtiniemi & Ruckenstein, 2019; O'Hara, 2019). We explored each of these, as well as both the commonplace, existing approach whereby digital services are responsible for managing data and an option to opt out. These approaches are not mutually exclusive, but we separated them out so jurors could deliberate distinguishing features and potential benefits and drawbacks. For each example, arguments for and against were presented. We have discussed these and other models extensively elsewhere (Hartman et al., 2020). In the interest of brevity, we sum them up here as follows:

1. Existing "terms of service" approach: digital services control people's data in exchange for providing them with a service.
2. Personal Data Store (PDS): a secure place where individuals can store and control data about them and who gets access to it.
3. Delegating responsibility to oversee data about you: this could be to an independent person, organisation or public body.
4. Data collectives: in which data is seen as a collective asset/public good which is managed collectively.
5. Opting out of online data collection, storage and use.

The question and answer session following this expert talk focused on the practicalities of how the models would work, for example, if people wanted to change from one model to another, the costs of the different models, and the types of data that would be covered under them. In their subsequent discussion of the models, these questions continued to be significant, and jurors identified potential benefits and risks for all models. In regard to the existing terms of service model, some jurors felt that "If you're thorough, and if it is clear, then [...] you should know exactly what you're signing up to" (Matthew), whereas others noted that people tend not to read terms and conditions in detail and that companies are aware of this and use it to their advantage.

Jurors asked lots of questions about other, less familiar models like the PDS. Some of them liked the idea of having all of their data in one place, and they felt that the control over their own data that this model enabled made it more trustworthy than other models. In contrast, some were concerned by the idea of all of their data being in one place, as this might make it less secure. Jurors were also concerned about how new models could be introduced: if the PDS model was adopted, would Facebook still control some historical personal data, they wondered.

For some jurors, the delegated responsibility model was trustworthy because it meant that data and related decision-making were in the hands of experts. Others were concerned that such an approach might be costly and impractical to introduce. One participant, Matthew, felt that delegated responsibility was less preferable to the PDS model for *some* types of data, but not all. He liked the idea of personal control over some of his data, but in the case of medical or health data, the delegated responsibility model felt more trustworthy than the PDS. Other jurors agreed with this view.

Jurors liked the democratic potential of the data collective model—Gillian described it as a "democratic way of working" and David felt it was more trustworthy than other models because "collective interest is involved". Others were concerned about whether it would be effective and wondered why introducing a DIY approach was preferable when the delegation model puts data-related decision-making in the hands of experts. Some jurors considered the idea of merging a collective model and a delegation model, so that both professionals and citizens are involved.

Finally, some jurors loved the idea of opting out of data collection: it sounded easy and enabled people to make their own decisions about whether their data is collected. Some were concerned that choosing to opt out of data collection might not erase historical data that had been gathered about people and they felt that this model was less trustworthy because what would happen to past and present data was unclear. Some jurors argued that there are benefits to data collection, for example, in relation to health or disease prevention as in the COVID-19 pandemic, and for this reason were not in favour of opting out.

After some discussion, jurors ranked models individually using the Mentimeter platform (an online tool for quizzes and ranking exercises). This enabled jurors to see how their rankings compared with those of other jurors. Combining these individual rankings, the option of opting out was preferred, followed by the PDS, and then the existing terms of service model. Delegating responsibility was the fourth preference, and the data collective approach was the least preferred of the five options. This is only partly consistent with the findings of our survey on the same topic (Hartman et al., 2020), in which opting out and personal control (which the PDS offers) were seen as desirable, but where the terms of service model were by far the least preferred approach. We suggest that these differences arise from the deliberative character of the citizen jury, and they confirm our argument that our methods shaped our findings. Jurors' deliberation focused on the practical difficulties of adopting new data management models and this made them doubt whether they could realistically be introduced. We suggest that this conclusion accounts for the relatively high ranking of the terms of service model. Furthermore, focusing only on combined rankings erases the nuance that was evident in jurors' deliberation, for example, about the possibility of combining models. This nuance was better captured in jurors' feelings notes about this exercise, in which they expressed caution about the models, which were seen to have "potential", but were yet to be "figured out".

After hearing from and questioning the expert witness and undertaking their deliberation, jurors came up with nine criteria in response to the question "What are your criteria for a trusted way of managing data?", which they then ranked in order from most to least important. As with the first session, every criterion that was suggested was added to the list using the facilitator's suggested phrasing. Jurors then ranked the criteria using Mentimeter and we identified the top five criteria after the jury was complete. These are shown in the top row of the middle column of Table 2.

Section 3: What Are the Most Important Criteria for the Design of Ethical, Just and Trusted Data-Driven Systems?

In the final session of the day, participants were asked to come up with their top five criteria to address the overarching question: *what are your most important criteria for the design of ethical, just and trusted data-driven systems?* Unlike in the previous sessions, in which we identified top criteria as part of our analysis, participants were tasked with collectively agreeing on the top five criteria and on their ranking. Whereas the first two sessions aimed to record the full spectrum of individual views on criteria, here we wanted the jurors to synthesise their views and come up with a collective recommendation. We anticipated that arriving at a consensus may be difficult, but in fact, jurors rapidly reached agreement on the five most important criteria for the design of ethical, just and trusted data-driven systems. Keen to ensure that the views of their co-jurors were reflected in the final list of criteria, they did so by sometimes combining multiple criteria and tweaking phrasing, as can be seen in Table 2. These criteria can be seen in the final column of Table 2.

Table 2 shows that the final criteria that jurors produced were very similar to those produced in the penultimate exercise of the day, which in turn built on criteria developed earlier. This suggests that the synthesis process was ongoing as it took place throughout the citizen jury. Some criteria were important from the beginning of the day, such as transparency and control. Some became more important as experts presented evidence, such as the need for human oversight of data-driven decisions, or as jurors deliberated, such as the need for regulation and sanctions. The final list of criteria, addressing the question: "What are the most important criteria for the design of ethical, just and trusted data-driven systems?" combines criteria which had previously been listed separately—transparency and accountability, regulation and sanctions—which suggests that jurors began to see a relationship between criteria as they deliberated them. Concluding feelings notes suggested that jurors could imagine a trustworthy data future based on the criteria that they produced. One wrote: "If most important criteria was implemented I would be confident in saying they were just and trusted data driven systems."

Reflections on Findings and on the Citizen Jury as Method for Researching Public Trust in Datafication

With this chapter, we add the citizen jury to critical data studies' methodological toolkit. We argue that with its aim of fostering informed public participation and civic engagement in data-related decision-making, it can facilitate data justice. In the citizen jury we describe here, criteria identified for the design of ethical, just and trustworthy data-driven systems echo some of the findings of previous research into public perceptions of data practices, including our own survey (Hartman et al., 2020). Like our jurors, participants in our survey preferred approaches that give people control over data about them, that enable people to opt out of data gathering or that include oversight from regulatory bodies. Other research has found that, like our jurors, people want transparency about and accountability in relation to data practices (e.g. Cabinet Office, Citizens Advice, 2016; see Kennedy et al., 2020 for more examples). Jurors' final criterion, that there are always human fail-safes, has not been identified as a preference in previous research, perhaps because it has not been considered as an option within the research process. Although the RSA (2018) found a desire for oversight over data-driven decision-making, the *human* dimensions of such oversight have not previously been prioritised.

Following Barnes (2008), and informed by the findings of our own prior research (e.g. Kennedy et al., 2020b), we tracked feelings throughout our jury to acknowledge their importance and include diverse groups. On the whole, jurors' thoughts and feelings changed throughout the day as they engaged in the process of deliberation. In the final feelings notes and as evidenced in the quote with which we end the previous section, five jurors explicitly referenced the criteria they had drafted. Their valuation of their own criteria is indicative of their understanding of the ambivalences of data power, at one and the same time potentially beneficial and potentially risky, and in need of careful governance. This also suggests that some jurors felt that one way to ensure that data-driven systems are ethical, just and trustworthy is to account for the views of citizens. The shifts that we saw in jurors' thoughts and feelings also highlight the democratic possibilities that citizen juries afford, which suggests that views can change when people are brought together and when they learn from one another, as well as from facilitators and experts.

The finding that human fail-safes matter provides evidence of our argument in this chapter that experts are not neutral and that methods shape findings. Helen took the decision to include "technological dependency"—or the belief that data-driven systems are accurate because they appear objective and scientific, which in turn causes people to defer to them—as the final risk that she discussed in her expert presentation. She did this because this has been identified as a problem by other expert commentators (such as Eubanks, 2017). Nonetheless, this decision shaped the subsequent question and answer session which was dominated by discussion of this issue, which, in turn, shaped the deliberation that followed, in which jurors started thinking about the *process* of data-driven decision-making. Prior to this expert presentation, jurors were principally concerned with *what* data-driven systems do. Afterwards, they became concerned with *how* data-driven systems do things.

The facilitator also shapes the outcomes of citizen jury research. As facilitator, Robin played a key role in generating the criteria, working to distil and translate the ideas of participants into words that could be added to criteria lists, often after a participant made a long statement or thought out loud. And yet, as Smith (2009) notes, facilitator style, values and philosophy are rarely acknowledged as a contributing factor in citizen jury literature. Participants also shape findings, and there is also little discussion of their role in the literature. The RSA's report on their Forum on Ethical AI is an exception, as it notes that juror selection is important in shaping how a citizen jury proceeds—like all research, the results of a citizen jury depends on who is in the room. Street et al.'s (2014) systematic review of citizen jury studies is another exception, as it recognises that both juror recruitment and moderation are important.

Finally, the jurors themselves also influenced findings. For example, one juror worked in financial services and was very familiar with credit scoring, which shaped the discussion about kinds of data scoring as other jurors listened attentively to what she had to say. While no juror was an expert on data-driven systems or data management models, their experiential and professional knowledge influenced their views and subsequently the course of their deliberations. We cannot say whether the demographic profile of participants influenced outcomes because our sample is small and, like all qualitative research, we do not consider participants to be representative of the demographic groups to which they belong. Other research has addressed this question of difference and inequality (e.g. Kennedy et al., 2020a), and there is more research to be done in this regard.

Implicit in Street et al.'s comment on juror recruitment and moderation is a suggestion that it is possible to do both of these things in ways that minimise juror and facilitator effects. We do not agree. We recognise the value of citizen juries both in centring citizens' experiential knowledge and in the deliberation and synthesising of collective views that they enable. But we also believe that all methods have effects, that they "bring into being what they also discover", as Law and Urry (2004, p. 393) put it. We have been researching public views and feelings about datafication for a number of years (see Hartman et al., 2020; Kennedy, 2016; Steedman et al., 2020) and we have used a variety of methods to do so, including interviews, focus groups, surveys, digital methods and now a citizen jury. Some of us have also carried out an extensive review of research into public understanding and perceptions of data practices (Kennedy et al., 2020a). In this review, we note that research methods, questions asked, how findings are interpreted and presented, the disciplinary background and political orientation of researchers all play a role in shaping research findings and the claims that are made. We argue that "[t]he wording of a survey question, the effect of interviewer presence, the framing of an issue and the impact of others in a focus group setting can all affect responses to research questions" (Ibid., p. 44).

The review discussed above and our own empirical research, including the citizen jury that we discuss in this chapter, lead us to conclude that all empirical research findings are shaped by their methods. Yet, as we state in our review, these well-known issues in social research are not widely acknowledged in research into public understanding and perceptions of data practices. This is not to argue that such research should be abandoned; rather, it is an argument that suggests the field might benefit from more reflection and greater transparency about positionality and approach. As Law et al. put it, it is important to think critically about methods, "about what it is that methods are doing, and the status of the data that they're making" (Law et al., 2011, p. 7). Neither the citizen jury nor research into public perceptions of datafication should be exempt from this kind of critical thinking. The contribution to critical data studies that we are trying to make in this chapter is to call for more critical attention to methods and to the role that researchers play in framing and shaping our research and the ways in which we understand data justice and civic engagement in datafied societies.

The review of research cited above (Kennedy et al., 2020a) also found that context influences perceptions of data practices. At the time of writing, exploring the effects of context on perceptions would mean researching whether and how the COVID-19 pandemic informs perceptions is necessary. Because this chapter focuses on research carried out before the pandemic, this important issue has not been discussed here. However, at the time of writing it is a central part of our current research projects, the results of which are forthcoming.

Acknowledgements This work was supported by grant reference number R/161466 from the Engineering and Physical Sciences Research Council's Human Data Interaction Network+.

References

Ada Lovelace Institute, understanding patient data & the Wellcome Trust. (2020). *The foundations of fairness for NHS health data sharing*. https://www.adalovelaceinstitute.org/the-foundations-of-fairness-for-nhs-health-data-sharing/

Barnes, M. (2008). Passionate participation: Emotional experiences and expressions in deliberative forums. *Critical Social Policy, 28*(4), 461–481. https://doi.org/10.1177/0261018308095280

Citizens Advice—Illuminas. (2016). *Consumer expectations for personal data management in the digital world*. https://www.citizensadvice.org.uk/Global/CitizensAdvice/Consumer%20publications/Personal%20data%20consumer%20expectations%20research.docx.pdf

Dodds, L. (2018, July 5). Who do we trust with personal data? https://theodi.org/article/who-do-we-trust-with-personal-data-odi-commissioned-survey-reveals-most-and-least-trusted-sectors-across-europe/

Doteveryone. (2018). People, power and technology: The 2018 digital attitudes report. Doteveryone. https://doteveryone.org.uk/report/digital-attitudes/

Escobar, O. (2011). *Public dialogue and deliberation. A communication perspective for public engagement practitioners*. Edinburgh Beltane – UK Beacons for Public Engagement. https://oliversdialogue.wordpress.com/2013/08/01/public-dialogue-and-deliberation-a-communication-perspective-for-public-engagement-practitioners/

Eubanks, V. (2017). *Automating inequality: How high-tech tools profile, police, and punish the poor*. St. Martin's Press.

Evans, R., & Kotchetkova, I. (2009). Qualitative research and deliberative methods: Promise or peril? *Qualitative Research, 9*(5), 625–643. https://doi.org/10.1177/1468794109343630

Geleta, S., Janmaat, J., Loomis, J., & Davies, S. (2018). Valuing environmental public goods: Deliberative citizen juries as a non-rational persuasion method. *Journal of Sustainable Development, 11*(3), 135. https://doi.org/10.5539/jsd.v11n3p135

Harding, S. (1992). After the neutrality ideal: Science, politics, and "strong objectivity"'. *Social Research, 59*(3), 567–587.

Hartman, T., Kennedy, H., Steedman, R., & Jones, R. (2020). Public perceptions of good data management: Findings from a UK-based survey. *Big Data & Society, 7*(1). https://doi.org/10.1177/2053951720935616

Hopkins Van Mil. (2015). *Big data: Public views on the use of private sector data for social research. A findings report for the Economic and Social Research Council*. https://esrc.ukri.org/files/public-engagement/public-dialogues/public-dialogues-on-the-re-use-of-private-sector-data-for-social-research-report/

ICO – Harris Interactive. (2019). *Information rights strategic plan: Trust and confidence*. https://ico.org.uk/media/about-the-ico/documents/2615515/ico-trust-and-confidence-report-20190626.pdf

ICO / Information Commissioner's Office. (2019). *When it comes to explaining AI decisions, context matters*. https://ico.org.uk/about-the-ico/news-and-events/news-and-blogs/2019/06/when-it-comes-to-explaining-ai-decisions-context-matters/

Ipsos Mori. (2018). *The state of the state 2017–2018: Austerity, government spending, social care and data*. https://www.ipsos.com/sites/default/files/ct/publication/documents/2017-10/thestate-of-the-state-2017-2018.pdf

Jasanoff, S. (2003). Breaking the waves in science studies: Comment on H.M. Collins and Robert Evans, 'The Third Wave of Science Studies'. *Social Studies of Science, 33*(3), 389–400. https://doi.org/10.1177/03063127030333004

Kennedy, H. (2016). *Post, mine, repeat: Social media data mining becomes ordinary*. Palgrave Macmillan. https://doi.org/10.1057/978-1-137-35398-6

Kennedy H., Oman, S., Taylor, M., Bates, J., & Steedman. R. (2020a). *Public understanding and perceptions of data practices: A review of existing research*. https://livingwithdata.org/project/wp-content/uploads/2020/05/living-with-data-2020-review-of-existing-research.pdf

Kennedy, H., Steedman, R., & Jones, R. (2020b). Approaching public perceptions of datafication through the lens of inequality: A case study in public service media. *Information, Communication & Society*. https://doi.org/10.1080/1369118X.2020.1736122

Law, J., Ruppert, E., & Savage, M. (2011). The double social life of methods. *CRESC Working Paper, 95*. http://www.open.ac.uk/researchprojects/iccm/library/164

Law, J., & Urry, J. (2004). Enacting the social. *Economy and Society, 33*(3), 390–410. https://doi.org/10.1080/0308514042000225716

Lehtiniemi, T., & Ruckenstein, M. (2019). The social imaginaries of data activism. *Big Data & Society, 6*(1), 1–12. https://doi.org/10.1177/2053951718821146

Moore, A. (2016). Deliberative elitism? Distributed deliberation and the organization of epistemic inequality. *Critical Policy Studies, 10*(2), 191–208. https://doi.org/10.1080/19460171.2016.1165126

O'Hara, K. (2019). *Data trusts: Ethics, architecture and governance for trustworthy data stewardship*. University of Southampton. https://doi.org/10.5258/SOTON/WSI-WP001

Pink, S., Lanzeni, D., & Horst, H. (2018). Data anxieties: Finding trust in everyday digital mess. *Big Data & Society, 5*(1), 1–14. https://doi.org/10.1177/2053951718756685

Roberts, J. J., Lightbody, R., Low, R., & Elstub, S. (2020). Experts and evidence in deliberation: Scrutinising the role of witnesses and evidence in mini-publics, a case study. *Policy Sciences, 53*(1), 3–32. https://doi.org/10.1007/s11077-019-09367-x

Robinson, G., & Dolk, H. (2015). Public attitudes to data sharing in Northern Ireland. *ARK Research Update*, No. 108. https://research.hscni.net/sites/default/files/0032-RESEARCH%20UPDATE%20108%20final.pdf

Royal Statistical Society. (2014). *Royal Statistical Society research on trust in data and attitudes toward data use / data sharing*. https://www.statslife.org.uk/images/pdf/rss-data-trust-data-sharing-attitudes-research-note.pdf

RSA / Royal Society for the Encouragement of Arts, Manufacture and Commerce. (2018). *Artificial intelligence: Real public engagement*. https://www.thersa.org/discover/publications-and-articles/reports/artificial-intelligence-real-public-engagement

Smith, G. (2009). *Democratic innovations: Designing institutions for citizen participation*. Cambridge University Press.

Steedman, R., Kennedy, H., & Jones, R. (2020). Complex ecologies of trust in data practices and data-driven systems. *Information, Communication & Society, 23*(6), 817–832. https://doi.org/10.1080/1369118X.2020.1748090

Street, J., Duszynski, K., Krawczyk, S., & Braunack-Mayer, A. (2014). The use of citizens' juries in health policy decision-making: A systematic review. *Social Science & Medicine, 109*, 1–9. https://doi.org/10.1016/j.socscimed.2014.03.005

Open Access This chapter is licensed under the terms of the Creative Commons Attribution 4.0 International License (http://creativecommons.org/licenses/by/4.0/), which permits use, sharing, adaptation, distribution and reproduction in any medium or format, as long as you give appropriate credit to the original author(s) and the source, provide a link to the Creative Commons licence and indicate if changes were made.

The images or other third party material in this chapter are included in the chapter's Creative Commons licence, unless indicated otherwise in a credit line to the material. If material is not included in the chapter's Creative Commons licence and your intended use is not permitted by statutory regulation or exceeds the permitted use, you will need to obtain permission directly from the copyright holder.

Worker Perspectives on Designs for a Crowdwork Co-operative

Jo Bates, Alessandro Checco, and Elli Gerakopoulou

Introduction

Crowdwork platforms such as Amazon Mechanical Turk (AMT) are a crucial infrastructural component of our global data assemblage. Through these platforms, low-paid crowdworkers perform the vital labour of manually labelling large-scale and complex datasets, labels that are needed to train machine learning and AI models (Tubaro et al., 2020) and which enable the functioning of much digital technology, from niche applications to global platforms such as Google, Amazon and Facebook.

J. Bates (✉)
University of Sheffield, Sheffield, UK
e-mail: jo.bates@sheffield.ac.uk

A. Checco
Department of Computer Science, University of Roma La Sapienza, Rome, Italy
e-mail: alessandro.checco@uniroma1.it

E. Gerakopoulou
University of Sheffield Information School, Sheffield, UK
e-mail: elli.gerakopoulou@gmail.com

© The Author(s) 2022

A. Hepp et al. (eds.), *New Perspectives in Critical Data Studies*, Transforming Communications – Studies in Cross-Media Research,
https://doi.org/10.1007/978-3-030-96180-0_18

While some social enterprise crowdwork platforms exist (Gray & Suri, 2019), the most popular platforms such as AMT are organised so that individual workers compete to be able to complete microtasks—or Human Intelligence Task (HITs)—advertised on the supposedly 'neutral' platform by requesters, most of whom pay far below minimum wage rates for the completion of tasks. The platform also takes a proportion of what the requester pays.

Crowdwork is often imagined as a solitary and isolating experience; however, researchers such as Gray et al. (2016) identify that crowdworkers often collaborate to meet their social and technical needs through, for example engaging in online forums. Beyond worker online forums, various socio-technical interventions in crowdwork infrastructures have been developed, including a number of popular browser plugins used by workers to help manage workflow. Most notable among these is Turkopticon, a browser plugin that enabled workers to review requesters, developed by Lili Irani and colleagues and actively maintained from 2008 to 2019.

In many ways, workers' efforts to adapt the dominant crowdwork infrastructure can be understood as them engaging in an act of bricolage aimed at re-constituting infrastructure to produce the most efficient workflow possible from the tools available. Nevertheless, despite such activity, significant barriers stand in the way of crowdworkers' efforts to collectively improve their labour conditions and Graham et al. (2017, p. 146) observe a "race to the bottom" in relation to worker pay and conditions.

Crowdworker co-operatives have been mentioned by a number of researchers and labour activists as a possible alternative, but have not yet been explored in detail (Gray & Suri, 2019; Graham et al., 2017; Scholz, 2016). In this chapter, we reflect on how a 'design justice' approach might be valuable to build on insights gained from a series of exploratory discussions we have engaged in with US-based crowdworkers about how a crowdworker co-operative might work in practice, and begin to sketch out a potential software architecture that could form the basis of future participative approaches to the design and development of a crowdworker co-operative.

We begin by discussing recent research on the possibility of 'platform co-operatives', including crowdwork co-operatives. We then go on to describe and reflect on our own evolving methodology and how it fits with the 'design justice' lens we propose for future work. Following this, we present findings from our discussions with crowdworkers about how a crowdwork co-operative might work in practice, including what values workers would like to see embedded in the design. We then finish with the outline of a prototype software architecture for a crowdworker co-operative that could be used as a starting point in future design work in collaboration with crowdworkers.

Platform Co-operativism

Different forms of co-operative, including worker and consumer co-operatives, have existed since the 1800s, as an alternative to strictly capitalist relations of labour and consumption. Many co-ops follow the Rochdale Seven Principles, originally formed by the Rochdale Society of Equitable Pioneers in the mid-1800s, and most recently updated in 1995. These are as follows:

1. Voluntary and Open Membership
2. Democratic member control
3. Member economic participation
4. Autonomy and independence
5. Education, training and information
6. Cooperation among co-operatives
7. Concern for community

More recently, in response to the advances of "platform capitalism" (Srnicek, 2017), people have begun to imagine how co-operative principles and practices might be adapted to address the forms of capitalist exploitation evident in the digital economy and which provide "an alternative to venture capital-funded and centralized platforms". Scholz (2016) and the Platform Co-operative Consortium, of which he is a member, have begun work to conceptualise different possible models of platform co-operative across the wider platform economy that includes crowdwork.

The Platform Co-operative Consortium proposes a "platform co-operative" should be based on principles including:

> Broad-based ownership of the platform, in which workers control the technological features, production processes, algorithms, data, and job structures of the online platform; Democratic governance, in which all stakeholders who own the platform collectively govern the platform; Co-design of the platform, in which all stakeholders are included in the design and creation of the platform ensuring that software grows out of their needs, capacities, and aspirations; An aspiration to open source development and open data, in which new platform co-ops can lay the algorithmic foundations for other co-ops. (https://platform.coop/)

Scholz (2016) proposes an initial typology for beginning to think through possible designs for platform co-operatives, identifying (1) Co-operatively Owned Online Labour Brokerages and Market Places, (2) City-Owned Platform Co-operatives, (3) Producer-owned Platforms, (4) Union-Backed Labour Platforms and (5) Co-operatives from Within.

A platform crowdworker co-operative offering decent work opportunities independent of existing platforms such as AMT—as in (1) or (4) above—has significant advantages, including the co-operative being able to use all income to pay workers and reinvest in the co-operative rather than channelling funds to Amazon. Nonetheless, challenges abound in relation to how a co-operative start up might compete with a platform such as AMT for tasks for workers to complete and hiring programmers to develop and maintain the platform. An alternative to setting up an entirely independent platform from scratch is the type Scholz refers to as "Platform from Within", which he describes as a form of "hostile takeover" in which "worker cooperatives form[] inside the belly of the sharing economy". The example Scholz uses is Uber drivers using Uber's technical infrastructure to run their own enterprise. Such a model is also possible with an infrastructure such as AMT—a worker co-operative could essentially 'plug in' to the AMT infrastructure to siphon off HITs, but have its own separate distribution and governance structures. Such an approach could be temporary while the co-operative develops the necessary expertise to disconnect from the AMT feed. However, despite the advantages of such an approach, depending on the specific actions of the co-operative, such activity may be against AMT terms and conditions, something which could be off putting for workers that do not want to risk being banned from AMT. These issues point to how the material conditions of workers and existing crowdwork infrastructures generate significant ambivalences around how co-operative models might be leveraged in efforts to address labour exploitation as a form of data power within the global AI assemblage.

Moving Towards a Critical Design Epistemology

This chapter, written in mid-2020, reflects on a moment in the unfolding of an interdisciplinary collaboration between the authors. How we are positioned as researchers in the field of crowdwork and how we came to work together is important for understanding the trajectory of our work and the nature of this particular contribution. Taking inspiration from Irani and Silberman's (2016) reflections on the "wider economic and cultural imaginaries of design as a social role", we here reflect on these challenges in relation to our own work.

While we have all long had interests in labour relations and capitalist modes of production, it was Alessandro that first became actively involved in research on paid crowdwork infrastructures. As a Computer Scientist (CS) working as a postdoctoral researcher on an EU funded project, Alessandro initially focused on experimental work with the goal of improving the efficiency and efficacy of crowdworker tasks, such as image recognition, labelling and annotation, text summarisation, translation, data cleaning and information retrieval. Unhappy with the surveillance-oriented methods of quality control, Alessandro used his CS expertise to develop a Gold Question (GQ) detector that if implemented could be used by crowdworkers to alert them to the existence of quality check questions within a task, and which with enough workers on board could potentially be used to initiate a digital strike. It was at this point that Alessandro invited Jo (a researcher in the field of Critical Data Studies [CDS]) to collaborate—in the first instance to help think through social theoretical lenses that could be used to frame this work on Gold Question detection.

After completing this initial work (Checco et al., 2018), we reflected on the lack of crowdworker voices in our nascent interdisciplinary collaboration and managed to secure a small amount of internal funding to pay a Research Assistant—Elli—to conduct a series of interviews with crowdworkers. These interviews explored a range of issues experienced by crowdworkers, as well as exploring their perceptions on the GQ detector and what ideas they had for improving crowdworker conditions. We decided to engage with US-based crowdworkers as we were particularly interested in how the increasingly global nature of the crowdworker labour market was being experienced by workers in countries with higher costs of living.

Emerging from some of our early interviews was the idea of a co-operative model for crowdwork. This idea of a crowdwork co-operative was also something that Alessandro had begun to explore from a Computer Science perspective, and we decided to begin actively asking crowdworkers about the co-operative idea in the second stage of interviews. Despite gathering significant insights from crowdworkers about the potential of different socio-technical interventions in the crowdwork infrastructure, our reflections on our evolving interdisciplinary approach raised questions about how the crowdworkers were included in the design of potential socio-technical interventions such as the design of a crowdwork co-operative.

Our research team discussions surfaced concerns about how we had begun by imagining the crowdworkers we spoke to as temporary participants in our research, rather than as co-designers of alternative digital labour infrastructures. While we had engaged with some crowdworker forum moderators and activists alongside conducting the interviews, the direction of the project remained relatively researcher-led.

This was in part due to how our own understanding of what we were trying to achieve unfolded over the years, but also the material constraints on our work. For example, we only had a small amount of internal funding which meant we could only compensate 20 crowdworkers for an hour of their time—an important ethical consideration when relying on the time and energy of very low-paid workers. Funding limitations also meant we were constrained in our ability to travel to meet crowdworkers and the number of languages we had at our disposal to interview crowdworkers. Other commitments in our work and lives also meant constraints in terms of the number of hours we as researchers could dedicate to engaging with the crowdworkers and on the project. Nonetheless, while we were cognisant of the need to recognise how the material realities of both the crowdworkers and researchers constrain practice, we still wanted to avoid our engagement with crowdworkers being 'tokenistic' (Arnstein, 1969) and instead try and foster a practice in any future work in which the locus of control would be with the crowdworkers.

Through our reflections on these issues, we began to explore ideas around critical design approaches. Critical design researchers have long considered how critical interventions within design might be undertaken. As Bardzell et al. observe, a key decision in any critical design project is understanding what it is about the world you are trying to "provoke" (2012, p. 290). For Bardzell et al. (2012) the answer was the gendering of space; for our research team it was what we perceived to be an exploitative capital labour relation at the core of the crowdwork infrastructure. While Bardzell et al.'s (2012) critical design process involved gathering reaction to 'provocative designs' that they embedded in gendered spaces (e.g. the home and locker room), our approach had been to garner the reaction of crowdworkers to ideas for 'provocative designs' (the Gold Question detector and crowdwork co-operative) that would disrupt the existing crowdwork infrastructure and the form of capitalist labour relation at its core.

In total, we spoke to 20 US-based crowdworkers over a period of 18 months. These interviews were conducted from the UK by Skype. We recruited participants via crowdworker forums and Slack channels, including seven who had previously taken part in experimental work conducted by Alex and a colleague in Computer Science on crowdwork quality control (three workers) and crowdworker cooperation (four workers). Of these, the idea for the crowdworker co-operative emerged in discussions with four out of the first ten interviews. We then actively asked about the crowdwork co-operative intervention in the final ten interviews.

Through this process of engaging with crowdworkers, we came to understand more about how US-based crowdworkers understood the crowdwork labour relation and their preference for constructive design interventions such as the 'worker co-operative' idea, rather than design ideas that some perceived as confrontational such as the 'Gold Question Detector' (Checco et al., 2018). However, what to do with this insight remained somewhat unclear until we considered emerging ideas relating to 'design justice' (Costanza-Chock, 2020).

Building on work in the fields of Participatory Action Research and co-design, and the ideas and practices of activists in the US-based Design Justice Network, Costanza-Chock argues the case for a design justice framework that extends beyond common framings of interventionist design paradigms such as 'social impact' and 'design for good'. Recognising "communities to be co-researchers and co-designers, rather than solely research subjects or test users", Costanza-Check presents a design justice framework that addresses questions of (1) values encoded in the design of systems and objects, (2) practices relating to who is engaged in and controls design processes, (3) narrative about how things are designed and how design problems are scoped and framed, (4) sites at which design takes place and how accessible they are by those most impacted by design processes and (5) pedagogies relating to teaching and learning about design justice (2020, pp. 35–36).

In writing this chapter, we were inspired by this framework in a number of ways:

- The writing style we decided to adopt was motivated by points (3) and (5) about narrative and pedagogies. We decided to produce an honest and reflective account of our own practices as an emergent interdisciplinary team of researchers and our intervention within the crowdwork space, both as an effort to reflect on our own narrative and also as a pedagogical intervention in terms of developing our

own learning about our practice while also producing a written artefact that could be used in other contexts of learning and teaching.
- The empirical work we present in the following section is guided primarily by (1). While our discussions with crowdworkers had confirmed our understanding of existing crowdwork infrastructures as embedding an exploitative capital-labour relation albeit experienced in different ways by different crowdworkers, we also took from these discussions a more detailed understanding of the values that US-based crowdworkers perceived as important to the design of an alternative and fairer crowdwork infrastructure.

We then began to think through how we might practically adopt these values to sketch out an initial prototype of what a crowdwork co-operative infrastructure might look like and which might be used as a starting point in future co-design activity. We began by incorporating lessons from Computer Supported Cooperative Work (CSCW), an approach that builds on 1980s Participatory Design, which stems from the Scandinavian tradition of trade union collaboration, and the second wave of Activity Theory. CSCW is an interdisciplinary field aimed at studying computer-assisted activities that involve people coordination; it therefore seemed appropriate for beginning to think through some practical concerns that may need to be considered in the ongoing co-design of a dynamic crowdwork co-operative infrastructure.

Work in the field of CSCW and cognate fields has recognised a number of important features of technology-assisted co-operative work that could be pertinent to consider in efforts to co-design a dynamic crowdwork co-operative prototype. These include the following:

- The recognition of the existence of multiple and conflicting incentives and disincentives in coordinated work, at the institutional, organisational and group level, and affecting differently the hierarchies embedded within each of these levels (e.g. competition might be more present amongst junior members, while collegial behaviour is more accepted at senior level) (Pratt et al., 2004).
- The recognition that co-construction is usually delegated to a higher class of workers, while lower-class workers are delegated to routine work, with a minimal input in the decision process (Bardram, 1998a). It is thus necessary to provide digital spaces to allow a democratic

co-construction phase that typically cannot happen during the routine work setting.
- The tension between work routine and the need for change for the sake of efficiency (Pratt et al., 2004).
- The cooperation breakdowns and exceptions are the salient points in which negotiation and the establishing of new work heuristics occurs—temporary failures of cooperation should be acknowledged as part of the learning and co-construction process (Bardram, 1998a; Bødker et al., 1988).
- Recognition of the importance of interrogating the dualism between competition and cooperation (Malone & Crowston, 1990).
- The importance of awareness (mutual or one-directional) of one individual's activity by other members, enhancing this awareness when needed is what makes collaborative systems successful (Carstensen & Schmidt, 1999; Pratt et al., 2004).
- The recognition that the design process of a manufacture-like process is relatively easy; however, the creative part of the interaction and the handling of interdependencies are characterised by an "overwhelming complexity" (Carstensen & Schmidt, 1999; Bødker et al., 1988): in our prototype we found this reflects difficulties in modelling recruitment, work division and meta-cognitive interactions.

In conclusion, emerging from critical design, CSCW and cognate fields is the clear understanding that we need to be aware of the risk of designers relegating workers to a non-creative routine, after only an initial phase when co-construction was informed by them (Bardram, 1998b). Similarly, design has a political effect that risks presenting the designers as "saviors descending to help others", shifting the focus away from the unjust system of value distribution of the platform economy (Irani & Silberman, 2016).

Inspired by our reading of critical design, design justice and CSCW approaches, the software architecture 'artefact' that we produced (Section "Crowdworker Perspectives on Crowdwork Co-operatives") is purposefully partial and contestable, and aims to embed the above insights from CSCW about the design of technology-assisted activities that involve the coordination of people, as well as the perspectives of the crowdworkers we spoke to as presented in the following section. We envisage the prototype as a possible starting point for future critical design practices that engage crowdworkers in different ways and across sites, in ways accessible in relation to material constraints such as time, resources and mobility.

Crowdworker Perspectives on Crowdwork Co-operatives

> MTurk somehow seems to promote the idea that ten cents a minute is standard pay. And I don't know who tried to live on ten cents a minute. (I7)

The crowdworkers we engaged with reported a wide range of challenges in their work, many of which reflected the findings of previous research (e.g. Hara et al., 2018; Martin et al., 2014; Ross et al., 2010; Kittur et al., 2013; Fort et al., 2011). In particular, they talked about the increasing number of workers on the platform resulting in reduced work availability and declining pay, their fear of having work rejected by the requester and therefore not receiving payment for work completed, unpaid time spent looking for good work on the platform, poor communication with requesters, and workers' lack of power to resolve issues and improve conditions.

All the crowdworkers that we spoke to directly about the crowdwork co-operative idea—both those that were positive about it and those that had some reservations—were enthusiastic and curious to explore what working in a crowdwork co-operative platform might be like.

Two participants (I13, I15) were very enthusiastic:

> Yes, that would, right there, solve so many problems, so many problems … the humanitarian in me, I'd be like, "Yeah, all over that". (I15)

Some perceived advantages around the possible culture shift that working in cooperation with other workers might engender within the currently hyper-competitive crowdwork culture:

> [I]t's really competitive and that puts people into a state of, you know, I have to protect myself over the next person. So when you take that threat away, when you give support where there was never any in certain areas, you're going to see a shift within that paradigm. (I15)

On the other hand, other workers did have reservations that would need to be addressed in the co-operative design. They perceived that the most pressing concerns that any crowdwork co-operative design would need to address were work availability, quality of work completed, dealing with issues of free riders and workers without appropriate skills, for example English language, and convincing the most successful workers and good requesters to join the co-operative.

Despite these reservations, as will be explored in the following sections, workers had many ideas for how challenges might be addressed and said that if they could be sure the work and payment distribution was fair and if there was enough work available to meet their income target they would be interested in trying to be part of a crowdwork co-operative.

The next section lays out the key issues that crowdworkers perceived would be fundamental to the design of a successful crowdworker co-operative.

Towards a Worker-Driven Design for a Crowdwork Co-operative

Co-operative Values from the Workers' Perspective

Across the different crowdworkers we spoke to, we identified a broad set of values that they imagined might be embedded within the design of a co-operative infrastructure for crowdwork. These included the following:

1. Fairness in payments—with their views reflecting notions of distributive fairness and procedural fairness as discussed by Fieseler et al. (2019)
2. Democratic or collective decision making, with an emphasis on equal representation (one person one vote)
3. Horizontal governance structure—they often recognised some coordination was required, but believed any 'manager' role should be equal to other workers in terms of power and reward
4. Transparent communication within the co-op
5. Open information and communication between workers and requesters
6. Accountability for both requesters and workers, including accountability for quality of work produced
7. Camaraderie and a sense of community, trust, mutual assistance and cooperation
8. Platform design based on workers' needs
9. Empowerment of workers in the co-operative to decide membership and rules
10. Security of work
11. Commitment to the co-operative

The broad set of values sit comfortably with the Rochdale principles (International Cooperative Alliance, 1995) mentioned in "Platform Co-operativism" section. While some of the values are more specific to the crowdwork context, for example, platform design based on workers' needs, the more general values of fairness in payment, democratic decision making, horizontal governance and empowerment of workers align with the Rochdale principles. The only Rochdale principles that are missing from what workers explored was 'co-operation among co-operatives' and the education and training aspects of 'education, training and information', although workers did discuss mutual support among workers for self-directed learning. They were not directly asked about these two issues, and it is something that could be explored further in the future.

Underlying this broad set of values, workers identified a number of specific ideas about what they imagined would be important practical components of a crowdwork co-operative design. These can be broken down into four thematic areas: (1) platform infrastructure, (2) payment, (3) quality control and (4) decision making and governance. Of these the first—platform infrastructures—is most specific to the material conditions of crowdwork, whereas the rest reflect the types of practical concerns that would likely be seen in any type of co-operative. Each of these themes was embedded within an overarching meta-theme of empowering workers. Table 1 identifies the different ideas generated by crowdworkers in relation to each of these themes. All ideas are detailed in the subsequent section.

Platform Infrastructure

Participants had a number of suggestions about how to optimise the platform infrastructure of the co-operative to help with the distribution and efficient completion of work, to empower workers and to help make working on the platform feel more personal and engaging.

A key suggestion made by four workers (I9, I10, I11, I13) was to have worker profiles integrated into the platform. This is in stark contrast to the existing AMT platform in which workers are anonymous and are only represented by a worker identity number. They thought that profile information such as basic demographics, level of education, skills and work history (types and number of HITs completed) could be useful for a number of reasons. Two workers thought that having such a profile which

Table 1 Key ideas for the design of a crowdwork co-operative from a worker perspective

Empowerment of crowdworkers			
Platform infrastructures	*Payment*	*Quality control*	*Decision making and governance*
– Worker profiles – Embedded scripts and tools – Communication channels with other workers and requesters embedded into the infrastructure – Requester profiles – Rating system for requesters – Ability to block requesters from platform	– Payment based on time worked tracked by the system – Potential for payment increase with experience and quality of work – Whether and how to pay worker benefits	– Ensuring high-quality work to attract enough high-quality requesters – Worker-controlled, transparent peer-review system for quality control of work	– Collaborative, possibly consensus, decision making – Democratic—one vote per person – Include manager/facilitator and other elected leadership roles – Possible application process to join (workers, requesters) – Collectively decide how to allocate work (e.g. potentially on the basis of skills, experience and geography)

listed their skill set would make the platform feel more humanised for them, rather than them just being a string of numbers.

> [S]omething where I'm not just—,[…], where I'm not just that worker ID number. (I13)

A number of workers also thought worker profiles would help the co-operative divide workers into subgroups based on their skill sets, help with filtering work in order to allocate it to the right workers and help workers quickly have a good sense of their role in the co-op [I9, I10, I13, I18]. Filtering work this way would make finding and completing tasks more time and cost effective and help avoid taking the same tasks twice, which is a significant inefficiency of the current infrastructure where all workers spend a lot of time searching for HITs independently. Two workers also thought worker profiles could help make sure malicious workers were not able to undertake tasks not appropriate for them [I9, I12].

As well as worker profiles, four participants [I5, I10, I11, I12] also recommended that requesters have a profile on the platform, both to make the process more personal and so workers would be able to find requesters in order to follow them if they like the work they provide and in order to check their duration on the platform. As well as requester profile, requester ratings were also suggested. While existing platforms such as AMT do not have a requester rating functionality many workers do use browser plugins such as Turkopticon to check requester ratings (Irani & Silberman, 2013); however, this plugin is no longer being supported by the developers. The most recommended [I10, I12, I16, I20] idea was to establish a rating system built directly into the platform similar to other apps, like Lyft and Uber, where the workers could rate the requesters and leave reviews based on the amount they pay and their task practices. Workers also recommended having a filter in the platform that would block requesters if they fall below a certain rating level or if they offer payment below a commonly agreed threshold [I4, I5].

> [I]t would be cool if on a site they could be basically removed at that rate [of pay] … for example, there should be like an override that says that if more than ten workers say that then we—Block the requester. (I4)

Beyond profiles, the workers [I4, I5, I12, I20] we spoke to suggested that the co-operative platform could directly integrate the various different scripts and tools that workers currently use as, for example browser plugins. This would help workers in the co-operative to find good work and work more efficiently from directly within the platform.

> [i]f you took all these features like you have Turkopticon and Hit Forker, and everything else designed to help workers find good work and help you out, you know if that was incorporated into the site to begin with. (I4)

They also suggested that a worker communication channel be embedded within the platform [I5, I9, I10, I15, I19, I20]. Currently workers use a variety of different forums and Slack channels to communicate with one another; however, they perceived that this communication would be better if integrated directly into the platform, creating a central hub of information and allowing requesters to participate in conversations.

> I would like it to have a more solidified community rather than the scattered forums. (I9)

I10 and I15 also discussed how these integrated communication channels could be extended to communications with requesters—for example, by including a chatbox next to tasks through which workers could report on any ambiguities. They perceived this would help decrease rejection rates and help requesters weed out malicious workers instead of rejecting workers in bulk.

Payment

Currently crowdworkers get paid a piece rate for work completed, with no payment for time spent searching for work or on tasks rejected by the requesters. Seven of the workers we spoke to thought the co-operative should shift to a payment model in which the amount paid reflects time spent working rather than tasks completed [I5, I12, I13, I16, I17, I18, I20], and that this would be a fairer reflection of the work undertaken to find and complete different types of HITs. They recommended having a system in place to log each worker's contribution. I12 suggested the payment to be calculated per minute in order to keep things as simple as possible.

> That would have to be logged somehow. So you would have to be able to either have it logged by the platform or have people check in and check out when they are working. But then at the end of any given project, let's say you'd have 20 workers who all had their varying amounts of time logged in, and then you would take whatever the fee was for that and divide it based on the time. (I17)

However, some workers also questioned if such standards for pay would be possible if work was scarce:

> [H]ow would you guarantee, you know, like $8 an hour if you have a bad day where there's no work for anybody, you know? (I19)

This issue of work scarcity on the ability of the co-operative to pay members was a concern for other workers. Some argued that the co-operative might struggle to attract requesters away from platforms such as

AMT, particular if the co-operative was perceived by requesters as a form of unionising [I20]. Others were concerned about how any scarcity issues might impact on the culture of cooperation a co-operative would be dependent upon:

> [T]hat that kind of scarcity makes people hungry. And when you're hungry you're not very cooperative, [laughs]. (I17)

Some workers also discussed how workers might be graded and whether requesters could pay a premium for high-quality, experienced workers [I5, I9, I10, I11, I13]. Some criticised the current 'Masters' system on AMT as lacking in transparency and argued that a transparent training and rating system in which workers could progress through different "quality tiers" [I15] would be beneficial to managing the co-operative and motivating workers.

There was no consensus among workers about whether higher quality or faster workers should receive more payment. However, a number of participants [I19, I11, I16, I17] recognised that the co-operative might not attract high-quality workers if they would have to share their income with the rest of the group and that it would therefore be challenging to have a standard hourly wage if the pace of work was not equal among members.

How to foster efficiency and trust within the co-operative was therefore another concern. Workers [I10, I12, I14, I16, I17] were concerned that some people might take a longer time to complete tasks perhaps because they were slower and less efficient at completing their work, or because they purposefully wanted to spread out the same amount of work over a longer period of time in order to increase the number of hours they received pay for.

> Any system you put in place, eventually somebody's going to try to find a way to abuse it, so you have to kind of find a way to safeguard before that happens, I guess. (I16)

Quality Control

One way to encourage both workers and requesters to an independent co-operative platform would be if the workers were trusted by requesters, and one another, to efficiently complete high-quality work. Workers recognised that some form of quality control process would be required by the co-operative. They suggested ideas for possible peer rating and review systems to try and identify poor quality work and people trying to abuse the system.

For example, I18 suggests to have a peer rating system among the workers where they would indicate if the other members were pulling their weight. The system would aggregate the reviews, and if the worker fell below a threshold they could be removed from the platform.

> I think if people had an anonymous way to, you know, rate everybody on the team, and if you had, you know, certain kind of thresholds, where somebody was doing not a good job. (I18)

I17 recommended having some form of statistic of how many units of the same project have other workers completed in a specific amount of time so as to identify if someone is misusing the system. Others discussed how starting small as a co-operative might also help with this issue. I15, for example, thought starting with a small group of workers and building it slowly would make it easier to find malicious workers, and others observed that starting small would also help to create a sense of community and help address governance issues in which a consensus was required [I15, I19, I20].

Beyond how to manage work quality in a way that would foster fair payment and encourage requesters to use the platform, we also explored whether a co-operative might be able to offer work benefits that are not currently available to crowdworkers. While some workers perceived themselves as self-employed and therefore responsible for their own pension and health- and sickness-related benefits, some did believe that it would be good to receive such benefits through their work on the platform. However, they thought given the downward trend in how much requesters are willing to pay for HITs this was unlikely.

> It's great to have the option. ... But, a lot of requesters have realised that there will always be people doing every hit regardless of how bad it's paid, regardless of anything. ... So, not only do I not expect for things like benefits and pension to come along, I actually expect this to be worse and worse paid compared to a normal job. (I20)

Given the co-operative would be competing against other platforms for requesters, it is likely that if the co-operative was to be able to pay a decent wage with benefits then regulatory action may also be required to enforce minimum standards. However, this is not something all crowdworkers are supportive of.

Decision Making and Governance

Workers thought that decision making in the co-operative should be done collaboratively, with each member of the co-operative having an equal vote [I10, I13, I16, I20]. However, one person did question what might happen if 'free riders' outnumbered productive workers in this model and thought something would need to be put in place to avoid that scenario:

> [S]ay there's five workers, John's one doing most of the work, but the other four would like to be, you know, paid the same amount as John, they—, again, they could outnumber him with a vote [laughs] and, you know, get paid the same. So you would … just have to find a way to make it fair, where people can't abuse it. (I16)

The question of how the co-operative might come to a consensus on issues was raised as potential concern by some, particularly if the co-operative grew beyond a small ground of workers:

> If it was a smaller [number of people] it would be easier to come to a consensus. Part of the problem with crowdwork is that you have so many opinions in the crowd that it's really hard to come to a consensus. (I19)

In relation to this issue, some workers thought some sort of responsibility structure would need to be put in place to help manage the co-operative:

> [I]f it's, let's say, 30–40 people, maybe it could be like a perfect democracy, and just everybody votes, but anything over that I would say, definitely there would need to be one or two people in charge of certain things. (I20)

Some sort of coordinator role was suggested by a number of workers [I15, I18, I19, I20]. However, they understood this role to be a facilitator role that would be equal with other workers in terms of power and pay, rather than being part of hierarchical management structure.

> [S]omebody like organising the whole thing is great, and making sure everybody's on the same page and stuff like that, but I wouldn't think that it would need to be a position by itself. (I18)

This role was imagined as involving helping to organise the work in the co-operative, ensuring work was completed, checking on the workers' progress, opening discussions about the rules in the co-op, promoting communication, helping resolve issues and being the voice of the co-operative.

One worker suggested that there could be more elected positions beyond a single coordinator:

> [A]s the group grows there could be like some elected positions maybe, like maybe a supervisor, maybe like a treasurer or something like that. (I20)

Key issues that would need to be decided through such decision-making processes include issues of membership of the co-operative and how to fairly allocate work.

The membership of the co-operative was recognised as a crucial aspect of decision making by a number of participants. Four participants [I9, I10, I13, I15] suggested the members of the co-operative would need a way to control membership. I9, for example, suggested a selective process to decide who could become a member of the platform, and I10 recognised the platform would need a system in place to deal with malicious workers. Requirements of membership suggested by the workers included things such as only taking workers above a particular approval rating or level of experience on existing platforms [I13, I15, I20], language proficiency tests to help with the allocation of tasks that require a particular language [I10, I11]. I10 also recommended that workers pay a fee for people to join the co-operative platform. I20 perceived that having some sort of application process to join the co-operative would help new workers demonstrate their commitment to the platform and create a stronger sense of community. Two workers [I4, I9] also suggested that requesters should have to apply to join the platform.

Another issue one worker [I9] raised related to decision making in task distribution related to economic geography. This worker argued that tasks should be allocated on the basis of geography, with requesters having to use workers from the country that they were located and at an appropriate rate for the living costs of that country:

> I think for a start having platforms that are dedicated to Europe and the US, where they get their workers. I mean part of me feels like that's not fair or just, but I feel like if a requester is in the US or Europe they should probably be getting European or US workers. It feels like the outsourcing is a little unfair. (I9)

This kind of issue would be the kind of question the co-operative would need to grapple with through the kind of democratic decision-making processes workers described earlier.

A Prototype Software Architecture for a Crowdwork Co-operative

As identified by the workers above, the first phase of a crowdwork co-op would face the problem acquiring a critical mass of requesters and workers to be able to be meaningful as an independent platform. For this reason, we will first focus on the intermediate "platform from within" (Scholz, 2016) idea of using an existing external platform (e.g. AMT) to acquire jobs, augmenting the worker experience with collaboration and cooperation tools, although with recognition that in some cases this may be against platform Terms and Conditions. Further, this approach also raises the challenge that the 'platform from within' can be disrupted by AMT (or other) platform architectural changes and therefore depends on external maintenance. This poses a challenge as the researchers and developers who could help maintain the platform, as they often cannot guarantee continued input to ensure sustainability, particularly if they do not have ongoing funding or support for the work from their institution or an external funder—an issue that Kristy Milland has drawn attention to in workshop discussions, for example Aroyo et al. (2018).

Building on the ideas of crowdworkers described in the above sections, we use Stanoevska-Slabeva's (2002) community-orientated design framework to identify the different components of a prototype software architecture that could be used as a starting point for future co-design practices with crowdworkers. This design framework is based on Schmid's Media Reference Model (1999), which identifies four 'views' that are critical to understanding a software architecture: community, implementation, transaction and infrastructure. Here we focus primarily on the community view as this will be a core aspect of any co-design process; we also highlight some of the key aspects of the implementation and transaction views.

Community View. This view captures the identity-shaping features and other static elements of the organisational structure.

- **Roles:** These are roles that a member of the co-op (together with automated tools) can undertake.
  - **Worker:** This is the basic role, in which a member is completing a HIT.
  - **Platform negotiator:** A member, with the aid of automatic tools, interacts with the platform before or after a HIT is accepted. This role is responsible for adding information in the worker role view (see Fig. 2) and in the shared knowledge database (as explained later in the knowledge service). Members could be assigned these roles through election by the membership. This role could be divided into three sub-roles:

Job seeking (human + tools): This is the role of scouting for suitable HITs in the platform. This is a task that is usually repeated by all workers, and having a dedicated role would save significant time. An experienced member would be able to locate promising jobs or jobs that are a good match to the members' skills. This role can also be assisted by automatic filters, with parameters decided by the members (e.g. payment threshold) and would be an improvement on the existing job-matching scripts, as suggested by the workers we spoke to. This role is important also to implement (even in a semi-automatic way) the boycott of requesters and batches.

Rule clarification: This is the role of contacting the requesters to clarify the rules of a batch. Currently, each worker will typically have to ask for some clarification from the requester before starting a batch, so having a dedicated role would save significant time. It would also increase the overall quality of the work, because all questions asked by the member assuming this role could be automatically propagated to all members working on the batch. Much of this role could be completely automated by a tool of rule clarification, where previously asked questions could be shared among members, but an experienced member could still have this role to catch important questions to ask when the job rules are ambiguous.

Rejection appeal (human + tools): This is a fundamental role, as a rejection of completed work can have a detrimental impact on workers. By monitoring the rejection rate and the workers' feedback, a member with this role can quickly identify unfair rejections, promptly contact the requesters to demand a reversal and flag the batch as unreliable to prevent additional HITs being completed by the co-op.

- **Training:** The training role could have two different ways of being implemented.

Indirect training: This is a role that any worker can have, even during the completion of a HIT: the co-op software could enable indirect messaging related to a batch while working so that each co-op member could leave notes on a batch for the rest of the members that encounter the batch.
Direct training: A worker could signal problematic tasks, and an experienced member could monitor these signals (together with workers' performance) and intervene by providing support via chat and screen sharing.

- **Coordinator:** A coordinator could interact with platform negotiators and workers, to dynamically select groups with similar skills/requirements, to create a critical mass of members working on the same batches. This would allow the members to increase the quality (training), efficiency (less job seeking) and the bargaining power of the co-op (rule clarification and rejection control).

Membership: A coordinator could also enforce the co-op rules and take care of new membership and revocation.

- **Metarole:** This is a role necessary to decide co-op rules and policies, and change the software itself. All members could vote to assign roles and change the structure of the co-op, for example changing the co-op rules (see Table 1, column 2). This can be achieved by a collaborative decision-making mechanism that can include coordinator roles. Important examples of rules that require an agreement are membership rules, job allocation rules, worker/requester exclusion rules and payment distribution rules.

- **Valid rules:** A clear set of rules and their enforcement are necessary to establish trust (Preece, 2000). It would be necessary that the co-op members agree on a set of policies for the governance of the co-op, together with the parameters of the software itself. These rules could change over time with a voting process. A rule with a high impact on the shape of the co-op is the payment distribution scheme, which as discussed above would require deliberation and agreements among co-op members given the different ideas suggested by the workers we spoke to (see Table 1).

- **Members' identity:** Members could have a profile that would help them present an identity to the community, as suggested by some of the workers we spoke to. This profile could include statistics, skills and preferences, and would help a member assuming the coordinator role to facilitate the job matching and training.
- **Common language:** Members use a common language and slang, originally inherited from the existing forums, and share a common vocabulary of terms related to the target platform, for example AMT.

Implementation View: This view contains the community processes that are the set of activities that can be performed by the co-op. We cannot list all processes here, but some of them are the membership process, the discussion process, the job flagging and training annotation process and so on.

Transaction View: This view describes the coordination and communication processes available in the co-op. They can be divided into the following:

- **Knowledge services:** To manage and use knowledge in the co-op. Some of this knowledge is obtained from workers signals (e.g. flag a job) and others automatically obtained from the platform (job search) and from the workers (aggregate job difficulty, worker performance). It is necessary to maintain a database for this.
- **Intention services:** To signal a member's intention or need, like the request for training, the need to abandon a batch and so on.
- **Negotiation services:** To make decisions about membership, new policies, jobs to target, how to allocate work and so on. A notable concept that needs to be negotiated is the potential *pay redistribution*: this would allow workers to share and thus mitigate the risk of having an underpaid batch affecting one worker's hourly wage. Similarly, *solidarity tools* like paid leave could be implemented.

We summarise this model in Fig. 1.

Example of the Worker Seeking Role and Worker Role Views

In Fig. 2, we summarise the view of the main views a worker would use. These would need to be implemented as a browser plugin. The worker seeking view will visualise a (re)ranked list of batches based on the job allocation rules decided by the co-op: this ranked list is potentially different for each worker. In this view the workers will be able to visualise requester statistics and other information on the batch (e.g. required

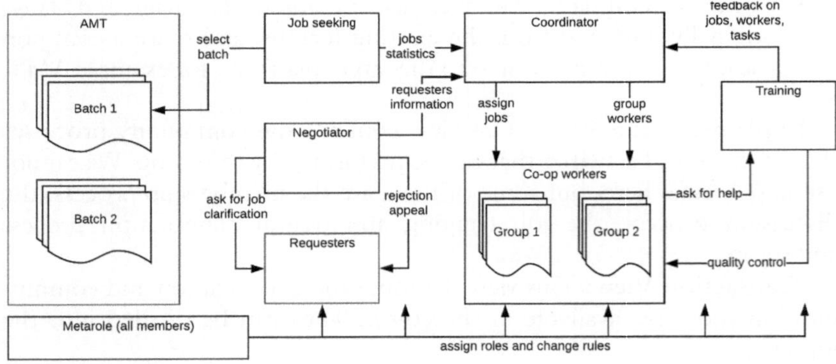

Fig. 1 Co-op software architecture model

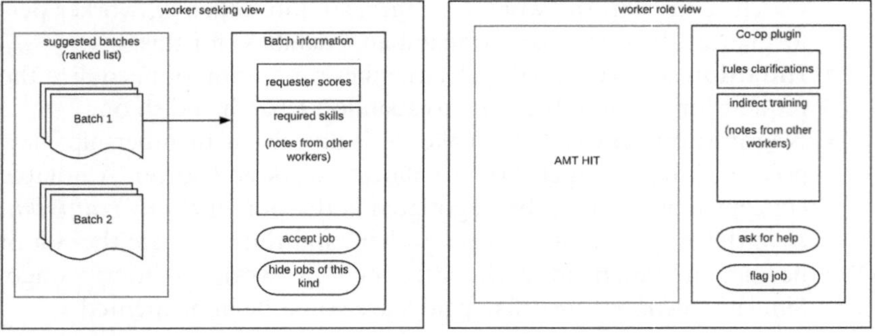

Fig. 2 Worker seeking role and worker role views

skills) obtained by workers that already selected this job, allowing an informed decision.

The worker role view will allow the worker to access, in addition to the external platform view (in red), to rule clarifications obtained by the corresponding role. Similarly, notes from other workers that already completed this batch will be displayed there. Moreover, the worker can communicate with other co-op members directly (ask for help) to obtain direct training. Finally, the worker can notify the co-op of potential problems with the job by flagging it.

Conclusion: Towards a Crowdwork Co-operative Prototype?

The above exploration of worker perspectives on a co-operative model for crowdwork platform design, and the resulting ideas for a prototype software architecture, aim to be a partial and contestable early step in exploring how workers and researchers might work together to re-imagine the organisation of the 'crowd' labour that contemporary AI systems are dependent upon.

The contribution draws upon insights from critical design and digital labour studies, to bring into focus the relevance of crowdwork platforms to Critical Data Studies. Emphasis in the Critical Data Studies field has tended to be on the expansion of datafication and outcomes for data subjects. Here, we draw attention to those people that labour within the infrastructures that support datafication processes, illuminate structures of labour exploitation that many contemporary AI systems are dependent upon and ask—with workers—how might these labour conditions be improved. In doing so, we also highlight the value of critical design studies and of interdisciplinary collaboration between the social and computer sciences, particularly as the focus of CDS expands from identifying instances of domination and exploitation resulting from deepening datafication, towards addressing the question of what can be done about it.

Through drawing on insights from different disciplinary perspectives as well as from the workers themselves, the ambivalences of data power and how to address it come into clearer focus. The material realities of workers' economic needs combined with the constraints baked into the existing capitalist crowdwork infrastructure, as well as the limitations on researchers' ability to guarantee sustainability of contributions within existing institutional arrangements, all interact to reinforce the complexity of the challenge. It is too early to understand how the recent COVID-19 pandemic will impact these issues. Certainly, the potential for sustainable contributions from university-based researchers in some countries will be further impacted by shifting priorities and reduced budgets. It is likely that the sharp increase in unemployment resulting from the pandemic may mean more people seeking work through crowdwork platforms (Moss et al., 2020), which could further push down wages if there is an increased supply of labour. Yet, on the other hand, it is also possible that with an increase in remote working more generally, researchers may make more use of platforms such as AMT to source research participants, thus increasing the demand on the platforms which may counter some of this effect. Any future research should therefore remain mindful of possible impacts of the pandemic on workers, requesters and researchers.

While our focus has been on co-operatives as a new way to organise 'crowdwork'—whether independent of or 'from within' existing infrastructures—we conclude by adding that we do not envisage a crowdworker co-operative as a standalone solution to worker exploitation in crowdwork markets. Clearly, in a 'platform from within' model the co-operative would still be tightly—and potentially vulnerably—embedded within the capitalist data economy, and an independent platform would likely not have the economies of scale to compete successfully with AMT. Neither of these models addresses the systemic low-pay issues in crowdwork that would make it difficult for a co-operative to pay a living—or even minimum—wage. Also, we recognise that much of the work being undertaken by crowdworkers, such as labelling of machine learning datasets, contributes to a complex set of challenges around the adoption of machine learning and AI across various sectors. It is important that any future work on crowdwork co-operatives remains mindful of this context. Nonetheless, we perceive that a co-operative model for crowdwork could be a progressive intervention in the context of broader developments involving labour market regulation in the interests of workers, and an AI sector regulated in line with egalitarian and democratic values.

References

Arnstein, S. R. (1969). A ladder of citizen participation. *Journal of the American Institute of Planners, 35*(4), 216–224.

Aroyo, L., Dumitrache, A., Paritosh, P., Quinn, A., Welty, C., Checco, A., Demartini, G., Gadiraju, U., & Sarasua, C. (eds.). (2018, July 5). *Proceedings of the 1st workshop on subjectivity, ambiguity and disagreement in crowdsourcing, and short paper proceedings of the 1st workshop on disentangling the relation between crowdsourcing and bias management (SAD 2018 and CrowdBias 2018)*, Zürich, Switzerland. http://ceur-ws.org/Vol-2276/

Bardram, J. (1998a). Designing for the dynamics of cooperative work activities. *CSCW '98: Proceedings of the 1998 ACM conference on computer supported cooperative work*, 89–98. https://doi.org/10.1145/289444.289483

Bardram, J. E. (1998b). *Collaboration, coordination, and computer support*. DAIMI.

Bardzell, S., et al. (2012, June). Critical design and critical theory: The challenge of designing for provocation. *DIS '12: Proceedings of the designing interactive systems conference*. https://doi.org/10.1145/2317956.2318001

Bødker, S., Ehn, P., Knudsen, J., Kyng, M., & Madsen, K. (1988). Computer support for cooperative design. *CSCW '88: Proceedings of the 1988 ACM conference on computer-supported cooperative work*, 377–394. https://doi.org/10.1145/62266.62296

Carstensen, P. H., & Schmidt, K. (1999). Computer supported cooperative work: New challenges to systems design. K. Itoh (ed.). *Handbook of Human Factors*. https://www.semanticscholar.org/paper/Computer-Supported-Cooperative-Work%3A-New-challenges-Carstensen-Schmidt/0ed4a37a0001169901ec31e31259f1af806fff62

Checco, A., Bates, J., & Demartini, G. (2018, July 5–8). All that glitters is gold – An attack scheme on gold questions in crowdsourcing. *Proceedings of the 6th AAAI Conference on Human Computation and Crowdsourcing (HCOMP 2018). The 6th AAAI Conference on Human Computation and Crowdsourcing (HCOMP 2018)*, Zurich, Switzerland. AAAI Press. https://www.aaai.org/ocs/index.php/HCOMP/HCOMP18/paper/viewFile/17925/16904

Costanza-Chock, S. (2020). *Design justice: community-led practices to build the worlds we need*. MIT Press.

Fieseler, C., Bucher, E., & Hoffmann, C. P. (2019). Unfairness by design? the perceived fairness of digital labor on crowdworking platforms. *Journal of Business Ethics, 156*, 987–1005. https://doi.org/10.1007/s10551-017-3607-2

Fort, K., Adda, G., & Cohen, K. B. (2011). Amazon mechanical turk: Gold mine or coal mine? *Computational Linguistics, 37*(2), 413–420.

Graham, M., Hjorth, I., & Lehdonvirta, V. (2017). Digital labour and development: Impacts of global digital labour platforms and the gig economy on worker livelihoods. *Transfer: European Review of Labour and Research, 23*(2), 135–162.

Gray, M., & Suri, S. (2019). *Ghost work: How to stop Silicon Valley building a new global underclass*. Houghton Mifflin.

Gray, M., Suri, S., Ali, S., & Kulkarni, D. (2016). The crowd is a collaborative network. *CSCW '16 Proceedings of the 19th ACM Conference on Computer-Supported Cooperative Work & Social Computing*. https://dl.acm.org/citation.cfm?id=2819942

Hara, K., Adams, A., Milland, K., Savage, S., Callison-Burch, C., & Bigham, J. P. (2018). A data-driven analysis of workers' earnings on amazon mechanical turk. *CHI '18: Proceedings of the 2018 CHI Conference on Human Factors in Computing Systems*, Paper No.: 449, 1–14. https://doi.org/10.1145/3173574.3174023

International Cooperative Alliance. (1995). Statement of Co-op Identity: The Rochdale Principles. Nasco. https://www.nasco.coop/resources/rochdale-principles

Irani, L., & Silberman, M. (2013). Turkopticon: Interrupting worker invisibility in amazon mechanical turk. *Proceedings of the SIGCHI conference on human factors in computing systems (CHI '13)*. Association for Computing Machinery, New York, NY, USA, 611–620. https://doi.org/10.1145/2470654.2470742

Irani, L., & Silberman, M. (2016). Stories we tell about labor: Turkopticon and the trouble with "design". *CHI '16: Proceedings of the 2016 CHI Conference on Human Factors in Computing*, 4573–4586. https://doi.org/10.1145/2858036.2858592

Kittur, A., Nickerson, J. V., Bernstein, M., Gerber, E., Shaw, A., Zimmerman, J., & Horton, J. (2013, February). The future of crowd work. *Proceedings of the 2013 conference on Computer supported cooperative work*, 1301–1318. https://doi.org/10.1145/2441776.2441923

Malone, T. W., & Crowston, K. (1990). What is coordination theory and how can it help design cooperative work systems? *Proceedings of the 1990 ACM conference on Computer-supported cooperative work*, 357–370. https://doi.org/10.1145/99332.99367

Martin, D., Hanrahan, B. V., O'Neill, J., & Gupta, N. (2014, February). Being a turker. *Proceedings of the 17th ACM conference on Computer supported cooperative work & social computing*, 224–235. https://doi.org/10.1145/2531602.2531663

Moss, A. J., Rosenzweig, C., Robinson, J., & Litman, L. (2020). Demographic stability on Mechanical Turk despite COVID-19. *Trends in cognitive sciences, 24*(9), 678–680

Pratt, W., Reddy, M. C., McDonald, D. W., Tarczy-Hornoch, P., & Gennari, J. H. (2004). Incorporating ideas from computer-supported cooperative work. *Journal of Biomedical Informatics, 37*(2), 128–137.

Preece, J. (2000). Online Communities: Designing Usability, Supporting Sociability. Chichester, UK: John Wiley & Sons.

Ross, J., Irani, L., Silberman, M., Zaldivar, A., & Tomlinson, B. (2010, April). Who are the crowdworkers?: Shifting demographics in mechanical turk. *CHI'10 extended abstracts on Human factors in computing systems, 2863–2872.* https://doi.org/10.1145/1753846.1753873

Schmid, B. (1999). *Elektronische Märkte. Merkmale, Organisation und Potentiale.* NetAcademy.

Scholz, T. (2016). *Platform cooperativism: Challenging the corporate sharing economy*. Rosa Luxemburg Stiftung.

Srnicek, N. (2017). *Platform capitalism*. Polity Press.

Stanoevska-Slabeva, K. (2002). Toward a community-oriented design of internet platforms. *International Journal of Electronic Commerce, 6*(3), 71–95.

Tubaro, P., Casilli, A. A., & Coville, M. (2020). The trainer, the verifier, the imitator: Three ways in which human platform workers support artificial intelligence. *Big Data & Society, 7*(1). https://doi.org/10.1177/2053951720919776

Open Access This chapter is licensed under the terms of the Creative Commons Attribution 4.0 International License (http://creativecommons.org/licenses/by/4.0/), which permits use, sharing, adaptation, distribution and reproduction in any medium or format, as long as you give appropriate credit to the original author(s) and the source, provide a link to the Creative Commons licence and indicate if changes were made.

The images or other third party material in this chapter are included in the chapter's Creative Commons licence, unless indicated otherwise in a credit line to the material. If material is not included in the chapter's Creative Commons licence and your intended use is not permitted by statutory regulation or exceeds the permitted use, you will need to obtain permission directly from the copyright holder.

Counting, Debunking, Making, Witnessing, Shielding: What Critical Data Studies Can Learn from Data Activism During the Pandemic

Stefania Milan

Introduction

The more than 70 per cent of the Peruvian labour force employed in the informal economy has been severely impacted by the lockdown imposed to curb COVID-19 diffusion. But government efforts to deploy algorithms to aid in the distribution of welfare subsidies relied on official yet inaccurate databases and technology designed with other purposes in mind—repeatedly failing to reach vulnerable households. As Cerna Aragon has warned, "[I]n a state that barely knows its population", "the technocratic asset of a rigorous algorithmic system brought woe for those in need. These technologies, by design and implementation, render some people invisible" (2021, p. 123). In India the biometric welfare system has

S. Milan (✉)
University of Amsterdam, Amsterdam, Netherlands
e-mail: s.milan@uva.nl

© The Author(s) 2022

A. Hepp et al. (eds.), *New Perspectives in Critical Data Studies*,
Transforming Communications – Studies in Cross-Media Research,
https://doi.org/10.1007/978-3-030-96180-0_19

come to a standstill due to pandemic-induced hygiene rules (Masiero, 2021), platform delivery workers in Barcelona (Spain) face a number of lose-lose dilemmas between survival and safety (Vieira, 2020), and in Ghana the government exploited the emergency to pass permanent legislation increasing state control over the national telecommunication system (Oduro-Marfo, 2020). Evidently, the COVID-19 crisis has exposed the open wounds of "the first pandemic of the datafied society" (Milan & Di Salvo, 2020), with the most vulnerable individuals and communities often paying the highest price. It has also laid bare how data power, understood as the variety of "problematic consequences of widespread datafication" (Kennedy & Hill, 2016, p. 775), evolves under the pressures of the pandemic, in ways that might undermine citizen agency even further.

The pandemic has considerably changed our lifestyles while also having effects on the information domain. We can identify as many as five significant adjustments. First, many of our daily activities have moved online and now unfold through a myriad of old and new e-commerce platforms, cloud computing, and videoconferencing facilities, exposing our tremendous dependence on for-profit digital infrastructures. Second, the increase in personal insecurity—for example, unemployment, poor access to health care, reduced mobility, and the suspension of school activities—has augmented social inequalities. It has paved the way for new forms of invisibility and exclusion to emerge, often propelled by algorithmic decision-making as in the case of Peru. Third, the uncertainties surrounding the virus as well as its related corrective measures have contributed to rising doubts among the populace, accelerating the spread of conspiracy theories and anti-scientific attitudes. Fourth, the techno-solutionism associated with the pandemic—see, for example, the governmental faith in contact tracing apps—has uncovered the tension between privacy and safety, and between individual and collective rights, often presented as irreconcilable dichotomies. Last but not least, the extended lockdowns have curbed the ability to mobilise social movements and other forms of aggregation in the public sphere, relegating the formation and expression of political opinion to the web.

Against this backdrop, data activism has stood out as a solid response to many of these problems, further consolidating its role within the social movement ecosystem. Emerging within the civil society realm, data activism embraces initiatives and mobilisations that take a critical approach to information and software and seek to marshal them for the social good—be it protecting online dissent and people's privacy, "translating" numbers into accessible stories, making the state more transparent, or mobilising for data justice. As we shall see, data activists have sought to meet the information and care needs of the citizenry during the pandemic. They have helped the general public to make sense of the tensions associated with the "governance by indicators" (Davis et al., 2012) that have characterised the government response to the crisis across the world. Among others, they have contributed to generate knowledge and alternative narratives of the emergency. Conversing with critical data studies and the sociology of social movements, this chapter analyses the evolution of data activism under the pressure of the first pandemic of the datafied society. In particular, it explores how citizens, advocates, and variably skilled users have engaged with data and technology in the wake of the COVID-19 crisis, surveying emerging practices of data activism as well as the challenges activists are likely to face in the post-pandemic world. It also derives lessons learnt that might inform critical data studies as the discipline further consolidates its role as interpreter of the datafied society.

The chapter is organised as follows. It starts by identifying two major shifts in data power that have been significantly accelerated by the global health crisis, namely the shift from state to corporate data infrastructure, and from private control over one's own data to the monopoly of digital platforms. It is in this complex environment that data activists intervene. The chapter then reviews the burgeoning literature on data activism, positioning the role of data activists as an emerging counterpower intercepting the above-mentioned shifts in data power. Next, five main tactics adopted by data activists during the pandemic are identified and described, followed by an analysis of three questions that data activists will have to face in the post-pandemic world. Finally, the chapter concludes by reflecting on new perspectives in critical data studies opened up by the evolution of data activism. The analysis is based on news sources and participant observation data assembled since early 2020, and with examples collected in the framework of the *COVID-19 from the margins* blog and book project (Milan et al., 2021).

Two Shifts of Data Power

As the world battled the COVID-19 pandemic with its corollary of insecurity and fear, power-holders and laypersons alike nurture hopes for "silver bullet" solutions such as smart applications that might help to win over the virus. Data have become a fundamental ingredient of any reporting on disease diffusion, betraying a positivist belief on the power of information to solve the most pressing problems of our times, on the grounds that "with enough data, the numbers speak for themselves" (Anderson, 2008). But these developments are not free of contradictions. Meanwhile, an "epic battle against coronavirus misinformation" (Ball & Maxmen, 2020) goes hand in hand with the normalisation of large-scale surveillance, putting civil societies under strain. These tensions are typical of what has been termed "surveillance capitalism", a system of power and trade grounded on the transformation of human actions and interactions into data points which can be quantified, analysed, and monetised (Zuboff, 2019). They have, however, been considerably amplified by the pandemic.

Since the inception of the COVID-19 crisis, the tech industry has assumed an ever more important role in providing crucial technical solutions to daily needs and activities. As a result, it has seen its profit margins rise massively, strengthening its quasi-monopoly in sectors like e-commerce, cloud computing, and content streaming. By way of example, Amazon doubled its revenues in the first quarter of 2020 (Faulkner, 2020), while the returns of Azure, Microsoft's cloud computing services, have increased by 48 per cent since the global explosion of the pandemic (Tilley, 2020). In the meantime, governments increasingly look at the possibilities offered by digital services in the response to the virus, verging on a one-size-fits-all techno-solutionism (Milan, 2020). The launch of questionable "immunity passports" based on "global" digital standards have raised concerns amongst privacy advocates and medical experts alike (Voo et al., 2020). State sovereignty appears increasingly at risk as many strategic infrastructures such as health care data or border control technology move into corporate hands (Latonero & Kift, 2018; Charitsis, 2019), while many human beings such as undocumented migrants are "invisibilized" by exclusionary data infrastructure and policies (Pelizza et al., 2021) In other words, data power—that is, the power of data actors and structures as well as the power exerted by data on social life—is rapidly evolving, not necessarily in the direction of progress or social justice.

Data power is shifting in two, worrying directions. On the one hand, state functions, prerogatives, and infrastructures are slowly but steadily moving towards the corporate world. This comes at a high cost: state oversight and sovereignty lose ground, while state powers (e.g., in the realm of repression and control) are augmented. Think, for example, of the involvement in the operations of the United States Immigration and Customs Enforcement agency of Palantir Technologies, a Silicon Valley company specialised in case management and data software. The partnership resulted in a "cruel new era of data-driven deportation", allowing authorities to cross-reference datasets more efficiently with the goal of identifying and expatriating migrants living illegally within national borders (Bedoya, 2020). On the other hand, we observe a shift from individual control over private information such as political preferences and biographical and biomedical data away from individuals themselves and into the grey area of corporate data infrastructure. For instance, in many countries, digital identity systems centralise medical and tax information in a single domain, often permeable to disparate state agencies and their commercial data management partners; biometrical systems track potential recipients of state welfare in countries as diverse as India and Colombia (cf. López, 2020).

To be sure, the two shifts in data power we have identified are the result of a complex process of rethinking access to information in the digital age—the so-called computational turn (Berry, 2012)—which started over half a century ago while the pandemic has played a part in dramatically accelerating. Tech solutions and functionalities such as location tracking (e.g., in contact tracing applications) and remote video surveillance (the infamous "proctoring" in university exams, see Maalsen & Dowling, 2020) have been introduced to facilitate activities otherwise paused by the rapid diffusion of the severe acute respiratory syndrome coronavirus 2. These developments have resulted in the fast-tracked legitimisation of large-scale data surveillance, with seemingly no end in sight. "Largely without public debate—and absent any new safeguards, we've become even more dependent on a technological ecosystem that is notoriously insecure, poorly regulated, highly invasive and prone to serial abuse", cautioned Canadian cybersecurity scholar Ronald J. Deibert (2020).

In addition, the imperative to identify solutions as fast as possible, typical of emergency situations like a global pandemic (cf. Calhoun, 2010), has encouraged governments to rely on ad hoc groups of experts as the central feature of crisis governance. "Task forces", "scientific councils", crisis managers, and special advisors have become a central cog in the machine of the problem-solving infrastructure—but at a high price in terms of democratic oversight. These technocrats are usually removed from existing mechanisms of democratic accountability, such as elections, and criteria for their selection are rarely made transparent. These moves have the added value of deflecting attention from broader systemic failures (such as the continued budget cuts affecting public health care systems in

Western Europe since the 1990s) and are expected to increase citizen confidence in their governments. However, this strategy seems to have achieved mixed results, as the increasing frequency of anti-lockdown or no-vax protests across the globe seem to signal (cf. Schradie, 2020). It is in this complex scenario that data activists operate.

Data Activism as an Alternative to Dominant Data Power

Surveillance capitalism has long been met by growing user concern about the aggressive intermediation of the industry, including social media platforms (Brown, 2020a). Also, state snooping, perpetrated, for example, through "smart city" projects, has been increasingly countered by grassroots resistance and attempts to create viable alternatives (Lynch, 2020). Over time, datafication and surveillance have become a target of contentious politics, permeating the agenda of social movements worldwide: think, for example, of Hong Kong pro-democracy protesters taking down "smart lampposts" suspected to deploy facial recognition technology (*Hong Kong: Anti-Surveillance Protesters Tear down "smart" Lamp-Post*, 2019). Revealing our dependence on the tech industry, the COVID-19 pandemic has registered a renewed interest in questions of data power.

As civil society's response to datafication, data activism is simultaneously a by-product of the datafied society and one of its most fascinating manifestations. Broadly speaking, it questions the role of data and data infrastructure (such as datasets, dashboards, apps, monitoring devices) in promoting or undermining social justice. It comprises a range of autonomous and rebellious actions that leverage technology and information to exert social change and to promote citizen agency and data justice.[1] It represents a practical counterpower to data power as described above, in that it systematically seeks to keep in check and to offset the consequences of a ubiquitous surveillance capitalism, while trying to exploit technological innovation for the social good.

[1] Data activism subsumes under the same label a number of distinct politically engaged identities, including but not limited to digital rights activism (cf. Maréchal, 2015), civic hacking (Schrock, 2016), transparency activism (Rajão & Jarke, 2018), and counting. While not all of these groups might identify themselves under the sphere of "data activism", they share a similar understanding of the role data and data infrastructure play in promoting justice and change.

Data activism is characterised by a distinctive action repertoire which focuses on the role of information and software in producing (or preventing) social change. Social movement scholars have characterised action repertoires as "sites of contestation in which bodies, symbols, identities, practices, and discourses are used to pursue or prevent changes in institutionalized power relations" (Taylor & van Dyke, 2004, p. 268). An action repertoire may embrace a number of distinct tactics, that is to say "alternative means of acting together on shared interest" in order to "make a statement of some kind" (Tilly, 1983, pp. 463–464). It is worth noting that tactics are not neutral means; rather, they "represent important routines, emotionally and morally salient in these people's lives. Just as their ideologies do, their activities express protestors' political identities and moral visions" (Jasper, 1997, p. 237).

If we look at tactical preferences of data activists as they fight a variety of manifestations of data power—most notably mass surveillance and the poor transparency practices of public administrations—we can identify at the bare minimum two ideal types of data activism (Milan & Gutierrez, 2015; Milan & van der Velden, 2016). These ideal types reflect distinct interpretations of the role of information in society, but share an overall "data justice" (Dencik et al., 2019) agenda. As is often the case with ideal types, there are not necessarily clear-cut boundaries between the two. However, they represent two distinct tactical preferences which typically reflect diverse identities. We can understand the two ideal types as positioned along the continuum of citizen engagement with data.

At one end of the spectrum, we find re-active data activism, which voices concerns over the social costs of "big data" and artificial intelligence technology in matters of surveillance and repression and exposes the consequent depletion of political agency. Examples of re-active data activism include the development of software able to offset privacy risks (Gürses et al., 2016), the promotion of security training to encourage human rights defenders to encrypt their communications (Daskal, 2018), and the forging of alternative imaginaries in an attempt to make sense of the complexity of our digital environment (Kazansky & Milan, 2021). In summary, re-active data activists seek to thwart the diffusion of "surveillance realism", whereby citizens can no longer imagine a society without mass surveillance (Dencik, 2018).

At the opposite end of the spectrum, pro-active data activists see information in all its present denominations as a key currency in the fight for progressive social change (Gutierrez, 2018). They may, for example, use publicly available data or access to information requests to audit the state (Torres, 2019). They engage in data-based storytelling to support public journalism or advocacy goals (Baack, 2018). They might browse the internet to collect evidence of human rights violations (Deutch & Habal, 2018) and gather publicly available data to be used in a court of law (Heller et al., 2012).

Focusing on the role of data as mediators of social action is another way of approaching data activism, as proposed by Beraldo and Milan (2019). Broadly understood, data can indicate what is at stake in an instance of collective action, that is to say data become issues and/or objects of political struggle in their own right. The mobilisation against the dismantlement of evidence-based environmental governance by the Trump administration (United States, 2016) is a case in point: over 175 US volunteers, including technologists and activists, embarked to archive vulnerable federal data corroborating climate change and environmental injustice, in an act of "data resistance" (Vera et al., 2018). But data can also become part of a movement's action repertoire, turning into a modular tool for political struggle mobilised alongside "traditional" tactics such as street protest, campaigning, or civil disobedience. Think for instance of Amnesty International's Decoders project, re-interpreting for the digital age the established tactic of "witnessing" as a way to generate evidence of human rights violations. Witnessing injustices with data means gathering evidence of historical abuses from newly digitised documents, and collecting proof of online abuse through the classification of data from the microblogging platform Twitter (Gray, 2019).

A second important distinction advanced by Beraldo and Milan (2019) concerns data activism as an individual practice versus data activism as collective action *strictu sensu*. Like earlier forms of activism focusing on media and technology, such as open-source software development (Coleman, 2013) or anti-copyright actions (Postigo, 2012), data activism unfolds into a myriad of individual practices such as encryption or access to information requests. These individual practices, however, assume meaning and exert impact only in relation to a broader community of acting individuals. Encrypting one's digital communications is a typical example: it is implemented by individual users, but it can only work when at least two people exchange encryption keys. Similar to what has been observed about other social movements, "there is protest even when it is not part of an organized movement" (Jasper, 1997, p. 5). Rather, activism results in (shared, recurrent) practices questioning or critically deploying data infrastructures.

The next section surveys how data activist tactics have been deployed during the pandemic by data activists often in collaboration with others, including volunteer citizens, in the hope of exploiting the potential of information to offset the social costs of the COVID-19 crisis.

Data Activist Tactics During the COVID-19 Pandemic

We have seen how the first global health scare of the datafied society has exposed fundamental tensions between privacy and safety, and between civil liberties and public health (see also Kitchin, 2020). In the increasing complexity of our digital ecosystem, where previously offline activities and forms of social aggregation have resorted to the digital realm, data activists have positioned themselves as the "interpreters" of these tensions to the benefit of the citizenry at large. A number of initiatives have materialised across the globe to help mitigate the impact of the pandemic, for example, producing "alternative" knowledge about virus diffusion, monitoring state measures, or building health care aids to counter the scarcity of medical devices. To make sense of the variety of grassroots initiatives that emerged during the pandemic, we can distinguish five focal approaches that are implemented by data activists or inspired by and derived from data activism and neighbouring communities of practice, such as the hacker movement (see, e.g., Jordan, 2016; Maxigas, 2012): counting, debunking, making, witnessing, and shielding, which I explore below. Interestingly, these tactics are more generally available to civil society actors in the pursuit of a collective response to the socio-economic crisis brought about by the pandemic.

Counting

Quantification is particularly alluring in uncertain times like a pandemic. This is because indicators and "numbers convey an aura of objective truth and scientific authority" (Merry, 2016, p. 1). Predictably, numbers have been at the core of the governmental and journalistic narrative of the pandemic. However, criticism emerged about partial data, non-transparent governments, and poorly reported figures, revealing the ways in which new "data gaps" currently haunt marginal communities and less resourced countries in the Global South (Milan & Treré, 2021). The archetypical data activist tactic of counting has been employed in many corners of the globe to produce "alternative" evidence about the pandemic. In Indonesia, to counter the absence of reliable official statistics on virus diffusion and the inability of the government to provide testing kits, citizens teamed up to collectively generate an alternative dataset by registering online suspected unreported cases in their neighbourhoods. Digital "information hubs" emerged to improve data transparency and help raise awareness among the citizenry (Nadzir, 2020). In Brazil, data activism proved "essential to challenge the [coronavirus-denying] state narrative about the pandemic and to prevent more deaths from COVID-19". Data activists "assumed governmental functions" by providing reliable figures to substantiate decisions. In particular, the activist group Brasil.IO independently collected data on COVID-19 cases and deaths, often manually compiling datasets and tabulating hundreds of local epidemiological bulletins. It also made available open-source software to empower others to scrape the datasets to fit their needs and run their own analyses (Füssy, 2021).

Debunking

No matter how seductive, numbers and indicators, anthropologist Sally E. Merry reminds us, are deeply affected by "the extensive interpretative work that goes into their construction" (2016, p. 1). But data activists can help interpret data in view of generating "alternative data epistemologies" to help non-experts interpret complex realities described through data (Milan & van der Velden, 2016). This skillset was put to good use during the pandemic when individuals and groups promoted initiatives oriented towards opening up the data vaults of public institutions, enhancing transparency and advancing independent investigations. They also spearheaded projects designed to mediate data for larger, lay audiences. In South Africa, a Johannesburg-based data journalism team aptly called the Media Hack Collective launched an independent national COVID-19 data visualisation dashboard with the goal of complementing the official narrative by making data available to the public (Odendaal, 2021). "Data silences", meaning the lack of data on marginal communities such as migrants, have been countered by national advocacy and campaigning initiatives. In Scotland, the non-governmental organisation Coalition for Racial Equality and Rights protested the poor data available regarding the impact of COVID-19 on minority populations living within national borders (Daly, 2021). Further, across the globe, activists have been calling for pandemic data to be released in an "open" format as a key step towards publicly informed evidence-based policymaking (Zingales, 2021).

Making

On account of the virus' sudden mass diffusion and the scarcity of medical equipment to protect caregivers or support declining respiratory functions, data activists have intervened by mobilising their "maker" skills. Although the maker and the data activism communities differ substantively (with the former being highly curated and removed from the grassroots, as illustrated by Hepp, 2020), the two share an "ethos of creativity, experimental and experiential learning, and sharing" (Davies, 2017, p. 171). During the pandemic this ethos has been applied to do-it-yourself (DIY) digital fabrication as well as to open hardware/software development (cf. Söderberg & Delfanti, 2015). For instance, data activists and professionals alike have attempted to produce DIY responses to the shortage of personal protective equipment (Richterich, 2020). In Lombardia, the Italian region that was most severely hit by COVID-19, a coalition of citizens, makers, and local administrations teamed up to transform consumer snorkelling masks into respirators for hospitals facing equipment shortages. Using a 3D printer made available by a local school to manufacture adaptors and fittings, the group not only began producing the makeshift respirators but also made the design available for noncommercial use (Morandi, 2020). Similarly, the Kenyan Ushahidi open-source mapping software, particularly popular among data activists since its launch, has been deployed to various ends including geolocating local needs and resources in quarantined Spain, ensuring food and medicine supplies are distributed to vulnerable communities in Italy, and documenting the outbreak in Nigeria, with over 200 grassroots mapping projects started in mid-March 2020 alone (Lungati, 2020).

Witnessing

The *SARS-CoV-2* virus was first identified in mainland China—a country plagued by pervasive information censorship. The first public reports of the virus' aggressiveness and its diffusion in the urban area of Wuhan, in the populous Hubei region, were met by government attempts to filter social media such as Sina Weibo in attempts to covering up the outbreak (Brown, 2020b). Chinese data activists, however, sought to preserve the memory of the communities affected by the pandemic. Evading the prevailing internet censorship by using Western services that still escape censorship, such as the software repository GitHub, they rallied volunteers in a collective documentation project, giving birth to alternative media projects involving citizens as well as journalists (Merini, 2021). "Witnessing" through data in the pandemic comprised collecting evidence to act, giving voice to marginalised groups, and enabling collective memory to counter mainstream narratives denying the pandemic or other social problems exacerbated by the lockdowns, such as the increased incidence of domestic violence. In Mexico, when the lockdown prevented feminist groups from taking to the streets while President Andrés Manuel López Obrador denied the increase of domestic violence within quarantined families, women mobilised on social media *en masse* using the hashtag #NosotrasTenemosOtrosDatos ("we have other data") in their demands for transparency in the identification and release of official figures (Villaseñor, 2021).

Shielding

In times of COVID-19, thermal facial recognition technology is regularly deployed to regulate access to public space such as airports (Kitchin, 2020). In China first, and later across the European Union, citizens were required to scan QR codes when accessing public space to verify their infection or vaccination status (see, e.g., Zhao, 2020). Once rolled out during crisis situations, however, these technologies often stay in place (see also Deibert, 2020). In the attempt to navigate the tension between privacy and public health, data activists have raised their voice against biometric mass surveillance presented as the necessary deterrent against virus diffusion, in the hope to "shield" the citizenry from unnecessary privacy breaches. Contact tracing apps are a case in point: they have been variably met with resistance across the world, which resulted in generally low adoption rates in most Western countries. In the Netherlands, a coalition formed by the non-governmental organisations such as Bits of Freedom, Waag, Platform Burgerrechten, and Amnesty International analysed the government plans for a contact tracing app, identifying ten principles to ensure that it would safeguard individual freedoms and rights, social security and cohesion, and pressed the government to respect these principles—with some success (VeiligTegenCorona.nl, 2020). In the European Union, a wide coalition of civil society organisations launched the "Reclaim Your Face" campaign in October 2020, which maintained that facial recognition technology is "Secretive. Unlawful. Inhumane" (ReclaimYourFace, 2020). They also launched a European Citizens' Initiative in early 2021, with the aim of gathering 1 million signatures across Europe and petition the European Union for a new law on the issue.

All things considered, data activism tactics have proved crucial in mitigating the negative effects of the pandemic on the citizenry. They have expanded the toolbox available to civil society actors so that they might get to grips with data power in all its denominations, be them of an informational or an infrastructural nature. But what does the future bear for data activism? The following section delves into this question and reflects on the open challenges data activism might face in the post-pandemic world.

Data Activism Reloaded: Open Questions for the Post-pandemic World

Notwithstanding the popularity of data activism tactics during the pandemic, there are at least three open questions that activists might have to face in the coming years if they are to maintain their active role as the interpreters of and as a counterforce to dominant data power.

The first challenge has to do with infrastructure and the ambiguous attitude displayed by data activists when it comes to distinguishing ideals from practice. Today, social movements rely on commercial infrastructure to mobilise, organise, and campaign. They reach their potential audiences on commercial social media services such as Facebook, Instagram, or Twitter; they petition on Google Forms; and they organise gatherings on Zoom or Microsoft Teams. Contrary to their predecessors of the 1970–1990s, who postulated the value of autonomy and self-determination in the realm of communication infrastructure (see, amongst others, Couldry & Curran, 2003; Downing, 2001; Milan, 2013), contemporary movements seem to have given up their role of critics of capitalism and surveillance capitalism in particular. Data activists, too, embody contradictory positions surrounding the role of corporate digital infrastructure: on the one hand, they embody a fierce critique of platforms and other commercial services, but on the other hand, they fail to embrace or promote radical practices of self-organisation online. In other words, their critique of data power does not adequately translate into an equally critical technical practice. However, the time might be ripe for significant change to happen. The change in privacy policy of the chat application WhatsApp in early 2021 has been followed by a surge in Signal and Telegram users, two privacy-friendly alternatives, forcing the company to address user concerns (Statt, 2021)—revealing that users are increasingly sensitive to matters of data power and thirsty for alternatives.

The second open question data activists might have to address in the near future concerns the impending forms of "data poverty" (Milan & Treré, 2020) that the pandemic has revealed. Data poverty has to do with the invisibility of certain social groups and communities along the lines of the pandemic's digital governance. As the Peruvian example made apparent, vulnerable categories must be visible to the state to benefit from welfare support, with privacy concerns being somewhat of a luxury in extreme poverty situations. Because today "data is tied to peoples' visibility, survival, and care", data poverty exposes visibility as "a sine qua non condition of existence" in the datafied society, which "gets to the bottom of what it means to be human" (2020, p. 2). But data activists have long assumed that the human right to privacy is (and should be) of primary concern to everyone, indirectly disregarding the fact that "being visible" might sometimes be more important. Re-negotiating this potential clash of values and priorities, re-assessing the question of privilege in relation to questions of data power, and branching out to other social groups whose top concerns have not (yet) emerged in the realm of digital rights, could be transformative when it comes to societal understanding of the perils of mass surveillance.

The third major challenge that data activists are likely to face in the post-pandemic world calls into question the multifaceted problem of digital literacy. Digital literacy is still relatively low in society: in 2019, 54 per cent of the European Union population had low or basic digital skills (Eurostat, 2019). But digital literacy encounters other types of specialised knowledge, in times in which people are increasingly critical of scientific knowledge or might simply be willing to trade privacy and data protection for a return to "normality". Understanding the risks of discrimination associated with immunity passports rolled out on a global or regional scale, for example, requires an appreciation of the technicalities of technology standards alongside the explanatory value of serological tests or vaccines. Data activists should identify the promotion of digital (data) literacy at large scale as a fundamental condition of survival for their progressive agenda. Furthermore, any efforts in support of digital literacy should not underestimate the current breadth of the digital divide (Van Dijk, 2020), considering that only 53 per cent of the world's population has "some" access to the internet (International Telecommunication Union, 2019).

What Path for Critical Data Studies?

This chapter has analysed data activism during the era of the COVID-19 pandemic, including tactics and outstanding issues, with a view of establishing the critical lessons the field of critical data studies might learn in the years to come. It has exposed how the COVID-19 pandemic has hastened a process of data power's centralisation, with two main shifts exacerbated by the global health crisis: state functions are increasingly delegated to the tech industry and key personal information is stored on corporate platforms. In this complex scenario, data activists represent a counterforce to predominant data power dynamics. Data activism has adopted five main tactics—counting, debunking, making, witnessing, and shielding—to mitigate the social impact of the crisis, contributing to raise awareness within civil society of the role played by information and software in contemporary societies. However, three open questions have the potential to jeopardise the advancement of data activism's agenda in the post-pandemic world: the inconsistent critique of infrastructure, the increase of data poverty and the related tension between privacy and visibility, digital literacy and the digital divide.

What can these observations on data activism tell us about critical data studies' prospects going forward? What new perspectives emerge to future-proof the discipline in a post-pandemic world? Still in its infancy, the interdisciplinary field of critical data studies has been at the forefront of the critical analysis of the relationship between data (and data infrastructures) and society. Scholars have foregrounded everyday forms of engagement with data (Kennedy, 2018), technical practice (e.g., Evans et al., 2020), and data practices (e.g., Neff et al., 2017), as well as the role of the state (e.g., Dillon et al., 2019) and industry (Couldry & Mejias, 2018). But we can identify at least two interconnected blind spots in the sprawling agenda of critical data scholars—and scrutinising the blind spots of data activism can help bring them into focus.

First, the datafied society is a deeply unequal society, where access, knowledge, infrastructures (cf. digital divide), and rights are not evenly distributed. Critical data studies should embrace an explicit (in)equality agenda in its normative analysis of the consequences of datafication. Incorporating perspectives from de- and postcolonial studies (e.g., van Schie et al., 2020), for example, might help to close the gap, overcoming the use of notions such as colonialism as mere evocative metaphors. Second, in a world where privacy is still a luxury for many and data literacy largely a mirage, the field should engage with a critique of data universalism, that is, the tendency to interpret data practices and infrastructure through Western lenses, values, and lifestyles (Milan & Treré, 2019). Although data activism, to name just one social phenomenon of concern to critical data studies, appears in various sociocultural contexts, as testified by the examples illustrating this chapter, the bulk of the discipline is still disproportionally white and "Western". Calls for "decolonizing" the discipline (e.g., Arora, 2019) or attempts to inject critical race and intersectionality perspectives (e.g., Ruberg & Ruelos, 2020) certainly move in the right direction. However, the field should be mindful of the risks connected to the "depoliticized languages of de-westernizing, internationalizing, and decolonizing" (Dutta, 2020, p. 228) and ought to simultaneously engage with the metalevel of institutional politics that interrogates "the politics of what counts as knowledge and how such counting is carried out within hegemonic structures" (Dutta, 2020, p. 233). Whether data activism and critical data studies will stand the test of time will depend on how seriously and skilfully these challenges will be addressed.FundingThis project has received funding from the European Research Council (ERC) under the European Union's Horizon 2020 research and innovation programme (grant agreement No. 639379-DATACTIVE; https://data-activism.net).

References

Anderson, C. (2008). The end of theory: The data deluge makes the scientific method obsolete. *Wired*. https://www.wired.com/2008/06/pb-theory/

Arora, P. (2019). Decolonizing privacy studies. *Television & New Media*, *20*(4), 366–378. https://doi.org/10.1177/1527476418806092

Baack, S. (2018). Practically engaged. The entanglements between data journalism and civic tech. *Digital Journalism*, *6*(6), 673–692. https://doi.org/10.1080/21670811.2017.1375382

Ball, P., & Maxmen, A. (2020). The epic battle against coronavirus misinformation and conspiracy theories. *Nature*. https://www.nature.com/articles/d41586-020-01452-z

Bedoya, A. M. (2020). The cruel new era of data-driven deportation. *Slate*. https://slate.com/technology/2020/09/palantir-ice-deportation-immigrant-surveillance-big-data.html

Beraldo, D., & Milan, S. (2019, July/December). From data politics to the contentious politics of data. *Big Data & Society*, 1–11. https://doi.org/10.1177/2053951719885967

Berry, D. M. (2012). Introduction: Understanding the digital humanities. In D. M. Berry (Ed.), *Understanding digital humanities* (pp. 1–20). Palgrave Macmillan.

Brown, A. J. (2020a). "Should I stay or should I leave?": Exploring (dis)continued Facebook use after the Cambridge Analytica scandal. *Social Media + Society, 6*(1). https://doi.org/10.1177/2056305120913884

Brown, L. (2020b). Fighting for Covid information: Meet the people who are ensuring that even those in the most censored environments receive accurate information on the pandemic. *Index on Censorship, 49*(4), 102–104. https://doi.org/10.1177/0306422020981289

Calhoun, C. (2010). *The idea of emergency: Humanitarian action and global (dis) order*. Zone Books.

Cerna Aragon, D. (2021). On not being visible to the state: The case of Peru. In S. Milan, E. Treré, & S. Masiero (Eds.), *COVID-19 from the margins: Pandemic invisibilities, policies and resistance in the datafied society*. Institute of Network Cultures.

Charitsis, V. (2019). Survival of the (data) fit: Self-surveillance, corporate wellness, and the platformization of healthcare. *Surveillance & Society, 17*(1/2). https://doi.org/10.24908/ss.v17i1/2.12942

Coleman, G. (2013). *Coding freedom: The ethics and aesthetics of hacking*. Princeton University Press.

Couldry, N., & Curran, J. (Eds.). (2003). *Contesting media power: Alternative media in a networked world*. Rowman & Littlefield.

Couldry, N., & Mejias, U. A. (2018). Data colonialism: Rethinking big data's relation to the contemporary subject. *Television & New Media, 20*(4), 336–349. https://doi.org/10.1177/1527476418796632

Daly, A. (2021). COVID-19 data on the fringes: The Scottish story. In S. Milan, E. Treré, & S. Masiero (Eds.), *COVID-19 from the margins. Pandemic invisibilities, policies and resistance in the datafied society*. Institute of Network Cultures.

Daskal, E. (2018). Let's be careful out there ...: How digital rights advocates educate citizens in the digital age. *Information, Communication & Society, 21*(2), 241–256. https://doi.org/10.1080/1369118X.2016.1271903

Davies, S. R. (2017). Characterizing hacking: Mundane engagement in US hacker and makerspaces. *Science, Technology, & Human Values, 43*(2), 171–197. https://doi.org/10.1177/0162243917703464

Davis, K. E., Fisher, A., Kingsbury, B., & Merry, S. E. (Eds.). (2012). *Governance by indicators: Global power through quantification and rankings*. Oxford University Press.

Deibert, R. J. (2020). The pandemic has made us even more dependent on a highly invasive technological ecosystem. *The Globe and Mail*. https://www.theglobeandmail.com/opinion/article-the-pandemic-has-made-us-even-more-dependent-on-a-highly-invasive/

Dencik, L. (2018). Surveillance realism and the politics of imagination: Is there no alternative? *Krisis: Journal for Contemporary Philosophy, 1*, 31–43.

Dencik, L., Hintz, A., & Cable, J. (2019). Towards data justice: Bridging anti-surveillance and social justice activism. In D. Bigo, E. Isin, & E. Ruppert (Eds.), *Data politics. Worlds, subjects, rights* (pp. 167–185). Routledge.

Deutch, J., & Habal, H. (2018). The Syrian archive: A methodological case study of open-source investigation of state crime using video evidence from social media platforms. *State Crime, 7*(1), 46–76.

Dillon, L., Lave, R., Mansfield, B., Wylie, S., Shapiro, N., Chan, A. S., & Murphy, M. (2019). Situating data in a Trumpian era: The environmental data and governance initiative. *Annals of the American Association of Geographers, 109*(2), 545–555. https://doi.org/10.1080/24694452.2018.1511410

Downing, J. D. H. (2001). *Radical media: Rebellious communication and social movements*. Sage.

Dutta, M. J. (2020). Whiteness, internationalization, and erasure: Decolonizing futures from the Global South. *Communication and Critical/Cultural Studies, 17*(2), 228–235. https://doi.org/10.1080/14791420.2020.1770825

Eurostat. (2019). *Individuals' level of digital skills*. https://appsso.eurostat.ec.europa.eu/nui/submitViewTableAction.do

Evans, S. W., Leese, M., & Rychnovská, D. (2020). Science, technology, security: Towards critical collaboration. *Social Studies of Science*. https://doi.org/10.1177/0306312720953515

Faulkner, C. (2020). Amazon doubled its profit during a pandemic. *The Verge*. https://www.theverge.com/2020/7/30/21348368/amazon-q2-2020-earnings-covid-19-coronavirus-jeff-bezos

Füssy, P. (2021). Liberating COVID-19 data with volunteers in Brazil. In S. Milan, E. Treré, & S. Masiero (Eds.), *COVID-19 from the margins. Pandemic invisibilities, policies and resistance in the datafied society*. Institute of Network Cultures.

Gray, J. (2019). Data witnessing: Attending to injustice with data in Amnesty International's Decoders project. *Information, Communication & Society*, 971–991. https://doi.org/10.1080/1369118X.2019.1573915

Gürses, S., Kundnani, A., & van Hoboken, J. (2016). Crypto and empire: The contradictions of counter-surveillance advocacy. *Media, Culture & Society, 38*(4), 576–590.

Gutierrez, M. (2018). *Data activism and social change*. Palgrave Macmillan.

Heller, C., Pezzani, L., & Studio, S. (2012). *Forensic oceanography. Report on the "left-to-die boat"*. Goldsmiths, University of London. https://content.forensic-architecture.org/wp-content/uploads/2019/06/FO-report.pdf

Hepp, A. (2020). The fragility of curating a pioneer community: Deep mediatization and the spread of the quantified self and maker movements. *International Journal of Cultural Studies*. https://doi.org/10.1177/1367877920922867

Hong Kong: Anti-surveillance protesters tear down "smart" lamp-post. (2019, August 26). https://www.theguardian.com/world/video/2019/aug/26/hong-kong-anti-surveillance-protesters-tear-down-smart-lamp-post-video

International Telecommunication Union. (2019). *Individuals using the internet, 2015–2019*. https://www.itu.int/en/ITU-D/Statistics/Pages/stat/default.aspx

Jasper, J. M. (1997). *The Art of moral protest: Culture, biography, and creativity in social movements*. Chicago University Press.

Jordan, T. (2016). A genealogy of hacking. *Convergence, 23*(5), 528–544. https://doi.org/10.1177/1354856516640710

Kazansky, B., & Milan, S. (2021). Bodies not templates: Contesting mainstream algorithmic imaginaries. *New Media & Society, 23*(2), 363–381. https://doi.org/10.1177/1461444820929316

Kennedy, H. (2018). Living with data: Aligning data studies and data activism through a focus on everyday experiences of datafication. *Krisis: Journal for Contemporary Philosophy, 1*. http://krisis.eu/living-with-data/

Kennedy, H., & Hill, R. L. (2016). The pleasure and pain of visualizing data in times of data power. *Television & New Media, 18*(8), 769–782. https://doi.org/10.1177/1527476416667823

Kitchin, R. (2020). Civil liberties or public health, or civil liberties and public health? Using surveillance technologies to tackle the spread of COVID-19. *Space and Polity*, 1–20. https://doi.org/10.1080/13562576.2020.1770587

Latonero, M., & Kift, P. (2018). On digital passages and borders: Refugees and the new infrastructure for movement and control. *Social Media + Society, 4*(1). https://doi.org/10.1177/2056305118764432

López, J. (2020). *Experimentando con la pobreza: El Sisbén y los proyectos de analítica de datos*. Fundación Karisma. https://web.karisma.org.co/wp-content/uploads/download-manager-files/Experimentando%20con%20la%20pobreza.pdf

Lungati, A. O. (2020). *Ushahidi in the era of COVID 19*. Ushahidi. https://www.ushahidi.com/blog/2020/03/30/ushahidi-in-the-era-of-covid-19

Lynch, C. R. (2020). Contesting digital futures: Urban politics, alternative economies, and the movement for technological sovereignty in Barcelona. *Antipode, 52*(3), 660–680. https://doi.org/10.1111/anti.12522

Maalsen, S., & Dowling, R. (2020). Covid-19 and the accelerating smart home. *Big Data & Society, 7*(2). https://doi.org/10.1177/2053951720938073

Maréchal, N. (2015). Ranking digital rights: Human rights, the internet and the fifth estate. *International Journal of Communication, 9*, 3440–3449.

Masiero, S. (2021). *Beyond touchscreens: The perils of biometric social welfare in lockdown. COVID-19 from the margins: Pandemic invisibilities, policies and resistance in the datafied society*. Institute of Network Cultures.

Maxigas. (2012, June). Hacklabs and hackerspaces – Tracing two genealogies. *Journal of Peer Production, 2*. http://peerproduction.net/issues/issue-2/peer-reviewed-papers/hacklabs-and-hackerspaces/

Merini, K. (2021). In memory of Covid-19 in China: Forms of digital resistance towards censorship. In S. Milan, E. Treré, & S. Masiero (Eds.), *COVID-19 from the margins. Pandemic invisibilities, policies and resistance in the datafied society*. Institute of Network Cultures.

Merry, S. E. (2016). *The seductions of quantification: Measuring human rights, gender violence, and sex trafficking*. University of Chicago Press.

Milan, S. (2013). *Social movements and their technologies: Wiring social change*. Palgrave Macmillan.

Milan, S. (2020, July–December). Techno-solutionism and the standard human in the making of the COVID-19 pandemic. *Big Data & Society*, 1–7. https://doi.org/10.1177/2053951720966781

Milan, S., & Di Salvo, P. (2020, June 8). Four invisible enemies in the first pandemic of a "datafied society". *Open Democracy*. https://www.opendemocracy.net/en/can-europe-make-it/four-invisible-enemies-in-the-first-pandemic-of-a-datafied-society/

Milan, S., & Gutierrez, M. (2015). Citizens' media meets big data: The emergence of data activism. *Mediaciones, 14*, 120–133. https://doi.org/10.26620/uniminuto.mediaciones.11.14.2015.120-133

Milan, S., & Treré, E. (2019). Big data from the south(s): Beyond data universalism. *Television & New Media, 20*(4), 319–335. https://doi.org/10.1177/1527476419837739

Milan, S., & Treré, E. (2020, July–September). The rise of the data poor: The COVID-19 pandemic seen from the margins. *Social Media + Society*, 1–5. https://doi.org/10.1177/2056305120948233

Milan, S., & Treré, E. (2021). A widening data divide: COVID-19 in the Global South. In S. Milan, E. Treré, & S. Masiero (Eds.), *COVID-19 from the margins: Pandemic invisibilities, policies and resistance in the datafied society*. Institute of Network Cultures.

Milan, S., & van der Velden, L. (2016). The alternative epistemologies of data activism. *Digital Culture & Society, 2*, 57–74. https://doi.org/10.14361/dcs-2016-0205

Milan, S., Treré, E., & Masiero, S. (2021). *COVID-19 from the margins: Pandemic invisibilities, policies and resistance in the datafied society.* Institute of Network Cultures.

Morandi, R. (2020). La "officina" che assembla respiratori con le maschere Decathlon. *VareseNews.* https://www.varesenews.it/2020/04/la-officina-assembla-respiratori-le-maschere-decathlon/917394/

Nadzir, I. (2020). Data activism and COVID-19 in Indonesia. *Jurnal Kependudukan Indonesia.* https://doi.org/10.14203/jki.v0i0.578

Neff, G., Tanweer, A., Fiore-Gartland, B., & Osburn, L. (2017). Critique and contribute: A practice-based framework for improving critical data studies and data science. *Big Data, 5*(2), 85–97. https://doi.org/10.1089/big.2016.0050

Odendaal, A. (2021). The trouble of visualizing COVID-19 during a nation-wide lockdown in South Africa. In S. Milan, E. Treré, & S. Masiero (Eds.), *COVID-19 from the margins. Pandemic invisibilities, policies and resistance in the datafied society.* Institute of Network Cultures.

Oduro-Marfo, S. (2020). Ghana. Transient crisis, permanent registries. In L. Taylor, G. Sharma, A. Martin, & S. Jameson (Eds.), *Data justice and COVID-19: Global perspectives.* Meatspace Press.

Pelizza, A., Milan, S., & Lausberg, Y. (2021). Undocumented migrants in COVID-19 counting: Rethinking the data-(in)visibility nexus. *Data & Policy.* 3, e18. https://doi.org/10.1017/dap.2021.19

Postigo, H. (2012). *The digital rights movement: The role of technology in subverting digital copyright.* MIT Press.

Rajão, R., & Jarke, J. (2018). The materiality of data transparency and the (re) configuration of environmental activism in the Brazilian Amazon. *Social Movement Studies, 17*(3), 318–332. https://doi.org/10.1080/14742837.2018.1459297

ReclaimYourFace. (2020). *The problem. Secretive, Unlawful, Inhumane.* https://reclaimyourface.eu/the-problem/

Richterich, A. (2020). When open source design is vital: Critical making of DIY healthcare equipment during the COVID-19 pandemic. *Health Sociology Review, 29*(2), 158–167. https://doi.org/10.1080/14461242.2020.1784772

Ruberg, B., & Ruelos, S. (2020). Data for queer lives: How LGBTQ gender and sexuality identities challenge norms of demographics. *Big Data & Society, 7*(1). https://doi.org/10.1177/2053951720933286

Schradie, J. (2020). "Give me liberty or give me Covid-19": Anti-lockdown protesters were never Trump puppets. *Communication and the Public, 5*(3–4), 126–128. https://doi.org/10.1177/2057047320969433

Schrock, A. R. (2016). Civic hacking as data activism and advocacy: A history from publicity to open government data. *New Media & Society, 18*(4), 581–599. https://doi.org/10.1177/1461444816629469

Söderberg, J., & Delfanti, A. (2015). Hacking hacked! The life cycles of digital innovation. *Science, Technology & Human Values, 40*(5), 793–798. https://doi.org/10.1177/0162243915595091

Statt, N. (2021, January 12). WhatsApp clarifies it's not giving all your data to Facebook after surge in Signal and Telegram users. *The Verge*. https://www.theverge.com/2021/1/12/22226792/whatsapp-privacy-policy-response-signal-telegram-controversy-clarification

Taylor, V., & van Dyke, N. (2004). "Get up, stand up": Tactical repertoires of social movements. In D. A. Snow, S. A. Soule, & H. Kriesi (Eds.), *The Blackwell companion to social movements*. Blackwell.

Tilley, A. (2020, October 27). Microsoft's earnings continue to ride pandemic-fueled demand for cloud videogaming. *The Wall Street Journal*. https://www.wsj.com/articles/microsofts-earnings-continue-to-ride-pandemic-fueled-demand-for-cloud-videogaming-11603831078

Tilly, C. (1983). Speaking your mind without election, surveys, and social movements. *The Public Opinion Quarterly, 47*(4), 461–478.

Torres, G. (2019). Institutional resistance to transparency: The quest for public Sector Information in Mexico. *Journal of Resistance Studies, 5*(2), 68–96.

Van Dijk, J. (2020). *The digital divide*. Polity Press.

van Schie, G., Smit, A., & López Coombs, N. (2020). Racing through the Dutch Governmental data assemblage: A postcolonial data studies approach. *Global Perspectives, 1*(12779). https://doi.org/10.1525/gp.2020.12779

VeiligTegenCorona.nl. (2020, April 8). *Bescherm onze gezondheid, maar bescherm ook onze rechten*. https://www.veiligtegencorona.nl

Vera, L. A., Dillon, L., Wylie, S., Ohayon, J. L., Lemelin, A., Brown, P., Sellers, C., Walker, D., & The Environmental Data and Governance Initiative. (2018). Data resistance: A social movement organizational autoethnography of the environmental data and governance initiative. *Mobilization: An International Quarterly, 23*(4), 511–529. https://doi.org/10.17813/1086-671X-24-4-511

Vieira, T. (2020). The lose-lose dilemmas of Barcelona's platform delivery workers in the age of COVID-19. *Social Sciences & Humanities Open, 2*(1), 100059. https://doi.org/10.1016/j.ssaho.2020.100059

Villaseñor, T. (2021). La otra epidemia: Los feminicidios que no se cuentan. In S. Milan & Treré, e., Masiero, S. (Eds.), *COVID-19 from the margins. Pandemic invisibilities, policies and resistance in the datafied society*. Institute of Network Cultures.

Voo, T. C., Clapham, H., & Tam, C. C. (2020). Ethical implementation of immunity passports during the COVID-19 pandemic. *The Journal of Infectious Diseases, 222*(5), 715–718. https://doi.org/10.1093/infdis/jiaa352

Zhao, Y. (2020, October 16). When health code becomes health gradient: Safety or social control? *COVID-19 from the margins.* https://data-activism.net/2020/10/bigdatasur-covid-when-health-code-becomes-health-gradient-safety-or-social-control-2/

Zingales, N. (2021). A Brazilian cautionary tale on pandemic negationism: Open data is an essential safeguard for evidence-based policy-making. In S. Milan & Treré, e., Masiero, S. (Eds.), *COVID-19 from the margins. Pandemic invisibilities, policies and resistance in the datafied society.* Institute of Network Cultures.

Zuboff, S. (2019). *The age of surveillance capitalism.* Profile Books.

Open Access This chapter is licensed under the terms of the Creative Commons Attribution 4.0 International License (http://creativecommons.org/licenses/by/4.0/), which permits use, sharing, adaptation, distribution and reproduction in any medium or format, as long as you give appropriate credit to the original author(s) and the source, provide a link to the Creative Commons licence and indicate if changes were made.

The images or other third party material in this chapter are included in the chapter's Creative Commons licence, unless indicated otherwise in a credit line to the material. If material is not included in the chapter's Creative Commons licence and your intended use is not permitted by statutory regulation or exceeds the permitted use, you will need to obtain permission directly from the copyright holder.

Index[1]

[1] Note: Page numbers followed by 'n' refer to notes.

© The Author(s) 2022

A. Hepp et al. (eds.), *New Perspectives in Critical Data Studies*, Transforming Communications – Studies in Cross-Media Research,
https://doi.org/10.1007/978-3-030-96180-0